T0297343

Dento/Oro/Craniofacial Anomalies and Genetics

Agnès Bloch-Zupan
Professor, University of Strasbourg
Reference Centre for Orodental Manifestations of Rare Diseases
Hôpitaux Universitaires de Strasbourg, Strasbourg, France

Heddie O. Sedano
Emeritus Professor, University of Minnesota
Lecturer, University of California, Los Angeles, USA

Crispian Scully
Emeritus Professor
University College London
University of Bristol, UK

ELSEVIER

AMSTERDAM • BOSTON • HEIDELBERG • LONDON • NEW YORK • OXFORD
PARIS • SAN DIEGO • SAN FRANCISCO • SINGAPORE • SYDNEY • TOKYO

Elsevier
32 Jamestown Road, London NW1 7BY
225 Wyman Street, Waltham, MA 02451, USA

First edition 2012

Notices

Knowledge and best practice in this field are constantly changing. As new research and experience broaden our understanding, changes in research methods, professional practices, or medical treatment may become necessary.

Practitioners and researchers must always rely on their own experience and knowledge in evaluating and using any information, methods, compounds, or experiments described herein. In using such information or methods they should be mindful of their own safety and the safety of others, including parties for whom they have a professional responsibility.

To the fullest extent of the law, neither the Publisher nor the authors, contributors, or editors, assume any liability for any injury and/or damage to persons or property as a matter of products liability, negligence or otherwise, or from any use or operation of any methods, products, instructions, or ideas contained in the material herein.

British Library Cataloguing-in-Publication Data
A catalogue record for this book is available from the British Library

Library of Congress Cataloguing-in-Publication Data
A catalogue record for this book is available from the Library of Congress

ISBN: 978-0-323-28223-9

This book has been manufactured using Print On Demand technology. Each copy is produced to order and is limited to black ink. The online version of this book will show color figures where appropriate.

Contents

Foreword

Not many clinical geneticists will easily confess their knowledge about teeth is limited to say the least. The opposite is probably also true: not many dentists will know much about genetics, or will recognize that the patient they treat with unusual dental findings also has unusual manifestations elsewhere. There are always exceptions to this rule, and probably the best known exception is late Professor Robert ('Bob') J. Gorlin. His knowledge both of dental and other intra-oral signs and symptoms, and the morphological characteristics of a huge number of syndromes elsewhere in the body is still unequalled. Bob went all the way to try to recognize entities. Once, when evaluating a patient with a cleft palate, he noticed unusual skin folds in the neck and subsequently he had no problem in checking the genital region for the presence of skin folds as well. Even he had to acknowledge, however, that this was about as far as a dentist could go.

Professor Gorlin worked mainly in a time of syndrome recognition based on the external phenotype and organ malformations. But progress in our understanding of the genetic background of syndromes and the function of the causative genes has been remarkable. Nowadays many syndromes are characterized not only by their phenotype but also by the gene(s) that cause it. This allows us to start to understand how the changes in genes cause particular signs and symptoms. And this also involves dental signs and symptoms.

The present book by Professor Agnes Bloch-Zupan and her co-workers Crispian Scully and Heddie Sedano is based on this principle: the recognition of syndromes allows recognition of dental findings, and the recognition of genes causing syndromes allows recognition of genes causing specific dental manifestations. This offers us a tremendous insight in the genes involved in dental morphogenesis. The authors have done a splendid job: the book contains a wealth of details, it is up to date and it is very well illustrated. The book will serve clinical geneticists and dentists alike, and patients of both groups of specialists will benefit by the information it provides.

And Bob Gorlin? He would have loved this book!

December 2011
Raoul C.M. Hennekam, MD, PhD
Amsterdam

Acknowledgments

Dear Reader,

Writing such a book was not an easy task as I was split between doubts, a busy schedule and life and continuously and fast evolving knowledge. I would like to take advantage of this tribune to thank warmly my co-authors who over the months and years remained patient and supportive.

I would like also to pay a tribute to all my godmothers and fathers in paediatric dentistry, developmental biology and genetics: the late Pr Jeanne Sommermater, Pr Jean Victor Ruch, Pr Robert J Gorlin, Pr Robin Winter.

I owe my inspiring colleagues, fruitful and stimulating discussions. Many of them gave me permission to publish their rare disease cases and I would like to acknowledge them for their trust (AC Acevedo, Y Alembick, I. Baileul-Forestier, N Chassaing, F Clauss, S Dewhurst, D Droz, M Fichbach, AL Garret, M Harrison, M Holder-Espinasse, M-C Maniere, GJ Roberts, A. Verloes, N Wolf, J Zschocke).

Pr Raoul Hennekam has kindly accepted to write a foreword for this book. His generous advices, vision, enthusiasm, sharing and friendship are a continuous enlightenment.

Heartfelt thanks to Dr. Ana Maria Johnson and Mr. Oliv Fluck for their invaluable preparation of the illustrations and the meticulous revision of the original manuscript, as well as to Drs. Katrina Dipple, Yoshio Setoguchi, Henry Kawamoto and the rest of the Craniofacial Team at UCLA for helping one of us (HOS) to keep updated in the fast growing field of Craniofacial Dysmorphology.

I am privileged to walk aside talented students.

Dear G, H, B, M please accept my apologies for the stolen time.

A special thanks to the patients and their families!

Agnès Bloch-Zupan
Heddie O. Sedano
Crispian Scully

Acknowledgments

Introduction

Knowledge of aetiology of anatomical, cellular or metabolic disturbances may lead to the identification of developmental processes involved in normality. On the other hand, understanding of dental genetics and developmental biology will allow identification of processes responsible for genesis of specific dental anomalies [1–3].

The objectives of this book are to present the background of dental and orofacial anomalies from their clinical and biological perspectives – discussing genes, their encoded proteins and the role of these proteins in developmental signalling pathways, sometimes via the analysis of genetically modified mice exhibiting such defects.

The book attempts to clarify the bewildering array of factors involved in tooth development (odontogenesis). This insight into the pathology offers clinicians a modern biomedical view on human dental and orofacial defects found and proposes ways of understanding orofacial manifestations of these genetic diseases.

The book will also present and illustrate the various dental and orofacial anomalies by type, signalling pathways and syndrome families.

1 Odontogenesis, Anomalies and Genetics

1.1 Odontogenesis

1.1.1 Tooth Development

Tooth development is embedded within craniofacial development. It originates from pluripotential cephalic neural crest cells which subsequently migrate towards the first pharyngeal arch, there to trigger (in combination with mesodermal cells) the development of many elements of the craniofacial structures [4–6].

Mammalian embryonic tooth development (odontogenesis), especially the mouse dentition, is an interesting model. Odontogenesis leads to specific crown and root morphogenesis for each type of tooth (incisor, canine, premolar, molar), to enamel organ histomorphogenesis and to terminal cytodifferentiations of odontoblasts, ameloblasts and cementoblasts. Evolutionary study of mammals is often focused on detailed analyses of teeth shapes. Molecular patterning may influence dental evolution via differences in gene expressions correlated with morphological variations [7–9].

The continuous and progressive stages of odontogenesis have classically been divided into the dental lamina, bud, cap and bell stages, root formation and tooth eruption. Tooth development is a kinetic dependent process mediated via epithelio-mesenchymal interactions between ectomesenchymal cells originating from cephalic neural crest cells and the first pharyngeal arch ectoderm [10–14]. These cells contribute to the formation of the dental mesenchyme, the dental pulp, odontoblasts, dentine matrix, cement and periodontium [15,16]. Extracellular matrix (i.e. basement membrane, predentine, and dentine) participates in odontogenesis either as a substrate with temporospatial specificity in interaction with receptors of the plasma membrane or as a putative reservoir of endocrine, paracrine and autocrine factors such as peptide growth factors.

Tooth morphogenesis is under strict genetic control, as is general embryonic development, and the participating genes are being discovered at an increasing rate. By 2008, more than 300 of these genes had been listed in the database created by Pekka Nieminen from Helsinki University, Finland, gathering expression pattern at the various stages of odontogenesis obtained from the laboratories of experts worldwide (http://bite-it.helsinki.fi) [17].

1.1.2 Developing Through Epithelio-mesenchymal Interactions with Signalling Molecules and Transcription Factors

A similar language in cell communication, conserved during evolution, is used throughout embryonic and tooth development, respectively. Implicated **signalling molecules**

Dento/Oro/Craniofacial Anomalies and Genetics. DOI: 10.1016/B978-0-12-416038-5.00001-9

and growth factors include transforming growth factor beta (Tgfβ) family with bone morphogenetic proteins (Bmps), activins and follistatin; fibroblast growth factors (FGFs); hedgehog (only sonic hedgehog, shh, plays a role during odontogenesis) and wingless-related MMTV integration site (Wnts) [18–23]. These molecules transmit their message via **signalling pathways** from the cell surface receptors towards the nucleus via effectors [24]. As a matter of writing convention used in this book, *genes* are in italics and proteins are non-italics; and human *GENES* and PROTEINS, capitalized; and mouse *gene*s and proteins, non-capitalized. Protein is then not italicized, and the first letter is capitalized.

Transcription factors modulate the expression of target genes, thereby inducing a modification of response and cell behaviour. Most of the genes outlined in http://bite-it.helsinki.fi participate in this cell communication network, and mutations in some of these genes are responsible for dental and oro-facial anomalies [25] (Figure 1.1).

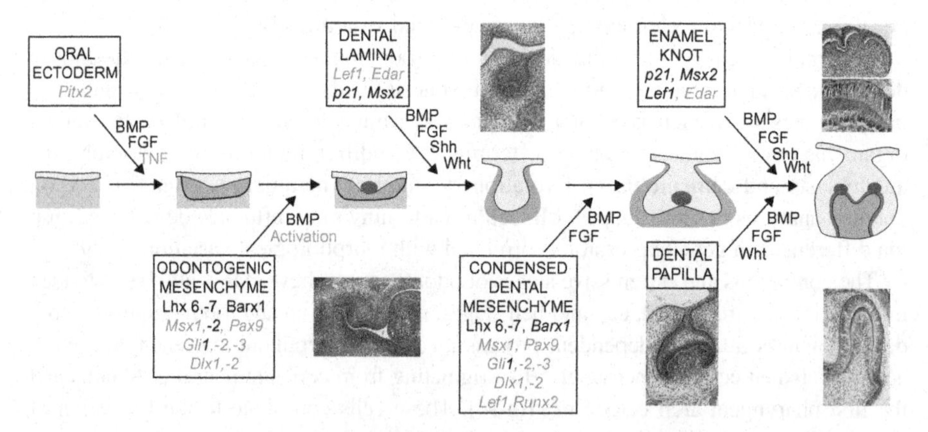

Figure 1.1 Signalling networks regulating tooth development. Genes displayed in green, if inactivated, are responsible for arrested tooth development. (For interpretation of the references to color in this figure legend, the reader is referred to the web version of this book.) *Source: From Ref. [26].*

Retinoids are also signalling molecules participating in dental and craniofacial development.

Other growth factors involved in odontogenesis include epidermal growth factor (EGF) family with EGF and transforming growth factor α (TGFα), hepatocyte growth factor (HGF), platelet-derived growth factor (PDGF), Midkine and pleiotrophin, neurotrophins nerve growth factor (NGF), brain-derived neurotrophic factor (BDNF), neurotrophin 3 (NT-3) and neurotrophin 4/5 (NT4/5) [27–33].

Teeth, then, are indeed organs that develop as epithelial ectodermal appendages and whose development is mediated by epithelio-mesenchymal interactions [22].

1.1.3 Signalling Pathways

Following is a description of the involvement of signalling pathways at different key stages of tooth development.

Initiation Stage

The localization, identity, shape and size of the teeth are determined during early stages of tooth development [34]. Early signals like Bmp4 (incisor region) and FGF8 (molar region) from oral ectodermal cells elicit the odontogenic potentialities of the underlying mesenchyme [35,36]. Numerous transcription factors expressed in the mesenchyme are then activated [13,37], including genes coding for several homeobox divergent transcription factors (e.g. *msx1 [muscle segment homeobox 1, alias hox-7], msx2, dlx1 [distal-less homeobox 1], dlx2, barx1 [BarH-like homeobox 1], lhx6 [LIM homeobox 6] and lhx7)* and involved in patterning (Figure 1.2). Many of these transcription factors, even within the same family, are coexpressed and participate to functional redundancy, allowing rescue mechanism in case of dysfunctions. Activation of these genes is necessary for the continuation of tooth development, as shown by the phenotype of genetically modified mice carrying inactivation of these genes. Tooth development is arrested at the initiation stage when *msx1* and *msx2* or *dlx1* and *dlx2* are inactivated [38,39].

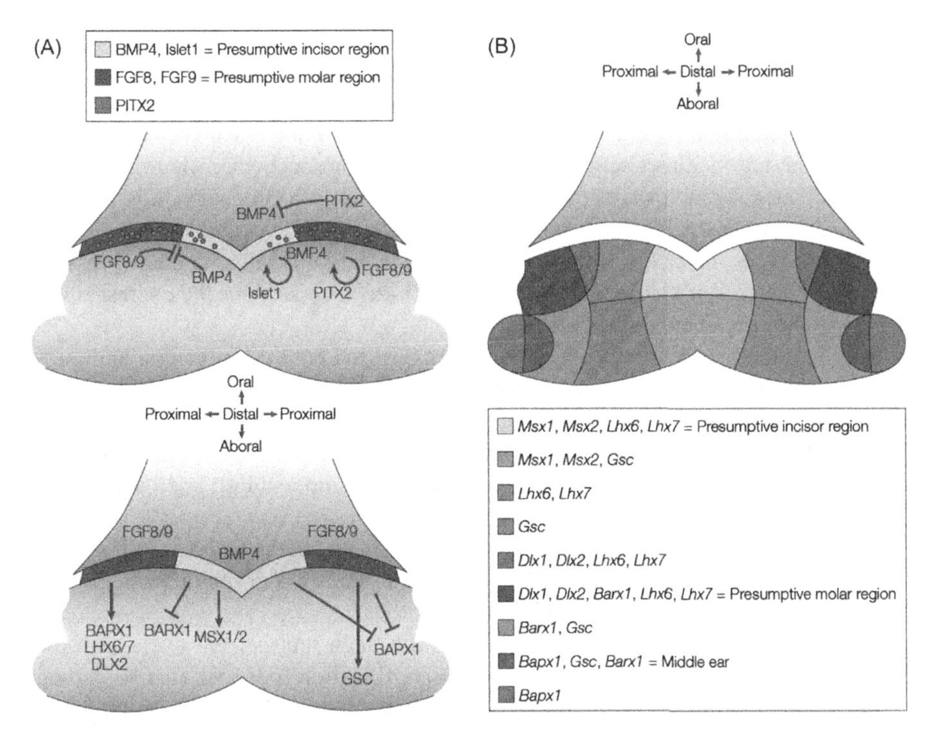

Figure 1.2 A divergent homeobox gene code within the maxilla and mandible.
Source: From Ref. [13] Reprinted by permission from Macmillan Publishers Ltd, copyright (2004).

Placode Formation

One of the key aspects of tooth development is the formation of ectodermal placodes, thickenings of the epithelium at the locations of each family of tooth. Similar placodes initiate the development of all organs developing as ectodermal appendages [22,40], such as the hair, nails, salivary, mammary and sebaceous glands.

Signalling molecules from the four well-known families previously listed (the Tgfβ, FGFs, hedgehog and Wnts families) participate in placode development. Fgfs and Wnts act as activators of placode formation in feather and hair follicle formation, and Bmps, as repressors [41,42], and this seems also to be the case during odontogenesis. Genes coding for molecules involved in placode formation participating in the TNF (tumour necrosis factor)/NF-kappaB signalling pathway, or the proto-oncogene p63, if mutated are responsible for ectodermal dysplasias, leading to diseases involving all these epithelial appendages and manifesting with hypodontia (less than six missing teeth), oligodontia (six or more missing teeth) and even anodontia (absence of all teeth) [43–46].

Bud-to-Cap Stage Transition

The underlying condensing mesenchyme controls the growth and folding of the epithelium. Mesenchymal signals induce, within the enamel organ, the formation of signalling centres called enamel knots – transitory structures which produce numerous signalling molecules at the cap stage [47]. The primary enamel knot is indispensable to crown development. These signals pass towards the mesenchyme, and within the epithelium regulate tooth shape. Shh is one signal essential for epithelial proliferation, but its direct action seems to target the mesenchyme, where it induces a feedback loop signal towards the epithelium [48]. Wnt and Bmp signalling regulate the formation of enamel knots.

Bmp4 induces the arrest of the cell cycle in enamel knot cells via the expression of a cyclin-dependent kinase inhibitor p21; Wnts are necessary for the expression of Fgf4 within the enamel knots [49–52].

Fgfs and their corresponding receptors are expressed both in epithelium and mesenchyme and reciprocally regulate proliferation in adjacent tissues [53,54]. Three transcription factors present at these stages in condensing mesenchyme are msx1, paired homeobox 9 (pax9) and runt-related homeobox 2 (runx2), whose expression is regulated by epithelial signals. *Msx1* is induced by Bmp and Fgf; *pax9* and *runx2*, by Fgf [55–58].

Runx2-deficient mice show arrested tooth development at the bud stage [58,59].

Signalling molecules from the TNF (Eda/Edar) family may participate in cusp formation via the enamel knot [61,62].

The Wnt signalling pathway is involved in the tooth replacement cycle with the development of the primary and then permanent dentition [63–65]. This signalling pathway also leads to the formation of supernumerary teeth through the multiplication of signalling centres via budding from the dental epithelium. The subsequent molars formed have a simplified cusp pattern.

Bell Stage

Molar cusp development is initiated by secondary enamel knots appearing within the inner dental epithelium at the site of the tips of future tooth cusps. They express numerous signalling molecules, amongst them FGF4, and stimulate cusps growth. The lineage relationships between primary and secondary enamel knots are not completely unravelled [66].

Proliferation outside the enamel knots area and non-proliferation in the knots coordinate foldings of the inner dental epithelium around the condensing mesenchyme [67,68]. Marked apoptosis within the knots will induce the progressive disappearance and the end of the signalling activity of these structures at the end of the cap stage and in the early bell stage [11,69].

Cell Differentiations

Odontoblasts are specialized ciliated cells [70], differentiating following a defined temporospatial gradient from the cusp tip towards the cervical area of the tooth, and induction by the inner dental epithelial cells. Odontoblasts synthesize dentine matrix proteins [71]: collagens (I, type I trimer, III, V, VI) and non-collagen proteins like osteonectin and osteocalcin, and SIBLINGs (Small Integrin-Binding LIgand, N-linked Glycoproteins) family proteins like osteopontin, MEPE (an extracellular matrix phosphoprotein), bone sialoprotein (BSP), dentine matrix protein 1 (DMP1), dentine sialophosphosproteins (DSPP or DSP and DPP) [72,73]. Other molecules which contribute to dentine formation include the proteoglycans; serum proteins such as albumin; enamel proteins such as amelogenins; matrix metalloproteinases (MMPs); tissue inhibitors of metalloproteinases (TIMPs); cathepsin and phospholipids [74–77]. SIBLINGs can bind to MMPs [78].

Ameloblasts differentiate following the same temporospatial gradient but with a different time frame. Amelogenesis proceeds in the presence of predentine/dentine and after disappearance of the basement membrane, when odontoblasts are functional. Ameloblast synthesize, participate in mineralization and resorb the enamel matrix proteins during the maturation stage via the proteinases enamelysine (MMP20) and kallikrein 4 (KLK4) [79]) leading to enamel, the most densely mineralized tissue of the body (95% in weight).

Structural proteins present in enamel matrix include amelogenins, ameloblastin (sheathline or amelin), enamelin, tuftelin and amelotin [80,81]. Other proteins like sulphated glycoproteins, calcium-binding proteins, DPP (dentine phosphoprotein), lipids and phospholipids also participate in enamel formation [77], as do enzymes, metalloproteinases, serine proteases and phosphatases.

The terminal cytodifferentiation of odontoblasts [82,83] and ameloblasts, as well as synthesis of dentine and enamel matrices, are regulated by signalling molecules from the Tgfβ family – especially Bmps and Fgfs [18,83–85]. The four main families of signalling molecules are involved in these processes [87]. These signals participating in the epithelio-mesenchymal interactions are synthesized by pre-odontoblasts and pre-ameloblasts and are carried by the basement membrane and the predentine/dentine.

Inhibiting Bmp4 signals (signals sent by the mesenchyme and then by pre- and functional odontoblasts to the inner dental epithelium and then pre-ameloblasts) by follistatin impairs terminal differentiation of ameloblasts from the mouse labial epithelial loop, a normal process occurring in the mouse lingual epithelial loop devoid of any enamel [88].

Stratum intermedium cells whose development is relayed by *msx2* would be involved in ameloblast differentiation via signalling through shh [89].

Odontoblasts and ameloblasts are vitamin D target cells [90].

Proteins synthesized by odontoblasts and ameloblasts could also play a signalling role, like DMP2 and amelogenin [87].

Dentine and enamel formation are interdependent. Odontoblasts secrete amelogenins [91], and ameloblasts transitory secrete DSP and DPP during enamel/dentine junction formation [92].

Eruption

As root and cement develop, eruption phenomena are regulated by tissue interactions mediated by various signals [93–96]. PtHrP (parathyroid hormone) seems to be involved in the regulation of surrounding bone resorption during tooth eruption [97,98]. Runx2 may participate in cementogenesis, formation of the periodontal ligament and in eruption *per se* [99].

1.2 Dental Anomalies

Developmental dental anomalies may exist in isolation or may be associated with extraoral clinical manifestations in syndromes: they can be of genetic origin or due to the action of teratogens [100–105].

Each dental anomaly can be classified into various categories, anomalies of number, shape or size (a continuum of anomalies); of structure for hard tissues formation; of root formation and eruption; and of resorption. They are correlated to specific genetic and developmental biology problems [106] such as the embryonic origins of dental cells, the patterning of the dentition, the defined location of tooth development, tooth identity, specific morphogenesis, histogenesis, terminal differentiation of odontoblasts and ameloblasts, dentine and enamel matrix synthesis followed by mineralization, root and periodontium formation and eruption of teeth [8,14,25,107].

Any interference with these developmental processes can lead to clinical anomalies and defects [107–110], and some may even lead to tumours arising from dental epithelial cells (odontogenic tumours) [111].

Acquired dental anomalies are relatively common and provide real markers of gene–environment relationships, pointing to pre- and post-natal pathological developmental pathology, easily accessible and readable via fixed teeth morphology and appearance throughout mineralization. This aspect is probably underestimated despite good evidence, for example, linking MIH (molar incisor hypomineralization) to dioxin exposure [100,104,112] or fluorosis to chronic fluoride intoxication [102,112], or anticancer treatments in children to their orodental repercussions.

Genetic dental anomalies have a prevalence of less than 200,000 affected individuals within the United States and are considered by NIH (National Institutes of Health, http://rarediseases.info.nih.gov/) as rare diseases. Prevalence of these anomalies is difficult to estimate as there are very few reliable epidemiological studies. Every type of inheritance is possible.

Missing teeth (anomalies of tooth number) concern tooth agenesis, hypodontia, oligodontia and anodontia [114] and are associated with mechanisms regulating tooth patterning as well as the transition from the bud to the cap stage. In the mouse, inactivation of different genes leads to the arrest of tooth development [115]. The first gene to be identified was *msx1* in 1994 [116]; knockout mice −/− for *msx1* demonstrate tooth developmental arrest at the bud stage. Subsequently, other genes shown to be involved in arrest of tooth development in mice include *dlx1, dlx2, lef1, pax9, pitx2, runx2/cbfa1* and the activin-coding gene [39,60,117–121]. In humans, mutations in these same genes, *MSX1, PAX9* and *PITX2*, also result in missing teeth according to specific patterns [122–127]. *AXIN2*, for example, is responsible for colorectal cancer susceptibility and dental agenesis [128,129].

Dental anomalies related to hard tissue formation appear to be related to genes coding for structural proteins contributing to the formation of dentine or enamel matrices or to the proteinases capable of degrading these matrices [72,130].

It is important to analyse and record the dental phenotype not solely tooth by tooth but also considering the whole dentitions. Very often in a case, it is possible to describe associations of dental anomalies. This is not surprising if one considers that the genes involved and the corresponding biological effectors that are the proteins may have similar or different functions throughout the entire course of teeth development from the sixth to the seventh week of gestation (even earlier, on day 22, if one considers neural crest cells) to 20–25 years, with the end of the apexogenesis of the third molars.

1.3 Syndromes and Dental Anomalies

Genetic dental anomalies are one of the phenotypical aspects of many rare diseases or syndromes.

Rare diseases, by definition, affect less than 1 in 2000 individuals within the population but include more than 7000 different entities and affect almost 25 million persons in Europe.

Dr Robert James Gorlin pioneered the area of genetics and dentistry [131–133]. Amongst more than 5000 known syndromes, more than 900 have a dento/oro/craniofacial counterpart [131], and more than 750 of them are associated with cleft lip and/or palate. Many are well described in Gorlin's publications.

Dental anomalies are often described in syndrome phenotypes [134–136] in association with other organs or system malformations [132], which is understandable because the same genes and signalling pathways regulate tooth development and the development of other organs such as the eye, for example. Dental anomalies are especially present in many syndromes affecting ectodermal derivatives as ectodermal dysplasias [22,45,46].

The achievement of the Human Genome Project should lead to an accelerated identification and understanding of important genes involved in craniofacial development and anomalies. New syndromes with an orodental phenotype are now described regularly [137,138].

Modern databases relevant to this field include

OMIM, Online Mendelian Inheritance in Man (http://www.ncbi.nlm.nih.gov/Omim/), which has 641 entries on teeth, 277 on teeth and genes and 201 on teeth and locus.

Orphanet, a Franco-European database dedicated to rare disease information (http://www.orpha.net/), lists 42 diseases after a search on 'dental anomalies' and 20 on the terminology 'anomalies of tongue, gingiva, oral mucosa' [139,140].

LDDB, the London Dysmorphology Database, outlines more than 919 syndromes with dental manifestations, 750 entities going along with clefting out of a total of 5978 syndromes [141–143].

It is important to remember that within the same syndrome different dental anomalies could be encountered, like hypodontia, enamel defects and delayed eruption, as the causal genes might be involved at different stages of tooth development.

2 Missing Teeth (Hypodontia and Oligodontia)

This chapter presents the genotypes and phenotypes associated with tooth agenesis.

2.1 Transcription Factors

Syndrome	Associated Features	Gene	Molecules	Inheritance	Locus
Transcription factors	**Divergent homeobox genes**				
Oligodontia #106600	Cleft palate	*MSX1*	Transcription factor	AD	4p16.1
Witkop syndrome #189500	Dysplasia of nails	*MSX1*	Transcription factor	AD	4p16.1
Wolf–Hirschhorn syndrome #194190	Growth retardation, microcephaly, dysmorphic features, intellectual disability, etc.	*MSX1*	Transcription factor	Chromosomal	4p
Oligodontia #604625 #106600		*PAX9*	Transcription factor	AD	14q12–q13
Rieger syndrome #180500	Anterior chamber anomalies, umbilical defects	*PITX2*	Transcription factor	AD	4q25–q26
#602482		*FOXC1*	Transcription factor	AD	6p25
#601499		?			13q14
Cleft palate CPX # 303400 [144]	X-linked cleft palate and ankyloglossia	*TBX22*	Transcription factor	X	Xq21.1

AD = autosomal dominant, AR = autosomal recessive, p = short arm of chromosome, q = long arm of chromosome. Missing teeth encountered in diseases caused by mutations in divergent homeobox genes usually follow a recognizable pattern.

Dento/Oro/Craniofacial Anomalies and Genetics. DOI: 10.1016/B978-0-12-416038-5.00002-0

2.1.1 MSX1 *Hypodontia and Oligodontia*

Definition

Hypodontia is an inherited condition characterized by developmentally missing teeth, although absent third molars are a "normal" variation and may not be considered to be part of hypodontia. It is very rare in the primary dentition and more common in females than males, with a 3:2 ratio. In addition to missing teeth, people with hypodontia may have rather small or very conical teeth [145].

- Hypodontia: less than six missing teeth (not taking into account the third molars)
- Oligodontia: six or more missing permanent teeth (not taking into account the third molars)
- Anodontia: no teeth present

OMIM Number

#106600

Prevalence

About 5% of the population have at least one tooth (other than a third molar) missing. Only about 0.3% are affected with oligodontia.

Inheritance

Autosomal dominant

Aetiology – Molecular Basis – Gene

The gene involved is *MSX1* (4p16.1). The Msx family of vertebrate homeobox divergent genes was originally isolated by homology to the msh (muscle segment homeobox) gene of the fly *Drosophila*. It encodes a transcription factor.

A mutation in the *MSX1* gene, detected in a single family, resulting in an Arg-Pro substitution in the homeodomain of the protein product of this gene, has previously been associated with the deficiency of second premolars and third molars [146].

Haploinsufficiency of *MSX1* is a possible mechanism for selective tooth agenesis [147]. A nonsense mutation in *MSX1* causes the Witkop or Tooth and Nail [TNS] syndrome (#189500) [148] (see next syndrome). *MSX1* mutation is also associated with orofacial clefting and tooth agenesis in humans [149].

A specific *MSX1*–R151S allele is a low-frequency, mildly deleterious allele for familial hypodontia that alone is insufficient to cause oral facial clefting. It may in fact contribute to the likelihood of common birth disorder phenotypes, such as partial tooth agenesis and oral facial clefting [150].

Animal Models/Main Features

Mice haploinsufficient for msx1 have a cleft palate and abnormalities of tooth development. The mandible and maxilla are underdeveloped, and the upper and lower incisors are absent. Molar development is arrested at the bud stage. There are abnormalities of the cranial bones, including Wormian bones. In the middle ear, the short process of malleus fails to form [116].

Clinical Description

Main features
The main features are the oro-dental features. The pattern of missing teeth is as follows:

2nd premolars > 3rd molars > 1st premolars > 1st molars

Management/Oral Health

Long-term pluridisciplinary management from paedodontics to orthodontics, prosthodontics, implantology and so on. Specialized hypodontia clinics might be available. Genetic counselling is important.

References

[116,127,145–152]

Pictures Illustrating the Oro-Dental Features

Figure 2.1

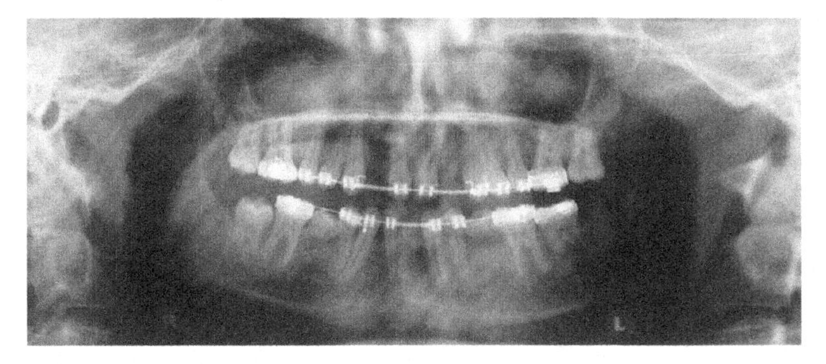

Figure 2.1 *MSX1*. Missing teeth in the spectrum of *MSX1* mutation (12, 22, 41, 31, 32, 45, 35, 48, 38).

2.1.2 Witkop Syndrome

Definition

This syndrome, also known as Tooth and Nail syndrome (TNS) or dysplasia of nails with hypodontia, is an ectodermal dysplasia affecting teeth, nails and hair.

OMIM Number

#189500

Prevalence

Several families have been reported from various countries and ethnic groups, but the prevalence is yet to be determined.

Inheritance

Autosomal dominant with variable expressivity and penetrance

Aetiology – Molecular Basis – Gene

The gene is *MSX1*, a transcription factor, which maps to the short arm of chromosome 4 (4p16.1).

Animal Models/Main Features

Histological analysis of *Msx1*-knockout mice, combined with a finding of *Msx1* expression in mesenchyme of developing nail beds, revealed that not only was tooth development disrupted in these mice, but nail development was affected as well. Nail plates in *Msx1*-null mice were defective and were thinner than those of their wild-type littermates [116].

Clinical Description

Main features

- Skull hair is sparse, thin and very friable.
- Nails of hands and feet are absent at birth and then they grow very slowly.
- Nails are soft, pliable and spoon-shaped (koilonychia).
- Nails are smaller towards the ulnar / fibular side.
- Nails can be absent in the smallest toe or the little finger.
- Patients cut their nails only once every 6 or 7 months.

Cranio-oro-dental features

- Absent teeth occur in both dentitions.
- The number of absent teeth is variable.
- Most frequently absent are mandibular incisors, second molars, maxillary canines.
- Teeth are conoid.
- Pouting lower lip occurs.

References

[148,153–157]

Pictures Illustrating the Oro-Dental Features

Figure 2.2

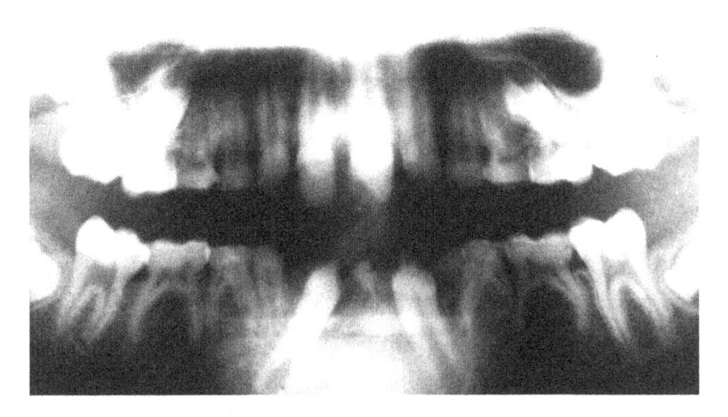

Figure 2.2 Witkop, or tooth and nail, syndrome. Panoramic radiograph from a 14-year-old boy showing partial anodontia, absence of teeth germs and conoid teeth.

2.1.3 Wolf–Hirschhorn Syndrome

Definition

The syndrome consists of learning disability, seizures and distinct facial appearance. The former Pitt–Rogers–Danks syndrome, is now considered part of Wolf–Hirschhorn syndrome (WHS) as it is caused by identical deletions of 4p.

OMIM Number

#194190

Prevalence

At birth, estimated at 2 in 100,000.

Inheritance

Isolated cases. In 87% of individuals, there is no family history of the disorder.

Aetiology – Molecular Basis – Gene (4p-)

Subtelomeric deletions in 4p between Huntington disease (*HD*) and *FGFR3* genes.

WHSC1 gene, which encodes a protein containing a SET domain that is found in proteins which are usually involved in embryonic development and which function as a transcription factor, is deleted in all cases. Another critical region, *WHSCR-2*, is also involved. The chromosome band 4p16.3 region also contains a gene called *DFNA6* for autosomal dominant non-syndromic hereditary hearing loss. The human *FGFRL1* gene, encoding a putative fibroblast growth factor (FGF) decoy receptor, has been implicated in the craniofacial phenotype of a WHS patient [158].

MSX1 gene might be deleted in some patients with WHS and oligodontia.

Animal Models/Main Features

Similar to WHS patients, these animals were growth retarded, susceptible to seizures and showed midline (palate closure, tail kinks) craniofacial and ocular anomalies (colobomas, corneal opacities) [159,160].

Targeted deletion of the mouse *Fgfrl1* gene recapitulates a broad array of WHS phenotypes, including abnormal craniofacial development, axial and appendicular skeletal anomalies and congenital heart defects [161].

Clinical Description

Main features

Zollino et al. [162] characterized the following minimal diagnostic criteria:

- Typical facial appearance
- Intellectual disability
- Growth delay
- Congenital hypotonia
- Seizures

Other frequent manifestations are:

- Microcephaly
- Cardiac defects
- Bones – short stature, fusion of ribs, anterior fusion of vertebrae, polydactyly, split hand
- Urinary and genital hypospadias – renal hypoplasia.

Cranio-oro-dental features

- Hypertelorism, prominent glabella, epicanthal folds (collectively described as 'Greek warrior helmet' features)
- Cleft lip or palate
- Hypodontia
- Occasional late dental development, taurodontism of the primary molars, microdontia

References

[162–172]

Pictures Illustrating the Oro-Dental Features

Figure 2.3

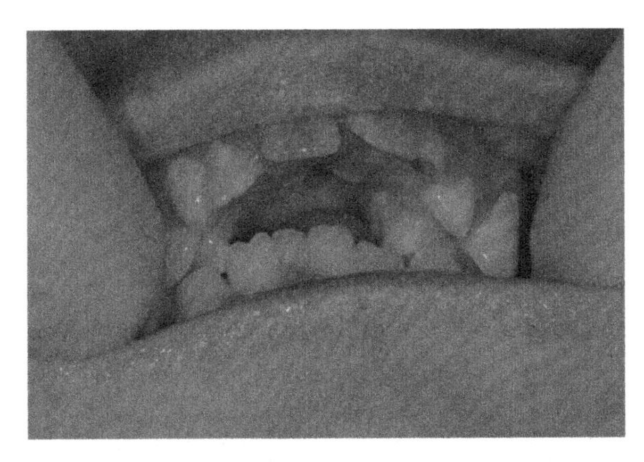

Figure 2.3 Wolf–Hirschorn syndrome. Reference Centre for Orodental Manifestations of Rare Diseases, Service de Médecine et Chirurgie Buccale, Hôpitaux Universitaires de Strasbourg, Université de Strasbourg, M-C Maniere.

2.1.4 PAX9 *Hypodontia and Oligodontia*

Definition

See *MSX1*.

OMIM Number

#106600, #604625

Prevalence

Not yet determined

Inheritance

Autosomal dominant

Aetiology – Molecular Basis – Gene

The gene involved is *PAX9*, 14q12–q13. Mutations in *PAX9* have been described for families in which inherited oligodontia characteristically involves permanent molars [173,174]. Primary molars can also be involved [124].

To various degrees, affected members lacked permanent first, second and third molars in all four quadrants. Several individuals with missing molars also lacked second premolars – most commonly, maxillary second premolars and mandibular central incisors [175,176].

Haploinsufficiency of *PAX9* is associated with autosomal dominant hypodontia [124].

In addition to permanent molars, some other teeth can be congenitally missing in the premolar, canine and incisor regions, and the teeth size is reduced [125].

A *de novo* mutation is found in *PAX9* [122] with third molars, second premolars and incisors missing.

More mutations and phenotype variability have recently been discovered [177–179].

The severity of tooth agenesis seems correlated to the DNA-binding capacity of the mutated *PAX9* proteins, supporting the hypothesis that DNA binding is responsible for the genetic defect [180].

Animal Models/Main Features

Homozygous *Pax9*-mutant mice die shortly after birth, most likely as a consequence of a cleft secondary palate. They lack a thymus, parathyroid glands and ultimobranchial bodies, organs which are derived from the pharyngeal pouches. In all limbs, a supernumerary preaxial digit is formed, but the flexor of the hindlimb toes is missing. Furthermore, craniofacial and visceral skeletogenesis is disturbed, and all teeth are absent. In *Pax9*-deficient embryos, tooth development is arrested at the bud stage [117].

Clinical description

Main features

The main features are the oro-dental features. The pattern of missing teeth is as follows:

3rd molars > 2nd molars > 1st molars > 2nd premolars

Management/Oral Health

Long-term pluridisciplinary management from paedodontics to orthodontics, prosthodontics, implantology is required. Specialized hypodontia clinics might be useful.

References

[122,124,125,127,129,173–181]

Pictures Illustrating the Oro-Dental Features

Figure 2.4

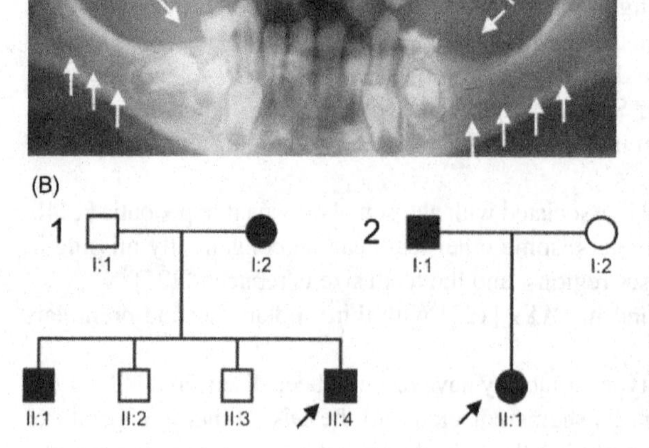

Figure 2.4 *PAX9*. Oligodontia phenotype and families
(A) Panoramic radiograph of proband of family 1 (II:4) at age 6. Missing teeth are depicted with arrows.
(B) Pedigrees of molar oligodontia families 1 and 2. Squares, males; circles, females; darkened, affected; arrows, probands. Identification of a nonsense mutation in the *PAX9* gene in molar oligodontia.
Source: From Ref. [173] Reprinted by permission from Macmillan Publishers Ltd, copyright (2001).

2.1.5 Axenfeld–Rieger Syndrome (Rieger Syndrome)

Definition

Rieger syndrome or Axenfeld–Rieger syndrome, can be divided into three types. These types present similar clinical findings, and here they will be discussed as a single entity.

OMIM Number

#180500	Type 1
#601499	Type 2
#602482	Type 3

Prevalence

0.5 in 100,000

Inheritance

Autosomal dominant with 95% penetrance and variable expressivity, especially of the dental abnormalities

Aetiology – Molecular Basis – Gene [182–188]

- Type 1 has been mapped to the long arm of chromosome 4 (4q24–q26), and it is considered to be the result of various mutations on the *PITX2* gene. *PITX2*, a bicoid-related homeobox gene (pituitary homeobox transcription factor), is involved in Axenfeld-Rieger syndrome and the left–right asymmetrical pattern formation in body plan.
- Type 2 has been mapped to the long arm of chromosome 13 (13q14). The gene involved is yet to be identified.
- Type 3 involves another transcription factor *FOXC1* (6p25.3).

Digenic inheritance of mutations in both *FOXC1* and *PITX2* has been reported [189].

In general, Axenfeld–Rieger patients who display defects in other organ systems, such as teeth or umbilicus, have mutations of the *PITX2* gene; whereas, in patients with isolated eye anterior segment dysgenesis, mainly *FOXC1* mutations are detected. However, patients with *FOXC1* mutations and dental anomalies have been described [187].

Animal Models/Main Features

In *Pitx2* gene-deleted mice, tooth development fails to progress past the full tooth bud stage [190,191].

Clinical Description

Main features [182,185]

- Cryptorchidism
- Foetal lobulation of kidney
- Congenital heart defects
- Anal stenosis
- Failure of involution of the periumbilical skin
- Growth hormone deficiency

Cranio-oro-dental features [192–195]
Facial features:

- Relatively underdeveloped premaxilla
- Small maxilla
- Relative mandibular prognathism
- Receding upper lip
- Sunken appearance of the mouth
- Broad nose
- Hypertelorism

Ocular findings:

- Mild hypertelorism with mild telecanthus
- Hypoplasia and flattening of the anterior surface of the iris
- Anterior synechiae running from the iris to the back of the cornea across the anterior chamber, with or without an abnormal iridiocorneal angle
- Dyscoria or slit-like pupils as a result from traction of these synechia
- Aniridia
- Optic atrophy
- Microcornea and corneal opacity
- Ectopic pupil
- Secondary juvenile glaucoma that may lead to total blindness

Sensory hearing loss
Dental findings:

- Marked hypodontia
- The teeth most commonly missing are the maxillary incisors
- Conical crown form
- Microdontia
- Enamel hypoplasia
- Teeth malposition

References

[182–189,192–196]

Pictures Illustrating the Oro-Dental Features

Figure 2.5

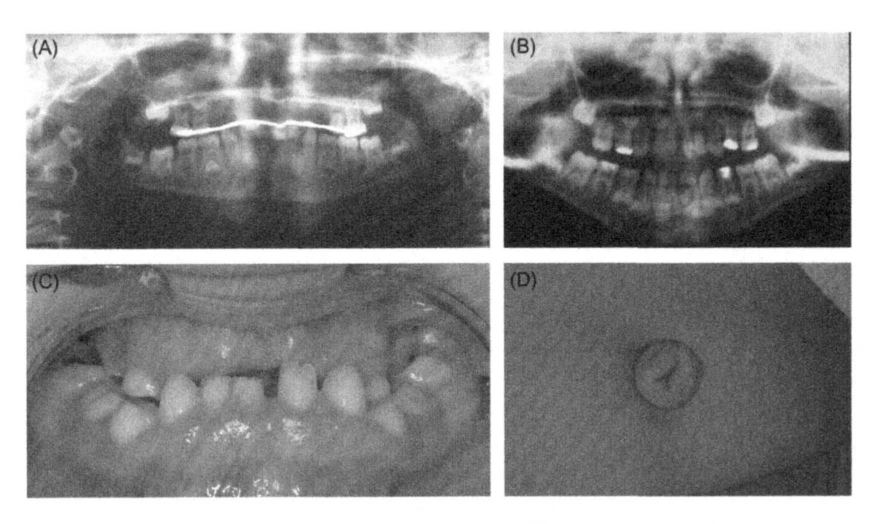

Figure 2.5 Rieger syndrome. (A) On the X-ray, the maxillary temporary and permanent incisors and canines are missing. 15, 24, 25, 17 and 27 tooth germs are also absent. The mandibular incisors (except 32) and canines, as well as 35 and 45, are also missing. (B) X-ray from another patient with oligodontia and absent permanent maxillary incisors and canines and second premolars. (C) Note the hypoplastic maxilla and the absent permanent upper and some lower incisors. (D) a protruding ombilic (A,C,D correspond to the same patient).

2.2 WNT Signalling Pathway

Syndrome	Associated Features	Gene	Molecules	Inheritance	Locus
WNT signalling pathway					
Oligodontia with neoplasia #608615	Colorectal neoplasia	*AXIN2*	WNT signalling regulator	AD	17q24
Odonto-onycho-dermal dysplasia (OODD; MIM #257980) [197,198]	Severe hypodontia, nail dystrophy, smooth tongue, dry skin, keratoderma and hyperhidrosis of palms and soles	*WNT10A*	Signalling molecule	AR	2q35
Schopf–Schulz–Passarge syndrome #224750	Numerous cysts along eyelid margins, benign and malignant skin tumours, palmoplantar keratosis and hypodontia	*WNT10A*	Signalling molecule	AR	2q35

(*Continued*)

(*Continued*)

Syndrome	Associated Features	Gene	Molecules	Inheritance	Locus
Isolated hypodontia Tooth agenesis selective 4 [199] #150400	Variable hypodontia involving the lateral incisors and premolar teeth	*WNT10A*	Signalling molecule	AD	2q35
Ectodermal dysplasia and oligodontia Bailleul-Forestier et al. personal communication 2011	Severe oligodontia, severe alveolar bone defect	*LEF1*	Transcription factor Involved in lef1/beta-catenin pathway with a negative feedback regulation on EDA/EDRA axis	?	4q25

The WNT gene family consists of structurally related genes encoding secreted signalling molecules that have been implicated in oncogenesis and in several developmental processes, including regulation of cell fate and patterning during embryogenesis.

Intracellular signalling of the WNT pathway (Figure 2.6) diversifies into at least three branches: (1) the beta-catenin pathway (canonical WNT pathway), which activates target genes in the nucleus; (2) the planar cell polarity pathway, which involves jun N-terminal kinase (*JNK*) and cytoskeletal rearrangements and (3) the WNT/Ca^{2+} pathway.

WNTs are secreted glycoproteins that bind to frizzled transmembrane receptors.

Canonical WNT signals are transduced through frizzled family receptors and LRP5/LRP6 coreceptor to the beta-catenin signalling and involve the Axin2 protein within the cytoplasm.

AXIN2 is an inhibitor of the WNT signalling pathway. It downregulates beta-catenin.

AXINs antagonize beta-catenin => loss-of-function mutations stimulate beta-catenin.

AXIN2 realizes a negative feedback loop.

Through *AXIN2* haploinsufficiency, failure of feedback regulation of WNT signalling induces failure of development of the permanent teeth.

Stimulation of WNT signalling as in Gardner syndrome and familial adenomatous polyposis (gene *APC* loss of function) will promote formation of extra teeth (see further).

WNT/beta-catenin and *EDA/EDAR/NF*-kappa B signalling pathways seem interdependent [200].

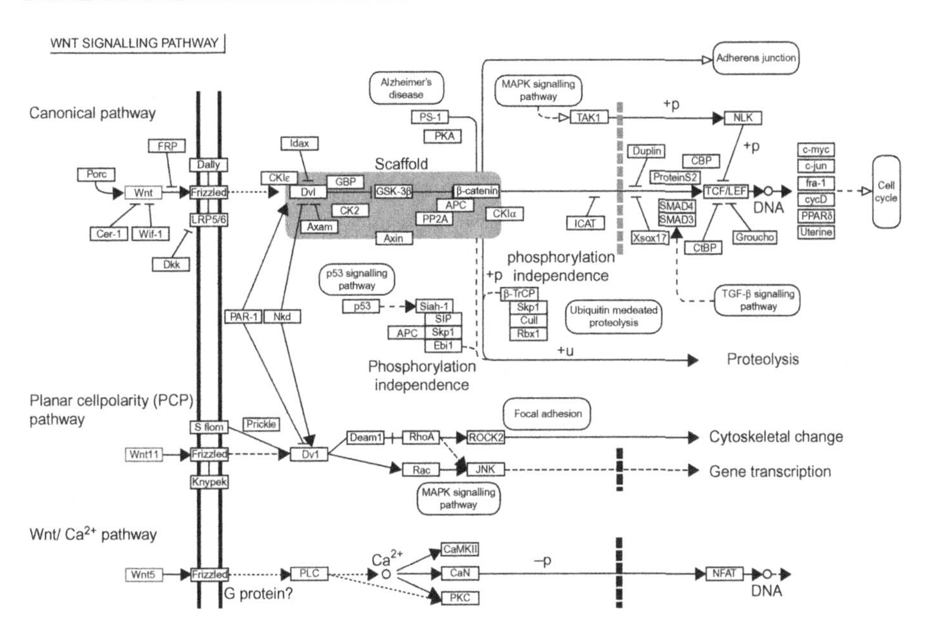

Figure 2.6 WNT signalling pathway KEGG database.

2.2.1 Oligodontia and Colorectal Cancer Syndrome

Definition

This disease combines susceptibility to colorectal cancer (with precancerous or existing lesions) and hypodontia.

OMIM Number

#608615

Prevalence

Not yet determined.

Inheritance

Autosomal dominant.

Aetiology – Molecular Basis – Gene

The gene involved, *AXIN2* (17q24), is a tumour suppressor (Conductin) protein negatively controlling the WNT signalling pathway. Mutations are expected to activate WNT signalling [128,201].

Animal Models/Main Features

Targeted disruption of *Axin2* in mice induces malformations of skull structures, a phenotype resembling craniosynostosis in humans [202].

Clinical Description

Lammi [128] identified a Finnish family in which severe permanent tooth agenesis (oligodontia) and colorectal neoplasia segregate with dominant inheritance. Eleven members of the family lacked at least eight permanent teeth, and two members developed only three permanent teeth. The teeth involved were permanent molars, premolars, lower incisors and upper lateral incisors. Three individuals lacked all canines. The teeth present were upper central incisors, usually canines, first premolars and first molars. Most affected persons had a normal primary dentition.

Colorectal cancer or precancerous lesions of variable types were found in eight of the patients with oligodontia. The entity was found to be caused by *AXIN2* mutations [203].

AXIN2 seems also involved in sporadic forms of common incisor agenesis [204] or isolated oligodontia [205].

Management/Oral Health

If there are missing teeth together with a family history of colorectal cancer, patients and their families should be referred to a geneticist for further investigation. Oral findings could lead to the diagnosis and considerably help and improve survival and quality of life for the affected individuals. Hypodontia should be managed through specialized clinics, and treatment involves paediatric dentistry, orthodontics, prosthetics, implantology and aesthetic dentistry.

References

[127,128,201,203–206]

Pictures Illustrating the Oro-Dental Features

Figure 2.7

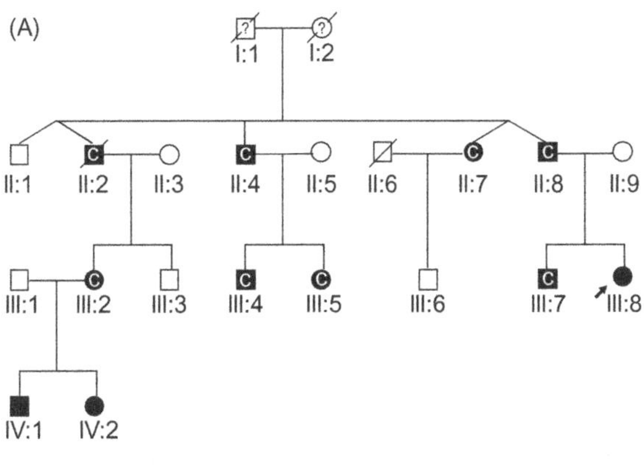

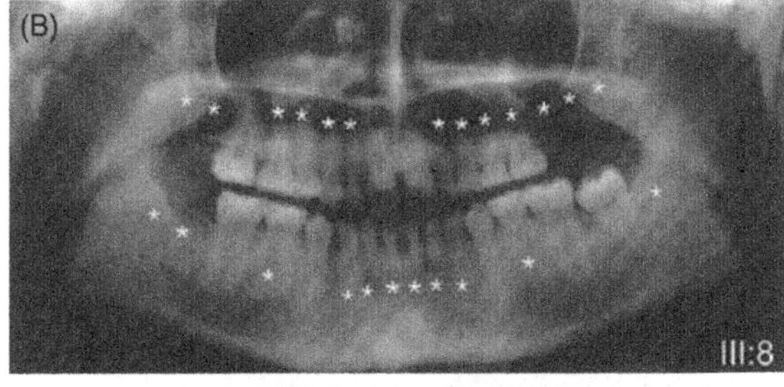

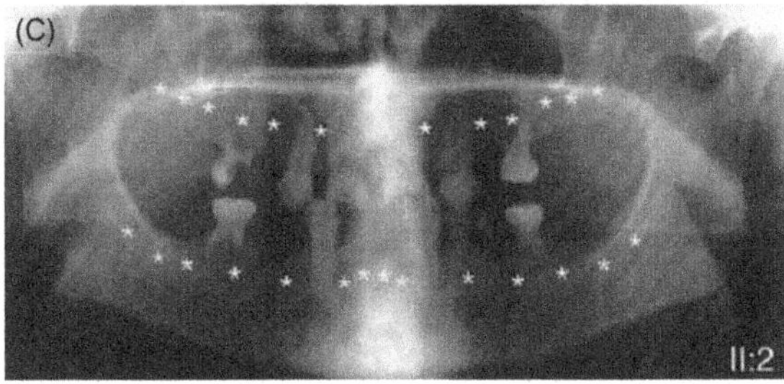

Figure 2.7 Pedigree of the four-generation family and oligodontia phenotypes. (A) Pedigree showing autosomal dominant inheritance of oligodontia and colorectal neoplasia. (B) and (C) Panoramic radiographs of the proband (III:8) at the age of 15 years (B) and her uncle (II:2) at the age of 34 years. (C) Stars indicate congenitally missing permanent teeth. Note that both family members have many persisting deciduous teeth. A persistent deciduous tooth is an often encountered finding when the corresponding permanent tooth has not developed. ★missing permanent tooth. Mutations in *AXIN2* cause familial tooth agenesis and predispose to colorectal cancer.
Source: From Ref. [128] Reprinted from Publication, Copyright (2004), with permission from Elsevier.

2.2.2 Odonto-Onycho-Dermal Dysplasia (OODD)

Definition

This syndrome is characterized by severe hypodontia, nail dystrophy ranging from onychodysplasia to total absence of nails, smooth tongue, dry skin, keratoderma and hyperhidrosis of palms and soles.

OMIM Number

#257980

Prevalence

Not yet determined.

Inheritance

Autosomal recessive.

Aetiology – Molecular Basis – Gene

The disorder can be caused by mutation in the *WNT10A* gene (2q35).

WNT10A mutations cause not only OODD but also other forms of ectodermal dysplasia, varying from apparently monosymptomatic severe oligodontia to Schopf–Schulz–Passarge syndrome (#224750) which associates numerous cysts along eyelid margins, increased risk to develop benign and malignant skin tumours, palmoplantar keratosis and hypodontia [207].

WNT10A can also cause isolated hypodontia [199].

Animal Models/Main Features

None reported in the literature. WNT/beta-catenin signalling is absolutely required for NF-kappa B activation, and Edar is a direct WNT target gene [200].

Clinical Description

Main features

- Absent hair
- Thin hair
- Sparse eyebrows
- Dystrophic to absent nails
- Palmar erythema
- Keratosis pilaris
- Keratoderma (palms and soles)
- Hyperhidrosis (palms and soles)
- Hyperkeratosis

Cranio-oro-dental features

- Severe hypodontia
- Smooth tongue with marked reduction of fungiform and filiform papillae

References

[197–200,207,208]

Pictures Illustrating the Oro-Dental Features

Figure 2.8 [787]

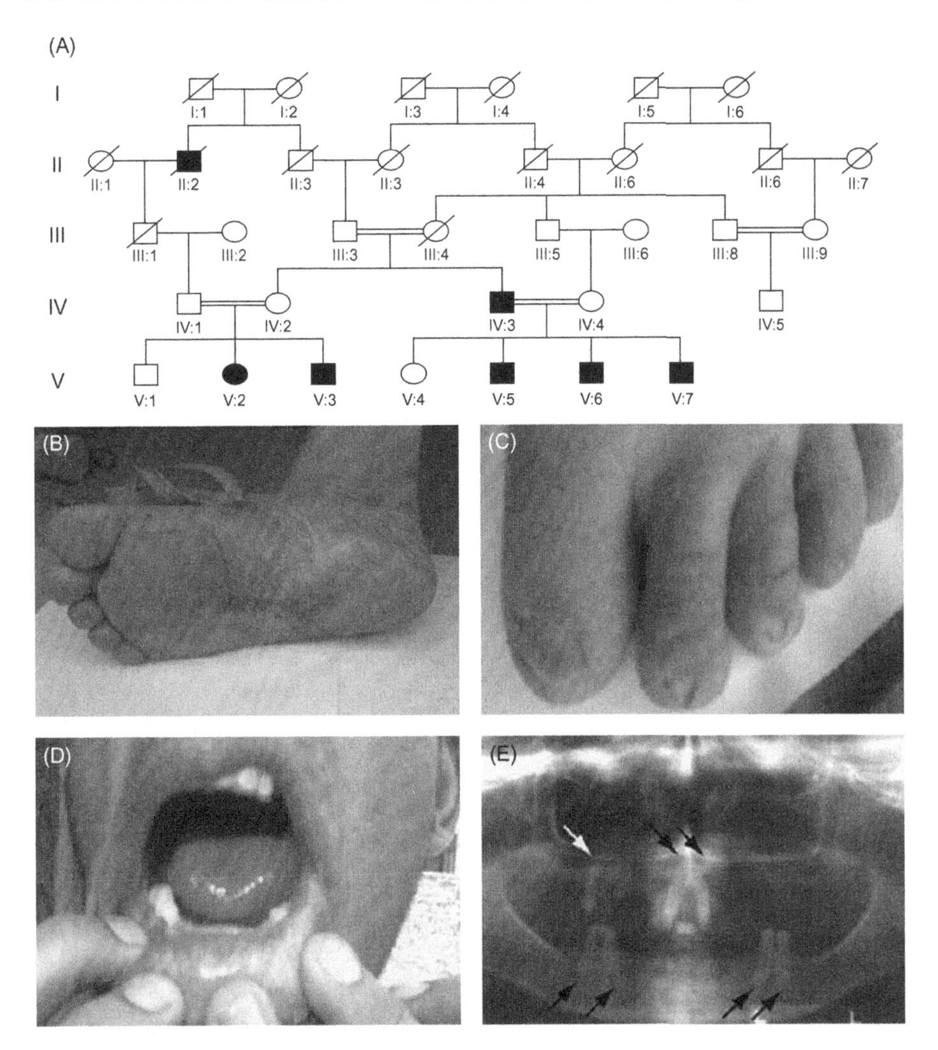

Figure 2.8 OODD. Pedigree segregating OODD and clinical features. (A) Several consanguinity loops are shown, and the OODD phenotypes appear in the two successive generations IV and V (right loop). Filled symbols denote affected individuals. (B) Sole of right foot in individual V:7 at the age of 37 years, illustrating plantar keratoderma. (C) Dystrophic toenails in individual V:2 at the age of 48 years. (D) Smooth tongue with reduced papillae in individual V:6 at 23 years of age. (E) Panoramic radiograph of the affected male V:3 at 39 years of age, showing agenesis of most permanent teeth, third molars excluded. In the upper jaw, there are two permanent central incisors and one permanent premolar with conical crown shape (upper arrows). In the lower jaw, there are four permanent premolars with reduced crown form (bottom arrows). WNT10A missense mutation associated with a complete odonto-onycho-dermal dysplasia syndrome.

Source: From Ref. [197] Reprinted by permission from Macmillan Publishers Ltd, copyright (2009).

2.3 TNF/NF-Kappa B Signalling Pathway

Syndrome	Associated Features	Gene	Molecules	Inheritance	Locus
TNF/NF-kappa B signalling pathway					
Hypohidrotic ectodermal dysplasia (HED) #305100	Sparse hair/absent sweating	*EDA* (ectodysplasin)	TNF signal	X	Xq12–q13.1
Non-syndromic oligodontia Selective tooth agenesis 1 #313500 [209]	Involving all classes of teeth Missing anterior teeth, lower and upper incisors	*EDA* (ectodysplasin)	TNF signal	X	Xq12–q13.1
Hypohidrotic ectodermal dysplasia #129490		*EDAR*	TNF-receptor, death domain protein	AD and AR	2q11–q13
Hypohidrotic ectodermal dysplasia #129490		*EDARADD* *Associated death*	Death domain adapter protein *domain*	AD	1q42.2–43
Non-syndromic oligodontia		*EDARADD*	Death domain adapter protein *domain*	AD	1q42.2–43
HED-ID #300291	Anhidrotic ectodermal dysplasia, immunodeficiency	*IKKgamma*	Activator of transcription factor	X	Xq28
HED-ID #300291	Anhidrotic ectodermal dysplasia, immunodeficiency	*NFKBIA, IKBA 164008*	NFKB complex is inhibited by I-kappa-B proteins (NFKBI A and B)	AD	14q13
Incontinentia pigmenti #308300	Increased skin pigmentation, eye, neurological problems	*IKKgamma, NEMO*		X	Xq28
Other ectodermal dysplasias					
EEC3 #604292	Ectodermal dysplasia, ectrodactyly, cleft palate	*P63*	Transcription factor	AD	3q27
Rapp Hodgkin #129400	Anhidrotic ED with cleft lip/palate	*P63*	Transcription factor	AD	3q27
Acro-dermato-ungual-lacrimal-tooth (ADULT) syndrome #103285 is probably the same as EEC	Ectodermal dysplasia Lacrimal duct obstruction Conjunctivitis Dysplastic nails	*P63*	Transcription factor	AD	3q27
Limb-mammary syndrome #603543	Lacrimal duct stenosis	*P63*	Transcription factor	AD	3q27
	Cleft palate bifid uvula Hypodontia				

	Small/absent breasts and nipples Hands and feet anomalies Ectrodactyly, etc.				
Ankyloblepharon-ectodermal defects-cleft lip/palate	Scalp erosion/infection	*P63*	Transcription factor	AD	3q27
Hay–Wells syndrome #106260	Ankyloblepharon filiforme adnatum; hypodontia; underdeveloped maxilla and cleft lip/palate				
CLPED1 #225060	Cleft lip/palate Ectodermal dysplasia	*PVRL1* (nectin-1)	Cell adhesion molecule	AR	11q23–q24
Ellis van Creveld syndrome #225500	Neonatal teeth, hypodontia, eruption, short limbs, short ribs, postaxial polydactyly and dysplastic nails and teeth [304]. Congenital cardiac defects	*EVC1, EVC2*	Cilia transmembrane proteins positive regulators of shh signalling	AR	4p16
Weyers acrofacial dysostosis #193530	Single central incisor, conical teeth Irregular, small, or absent permanent incisors Postaxial polydactyly, short limbs	*EVC1*		AD	4p16
Van der Woude syndrome #119300	Pits and/or sinuses of the lower lip, and cleft lip and/or cleft palate	IRF6	*Interferon regulatory factor-6*	*AD*	*1q32–q41*
OFDI #311200	Face, oral cavity, and digits brain, polycystic kidneys	*OFDI (Cxorf5/71-7a)*	Protein containing coiled-coil alpha-helical domains	X	Xp22
Williams (-Beuren) syndrome	Intellectual disability, unusual face, aortic stenosis, unusual behavior Hypodontia, microdontia, enamel hypoplasia	*ELN* *RFC2* *LIMK1* *GTF2IRD1/GTF21* *FKBP6*	Elastin LIM kinase	AD	7q11.23 Contiguous gene deletion syndrome
Oculodentodigital syndrome (ODD)	Unusual face, hearing loss, microphtalmia, cleft lip/palate, hypodontia, enamel defects, taurodontism, hyperkeratosis, fine hair	*GAJ1*	Connexin43	AD	6q21–q23.2

The transcription factor NF-kappa B regulates the expression of genes controlling the immune and stress responses, inflammatory reaction, cell adhesion and protection against apoptosis.

Members of the tumour-necrosis factor receptor (*TNFR*) family that contain an intracellular death domain initiate signalling by recruiting cytoplasmic death domain adapter proteins.

Ectodysplasin (*EDA*), a type II transmembrane protein with tumour necrosis factor alpha domains, binds to the membrane-bound TNF-receptor *EDAR* (ectodysplasin receptor), which recruits the intracellular adaptor *EDARADD* (*EDAR*-associated death domain). *EDARADD* binds *TRAF*, which lead to activation of the *IKK* complex (*NEMO* = *IKK*γ, *IKK*α, *IKK*β). The complex phosphorylates *IKB*, marking it for degradation, which liberates the NF-kappa B transcription factor, allowing it to move to the nucleus and activate its target genes.

The *NEMO-NF*kappa B signalling pathway is utilized by several groups of immune receptor: members of the tumour necrosis factor alpha (*TNF*-α) receptor superfamily (*TNFR*), represented by CD40 and *TNFR*, antigen receptors (*TCR* and *BCR*) and members of the Toll-interleukin-1 receptor (*TIR*) superfamily (IL-1Rs/TLRs). Toll-like receptors (TLRs) are engaged by well-defined chemical agonists that mimic microbial compounds, raising the possibility that human TLRs play a critical role in protective immunity *in vivo* [210].

The NF-kappa B pathway is essential for the development of hair follicles, teeth, exocrine glands, mammary glands and other ectodermal derivatives [43,45,62,211,212].

Dysfunction of this pathway, altering epithelio-mesenchymal interactions, causes ectodermal dysplasias that show sparse hair, a lack of sweat glands and malformation of teeth. Ectodermal dysplasias form a heterogeneous group of disorders with more than 170 different phenotypes described.

They can be classified in different groups [213–215] based on clinical and molecular findings:

Group 1 is associated with disruption of epithelio-mesenchymal interactions and either of the following:

- Major ectodermal implications (HED) NF-kappa B pathway associated with immune or neurological anomalies (incontinentia pigmenti, HED-ID [immune deficit]) or with skeletal manifestations EEC, p63 syndrome or TDO tricho-dento-osseous syndrome. p63 is a homologue of the tumour suppressor p53. The p53 family of transcription factors consists of three members: p53, p63 and p73. Ectodermal organs such as hairs, whiskers, teeth and several glands, including mammary, salivary and lacrimal glands, are lacking in p63-deficient mice. p63 regulates many genes in different signalling pathways involved in placode initiation [216,217].
- Endocrine abnormalities

Group 2 is associated with causative molecules involved in the cytoskeleton and cellular stability like connexins, adhesion molecules: Ectodermal dysplasias, with the following:

- Hyperkeratosis, Clouston syndrome (connexin)
- Hearing deficit (connexin)
- Cleft lip and palate
- Retinal dystrophy

2.3.1 Hypohidrotic Ectodermal Dysplasia (HED)

Definition

Patients with hypohidrotic ectodermal dysplasia (HED) present variable degrees of hypodontia or oligodontia, hypohidrosis (decreased ability to sweat) and hypotrichosis (sparse hair). There are three varieties which are clinically similar but with different modes of inheritance, as noted here.

OMIM Number

#305100	X-linked recessive, (Christ–Siemens–Touraine) syndrome, ED1
#224900	autosomal recessive form, ED2
#129490	autosomal dominant form, ED3

Prevalence

Estimated at 1 in 17,000 births. The X-linked form is the most common, concerning almost 80% of ED. The incidence in males is 1 in 100,000, and in carrier females, 17.3 in 100,000.

Inheritance [218–220]

- Genetic heterogeneity, in the majority of families it is inherited as an X-linked recessive disease, but in some families, as an autosomal dominant, and still in other families, as autosomal recessive; therefore, affected females do occur.
- Clinical manifestations are identical for the X-linked and the autosomal recessive forms but markedly variable for the autosomal dominant form.
- In the X-linked form, only males are affected. It is inherited through carrier females who display no to limited clinical manifestations.

Aetiology – Molecular Basis – Gene [221–225]

- The X-linked form has been mapped to the long arm of the X chromosome (Xq12–q13.1).
- The *ED1* or *EDA1* gene from that region encodes a transmembrane protein, Ectodysplasin A.
- Both the autosomal recessive and autosomal dominant forms are caused by mutations in the ectodysplasin 1 receptor (*EDAR* = *EDA3*) gene (2q11–q13) and the EDAR-associated death domain (*EDARADD*) gene (1q42.2–q43). *WNT10A* (2q35) is also involved in hypohidrotic/anhidrotic ED.
- *EDAR* is implicated in about 25% of non-*ED1 HED* [226].

Only four genes (*EDA1, EDAR, EDARADD* and *WNT10A*) account for no less than 90% of hypohidrotic/anhidrotic ectodermal dysplasia cases [227].

Animal Models/Main Features

The tabby mouse (*eda*) [228,229], the downless mouse (*edar*) and the crinkled mouse (*edaradd*) [62,230].

Clinical Description

Main features

- No hair at birth.
- The skin is dry and wrinkled because of decreased sweating.
- There is a total absence of body hair.
- Nails are slightly underdeveloped, mainly in the toes and pinky.
- The almost total absence of sweat glands is responsible for altered body temperature regulation, especially if these patients exercise actively or if they live in hot climates.
- There have been several reports of patients with HED dying as consequence of hyperthermia.
- Affected individuals have few sweat pores, which are abnormal both in distribution and size.
- Carrier females have fewer sweat pores, which are smaller in size than normal sweat pores.
- Lyonization is well demonstrated by the number and arrangement of sweat pores. Normally, the palms of the hands have anywhere from 20 to 40 sweat pores per linear millimetre. The pores are located in the flexion creases and are just about the same size and separated by about the same distance.

Cranio-oro-dental features [231–234]

- Almost total absence of scalp hair
- Prominent forehead
- Eyelashes and eyebrows are minimally formed or absent
- Depressed nasal bridge
- Periorbital hyperpigmentation
- Markedly protruded lips
- Narrow and underdeveloped maxilla
- Small palatal depth
- Almost total absence of teeth formation in both dentitions (oligodontia, anodontia)
- Teeth present might be conical
- Generally, poorly calcified enamel on the few teeth that are formed
- Taurodontism
- No alveolar bone formation
- Medullary bone hyperdensity, including in the mandibular symphysis area
- Loss of vertical dimension
- Salivary secretion is diminished in some patients as well as tear production.

Management/Oral Health [235–239]

- Early in life, either partial, complete or overlay dentures which are replaced as the patient grows should be constructed.
- Dental implants in the anterior mandible can be used as early as 6 to 7 years of age.
- Dental implants in other locations, to support dentures, can be placed as early as 14 or 15 years of age. The age for dental implants placement is, however, still debated in the literature.

Therapy for this disease, using recombinant replacement ectodysplasin, is being developed [240,241].

References

[218–227,231–237,239,241–245]

Pictures Illustrating the Oro-Dental Features

Figures 2.9 and 2.10

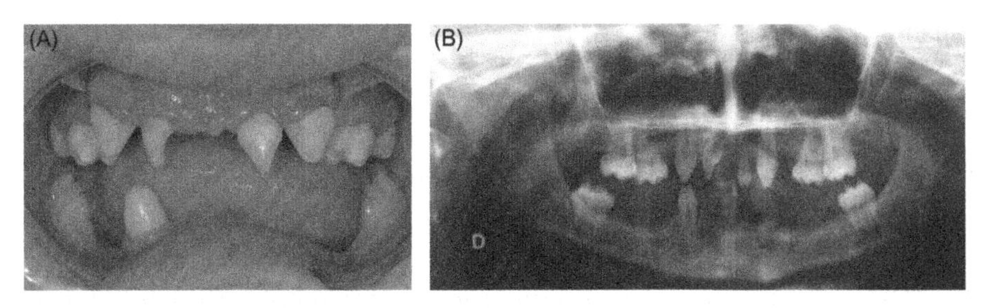

Figure 2.9 (A and B) Hypohydrotic ectodermal dysplasia, X-linked hypohydrotic ectodermal dysplasia with *EDA* gene mutation (exons 4–9 deletion). Reference Centre for Orodental Manifestations of Rare Diseases, Pôle de Médecine et Chirurgie Bucco-dentaires, Hôpitaux Universitaires de Strasbourg, Université de Strasbourg, F Clauss.

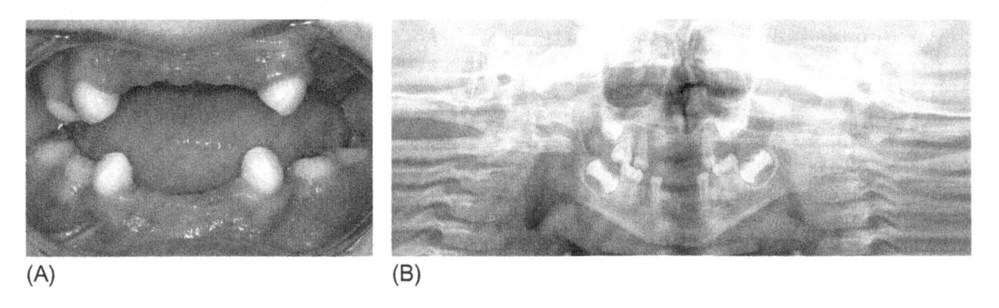

Figure 2.10 (A and B) Hypohydrotic ectodermal dysplasia with *EDAR* gene mutation. Reference Centre for Orodental Manifestations of Rare Diseases, Pôle de Médecine et Chirurgie Bucco-dentaires, Hôpitaux Universitaires de Strasbourg, Université de Strasbourg, F Clauss.

2.3.2 Hypohidrotic Ectodermal Dysplasia with Immune Deficiency (HED-ID)

Definition

The syndrome consists of abnormal development of ectoderm-derived structures such as skin, skin appendages and teeth, and a broad spectrum of infectious diseases [246].

OMIM Number

#300291	X linked
#612132	AD
	anhidrotic ectodermal dysplasia with T-cell immunodeficiency

Prevalence

Not yet determined.

Inheritance

X-linked recessive, AD

Aetiology – Molecular Basis – Gene

The gene involved is *NEMO*: Xq28, *IKBA*: 14q13) [210,247–252].

Mutations are found in *NEMO* (IKKgamma) (NF-kappa B essential modulator). This gene codes for a protein which plays a crucial role in the canonical nuclear factor-kappa B (NF-kappa B) signalling pathway as the regulatory component of the IKK (I-kappa B kinase) complex. This leads to impaired, but not abolished, NF-kappa B activation in response to a variety of stimuli, including Toll-like receptor agonists. There is impaired Toll-like receptor (TLR) signalling, resulting in a broad spectrum of infectious diseases.

There are either X-linked recessive hypomorphic mutations in *NEMO* or autosomal dominant hypermorphic mutations in *IKBA*(=*NFKBIA*, located at 14q13) #612132.

Animal Models/Main Features

Heterozygous female mice developing patchy skin lesions and male lethality are hallmarks of incontinentia pigmenti (*NEMO*). The animals also have an inflammatory-bowel-disease-like phenotype [253,254].

Clinical Description

Main features [255]

- Increased susceptibility to pyogenic bacteria, viruses and non-pathogenic mycobacterial infections
- Inflammatory bowel disease
- Thin hair/ underdeveloped glands

Cranio-oro-dental features

- Cone-shaped teeth and hypodontia, oligodontia

Management/Oral Health

Multidisciplinary management.

References

[210,246–252,255–258]

2.3.3 Incontinentia Pigmenti

Definition

This syndrome, also known as Bloch–Sulzberger syndrome is characterized by vesicles, subsequently verrucae and later on pigmented macules of the skin, and anomalies involving eyes, skeletal system, central nervous system and teeth.

OMIM Number

#308300

Prevalence

- Nearly all affected individuals are women (95%).
- A few affected males have been reported, but they died shortly after birth or within the first year of life.
- Prevalence is 0.2 in 100,000 [259].

Inheritance

X-linked dominant, usually lethal prenatally in male.

Aetiology – Molecular Basis – Gene

- The responsible gene is *NEMO* (also named *IKK*-gamma gene *IKBKG*, NF-kappa B essential modulator) which maps to the long arm of the X chromosome (Xq28).
- Incontinentia pigmenti is caused in 80% of cases by a single mutation in *NEMO*, resulting in a deletion of exon 4 through 10 of the *NEMO* gene.
- About 50% of IP cases have a positive family history.
- Ninety-five per cent of cases are females; affected males are probably the result of spontaneous mutations.

Animal Models/Main Features

Many animal models exist for this disease. They reproduce the phenotype with shorter life span, immune, haematopoietic, pigmentation and skin/coat/nails deficits. Male lethality is also observed in animal models [253,260–262].

Clinical Description

Main features [263,264]

- Physical growth is within normal limits.
- Mental handicap of variable degree has been observed in about 30% of affected patients.
- Linear or grouped vesicles containing brownish-coloured serum appear on the skin of the extremities during the first weeks of life.

- These vesicles disappear by the end of the first month.
- The vesicles may recur and they may be replaced by violaceous papules or inflammatory lesions or both.
- Pigmented macules, brownish-grey in colour, with a reticular or whorl pattern, generally present at birth, occur over the trunk and extremities.
- The pigmentation tends to disappear by the second year of life.
- Hyperkeratotic warty lesions usually appear on the dorsal surface of the digits, knuckles and joints by the second month of life, but they disappear after 3 or 4 months.
- Occasional findings are dystrophic fingernails, subungeal fibromas and breast asymmetry.
- Syndactyly, supernumerary ribs, hemiatrophy and shortening of the legs and arms are present in about 20% of patients.

Cranio-oro-dental features [265–267]

Craniofacial features:

- Hydrocephalus, spastic and lax paralysis, microcephaly and convulsive episodes in 20–25% of affected persons.
- Alopecia areata is present in 20–25% of affected persons.
- Narrow, highly arched palate

Ocular findings:

- Paresis of eye muscles is seen in 35–40% of patients.
- Cataracts, strabismus and optic atrophy are seen in 25–35% of patients.

Dental findings:

- Occur in more than 80% of all patients, 65% having major dental anomalies
- Delayed tooth eruption
- Hypodontia, oligodontia
- Supernumerary teeth
- Retained teeth
- Pegged teeth
- Teeth with conical crowns
- Elongated crowns in anterior teeth
- Malformed teeth
- Tulip-shaped maxillary permanent central incisors
- Supernumerary cusps
- Diastemas
- Short roots
- These teeth findings are present in both dentitions.

References

[249,256,263–272]

Pictures Illustrating the Oro-Dental Features

Figure 2.11

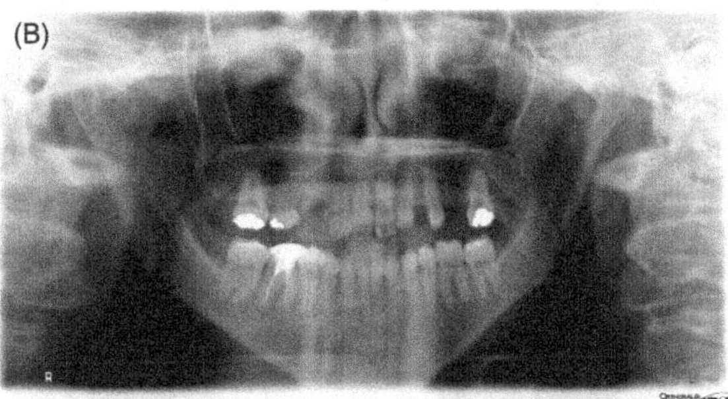

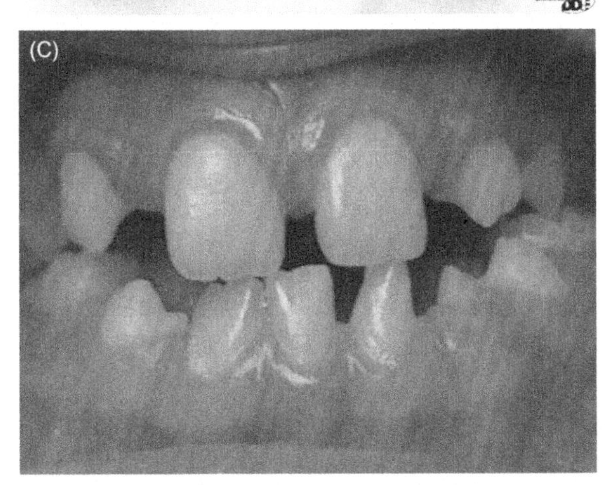

Figure 2.11 Incontinentia pigmenti. (A) Skin pigmented lesion encountered in IP. (B) OTP of patient affected with IP. The following teeth are missing: 13, 14, 15, 17, 18, 23, 25, 27, 28, 43, 45, 48, 33 and 38. (C) Intraoral view of the same patient at age 11.

2.3.4 Ectrodactyly, Ectodermal Dysplasia and Clefting (EEC)

Definition

This syndrome, characterized by hands and feet anomalies, various manifestations of ectodermal dysplasia and clefting of lip and palate.

OMIM Number

#129900	EEC1
#602077	EEC2
#604292	EEC3

Prevalence

1 in 50,000 live births.

Inheritance

- It is autosomal dominant, with reduced penetrance and variable expressivity.
- Fifty per cent of cases are isolated examples.

Aetiology – Molecular Basis – Gene

- EEC1 7q11.2–q21.3
- EEC2 19p13 (debated) This family has a p63 mutation as well
- EEC3 3q27, missense mutations of the p63 (TP73L) gene [46,273]
- Allelic with ADULT syndrome (#103285), split hand/foot malformation 4 (#605289), Rapp–Hodgkin syndrome (**#129400**), Hay–Wells syndrome (**#106260**) and limb-mammary syndrome (**#603543**). More and more it becomes clear that this distinction in subtypes is artificial.

Animal Models/Main Features

Animal models display limb anomalies and defects of ectodermally derived tissues [216,274–276]. The ectodermal placodes that mark early tooth and hair follicle morphogenesis do not form in *p63*-deficient embryos [217].

Clinical Description

Main Features

- Ectrodactyly, the congenital absence of one or more fingers, in 84% of patients; called 'lobster-claw anomaly' because the hands or feet of those affected can appear claw-like
- Ectodermal dysplasia with skin, hair, nails and teeth anomalies
- Small thin and malformed nails in hands and feet in 80% of cases
- Genitourinary tract anomalies
- Conductive hearing loss
- Growth hormone deficiency

Cranio-oro-dental features

Craniofacial features:

- Dry, hard hair, eyebrows and eye lashes
- Broad nasal tip
- Mild malar underdevelopment
- Choanal atresia
- Short philtrum

Ocular findings:

- Absent lacrimal puncta in 87% of cases
- Dacryocystitis

Oro-dental findings:

- Bilateral cleft lip/palate in 60–75% of patients
- Isolated cleft palate in 10% of cases
- Absent permanent teeth
- Conoid teeth
- Enamel hypoplasia
- Xerostomia due to atresia of the Stensen ducts

References

[46,216,273,275–285]

Pictures Illustrating the Oro-Dental Features

Figure 2.12

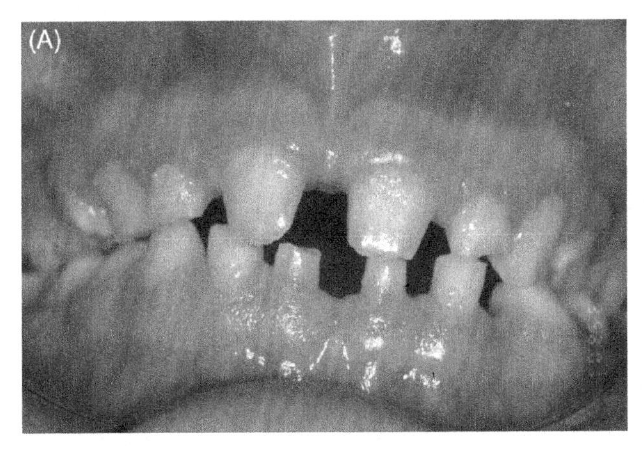

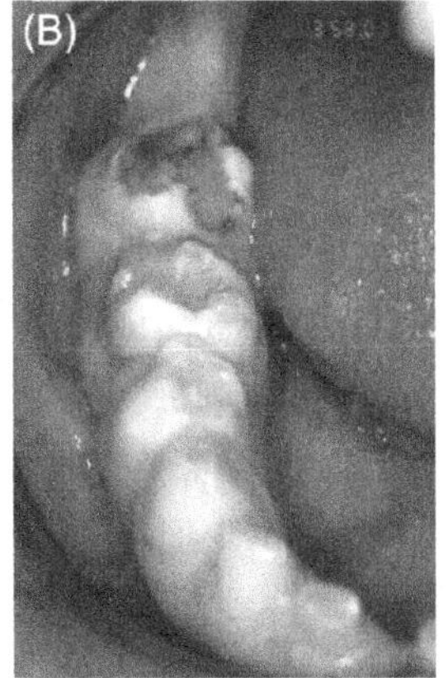

Figure 2.12 EEC. (A) Oligodontia, microdontia and teeth shape anomalies encountered in an EEC-affected patient. (B) Enamel defects/hypoplasia are also present.

2.3.5 CLPED1 Syndrome (Recessive Cleft Lip/Palate – Ectodermal Dysplasia; Zlotogora–Ogur Syndrome; Ectodermal Dysplasia, Cleft Lip and Palate, Learning Disability, and Syndactyly; Rosselli–Gulienetti Syndrome)

Definition

The syndrome consists of cleft lip and palate, ectodermal dysplasia, learning disability and syndactyly.

OMIM Number

#225000

Prevalence

Not yet determined.

Inheritance

Autosomal recessive.

Aetiology – Molecular Basis – Gene

PVRL1 (poliovirus receptor-like 1) gene (11q23–q24) encodes Nectin-1, an immunoglobulin (Ig)-related transmembrane cell–cell adhesion molecule that is a component of adherens junctions. Nectin-1 is also the principal cell surface receptor for alpha-herpes viruses.

PVRL1 variants contribute also to non-syndromic cleft lip and palate in multiple populations [286].

Animal Models/Main Features

Mice lacking nectin-1 exhibit defective enamel formation [287]. Cooperation of nectin-1 and nectin-3 is required for normal ameloblast function and crown shape development in mouse teeth [288].

Clinical Description

Main features

- Cleft lip and palate
- Ectodermal dysplasia
- Learning disability
- Syndactyly

Cranio-oro-dental features

- Micrognathia,
- Cleft lip/palate
- Hypodontia, anodontia
- Microdontia

References

[286–297]

Pictures Illustrating the Oro-Dental Features

Figure 2.13

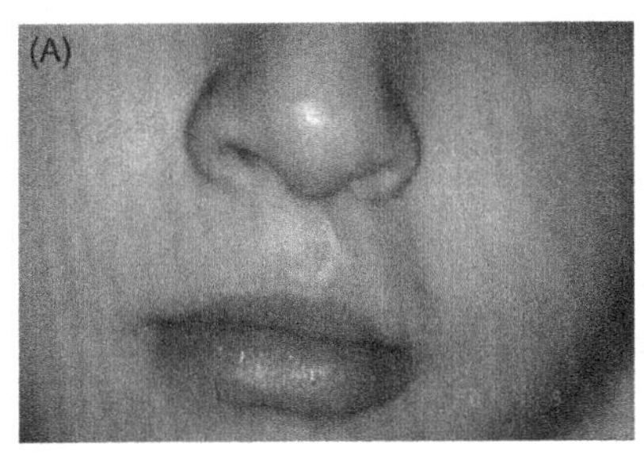

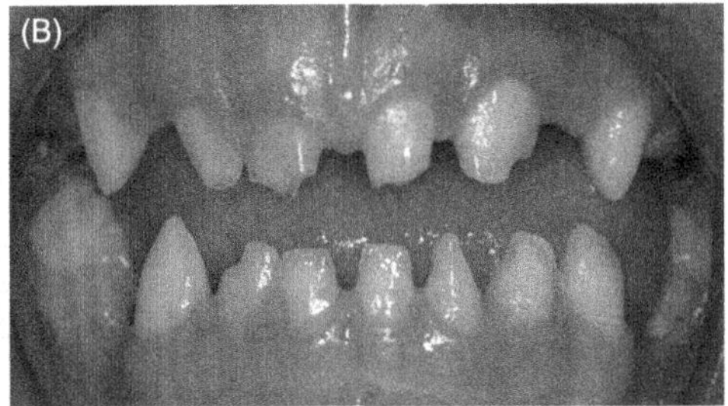

Figure 2.13 CLPED1. (A) Repaired CLP. (B) Intraoral pictures showing a clinical phenotype overlapping with oral manifestations of ectodermal dysplasia with oligodontia and size and shape tooth anomalies.

2.3.6 Ellis-van Creveld Syndrome (EVC)

Definition

Previously known as chondro-ectodermal dysplasia, EVC is a rare, inherited syndrome characterized by acromesomelic dwarfism, polydactyly and hidrotic ectodermal dysplasia symptoms, with malformed nails, hair defects and typical oro-dental findings.

OMIM Number

#225500

Prevalence

- Prevalence is estimated at 0.9 in 100,000 [298].
- There are over 150 cases reported worldwide [259].
- EVC is the most common form of dwarfism in the Amish population.

Inheritance

Autosomal recessive with variable expression.

Aetiology – Molecular Basis – Gene

- The syndrome is due to mutations in the *EVC1* gene (4p16 region) and also could be the result of mutations in the *EVC2* non-homologous gene (4p16) located near, but head to head, to the *EVC1* gene.

The *EVC2* gene encodes a protein predicted to have one transmembrane segment, three coiled-coil regions and one RhoGEF domain. The EVC2 protein (LIMBIN) has significant sequence homology with the tail domains of class IX non-muscle myosins.

- Mutations of the *EVC1* or *EVC2* gene are also responsible for Weyers acrodental dysostosis. The phenotype associated with the mutations in these two genes is indistinguishable.

- Weyers acrodental dysostosis (**#193530**) (autosomal dominant) and Ellis-van Creveld syndrome are allelic disorders.
- Both Evc1 and Evc2 proteins localize to the basal bodies of primary cilia. Hedgehog signalling is impaired in the absence or malfunction of Evc1 and Evc2 [299,300].

Animal Models/Main Features

Evc (−/−) mice develop an EVC-like syndrome, including short ribs, short limbs and dental abnormalities [301].

Clinical Description

Main features

- Dwarfism characterized by short forearms and legs; less marked shortening of the humerus and femur
- Polydactyly of hands (six fingers in each hand)
- Rarely polydactyly of toes
- Cardiac malformations present in up to 60% of affected children (single atrium, cor-biloculare or triloculare)
- Thin, spoon-shaped fingernails
- Cryptorchidism, hypospadias and other genitourinary anomalies sometimes present in males

Cranio-oro-dental features

Teeth:

- Natal or neonatal teeth
- Hypodontia
- Supernumerary teeth
- Microdontic and conoid teeth
- Shovel-shaped incisor
- Talon cusp
- Taurodontism
- Hypoplastic enamel
- Delayed teeth eruption
- Premature exfoliation
- Teeth transposition
- Absent maxillary mucobuccal fold due to fusion of the lip mucosa to the maxillary gingiva from canine to canine
- The anterior mandibular alveolar ridge is traversed by several thick frenula giving it a serrated appearance
- Accessory labiogingival frenula
- Gingival hypertrophy
- Partial cleft lip
- Malocclusion

References

[259,299–322]

Pictures Illustrating the Oro-Dental Features

Figure 2.14

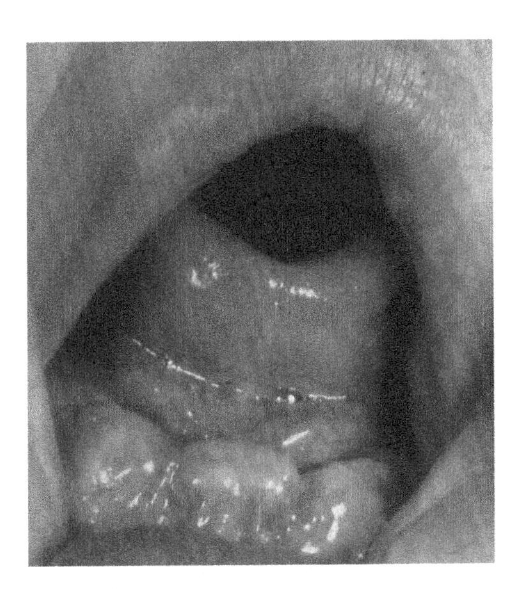

Figure 2.14 EVC. Serrated appearance of the anterior mandibular gingival ridge, produced by thick frenula in a child with the EVC syndrome, who also had bilateral hand polydactyly.

2.3.7 Weyers Acrofacial Dysostosis Syndrome (Curry–Hall Syndrome)

Definition

Weyers [330] described a syndrome of postaxial polydactyly with anomalies of the lower jaw, dentition and oral vestibule, similar to Ellis–van Creveld syndrome.

OMIM Number

#193530

Prevalence

Not yet determined.

Inheritance

Autosomal dominant with variable expression.

Aetiology – Molecular Basis – Gene (4p16)

Loss of function mutations in the *EVC1* gene or *EVC2*, which are also responsible for Ellis–van Creveld syndrome, cause Weyers acrofacial dysostosis.

Animal Models/Main Features

Evc (−/−) mice develop an EvC-like syndrome, including short ribs, short limbs and dental abnormalities [301].

Clinical Description

Main features

- Normal intelligence
- Mild short stature
- Short limbs
- Postaxial polydactyly (of toes is a frequent sign)
- Nail dystrophy

Cranio-oro-dental features

- Hypertelorism
- Prominent ears
- Mucosal bands obliterating the oral vestibule
- Bifid tip of the tongue
- Split epiglottis
- Irregular, small or absent permanent incisors
- Single central incisor
- Conical teeth
- Microdontia
- Taurodontism
- Premature eruption of teeth

References

[131,299,301,305,318,323–331]

2.3.8 Van der Woude Syndrome (VWS)

Definition

This syndrome, consisting of cleft lip/palate and paramedian sinuses or pits of the lower lip, has been classified as type 1 (with most reported cases VWS1 linked to chromosome 1q32–q41) and type 2 (VWS2 linked to another locus mapped to 1p34).

OMIM Number

#119300	Type 1
#606713	Type 2

Prevalence

- Over 900 cases have been reported from various countries and ethnic groups.

- The frequency of VWS is estimated between 1 in 35000 and 1 in 100000 in the Caucasian population; 2 in 100,000 is reported in [259].
- VWS represents around 2% of all patients with cleft lip/palate.

Inheritance

- Autosomal dominant with incomplete penetrance
- Penetrance has been estimated at 96.7% of cases
- New mutation in 30–50% of cases

Aetiology – Molecular Basis – Gene

- Type 1: gene at 1q32–q41; type 2: gene at 1p34.2.
- Mutations in the gene encoding *interferon regulatory factor-6* (*IRF6*) are responsible for VWS1. *IRF6* is a transcription factor that belongs to a family of nine transcription factors that share a highly conserved helix-turn-helix DNA-binding domain and a less conserved protein-binding domain. Most *IRFs* regulate the expression of interferon-alpha and -beta after viral infection, but the function of *IRF6* is related to epidermal development.
- Van der Woude and popliteal pterygium syndromes are allelic [332].
- In some pedigrees of families affected with VWS, hypodontia could be the only visible phenotype in some individuals [333].
- *IRF6* is involved in human non-syndromic cleft [334,335] and tooth agenesis [336].
- *IRF6* should be screened when any doubt rises about the normality of the lower lip and also if a non-syndromic cleft lip patient (with or without cleft palate) has a family history suggestive of autosomal dominant inheritance [337].
- *WDR65*, situated in the (VWS2) locus at 1p34.2, is the gene involved in Van der Woude syndrome and oral clefting (Type 2) [338].

Animal Models/Main Features

Mice deficient for *Irf6* have abnormal skin, limb and craniofacial development. Histological and gene expression analyses indicate that the primary defect is in keratinocyte differentiation and proliferation [339]. The cleft palate 1 (clft1) mutant mouse displays a mutation in *irf6* (Van der Woude syndrome mutation) [340].

Clinical Description

Main features

Occasionally reported:

- Talipes equinovarus
- Syndactyly of third and fourth fingers
- Ankyloblepharon (this can also be a mild expression of the pterygium popliteal syndrome)

Cranio-oro-dental features

- Lip pits present in 88% of affected patients
- Circular or oval sinuses or pits symmetrically located in each side of the lower lip midline
- Bilateral pits which can be asymmetric
- Occasionally, a single pit in the midline
- Lip pits that may be connected to minor salivary glands of the lip and secrete saliva spontaneously or under pressure

- Lower lip bulges (might represent a discrete sign)
- Lip pits and cleft lip/palate in 33% of cases
- Lip pits and cleft palate or submucous cleft palate in 33% of cases
- Lip pits without cleft in 64% of cases
- Cleft lip and palate
- Cleft or bifid uvula
- Syngnathia
- Narrow highly arched palate
- Ankyloglossia
- Hypodontia with preferential tooth agenesis of incisors and premolars

References

[332–350]

Pictures Illustrating the Oro-Dental Features

Figure 2.15

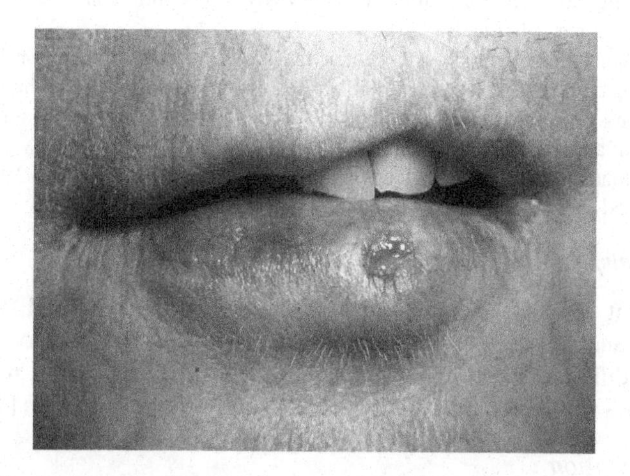

Figure 2.15 Van der Woude. A 50-year-old woman with Van der Woude syndrome. Note the paramedian lip pits and the scar from the correction of the cleft lip.
Reprinted from O.A.C. Ibsen, J.A. Phelan, Genetics (Chapter 6), in H.O. Sedano (Ed.), Oral Pathology for the Dental Hygienist. Figs. 6–29, p. 237, Copyright 2004, with permission from Elsevier.

2.3.9 Oral–Facial–Digital Syndrome Type I (OFDI)

Definition

This syndrome, also known as Papillon–Léage–Psaume syndrome, is a ciliopathy characterized by pseudocleft of the upper lip, multiple oral frenula and various digit anomalies.

OMIM Number

 # 311200

Prevalence

- OFDI is estimated at around 1 in 50,000 live births. There are over 250 cases reported in the literature.
- Seventy-five per cent of cases are sporadic.

Inheritance

X-linked dominant limited to women and lethal in heterozygotes men.

Aetiology – Molecular Basis – Gene

OFDI results from mutations in the *CXORF5* gene (chromosome X open reading frame 5) which maps to the short arm of the X chromosome, locus Xp22.3–p22.2, and encodes a protein containing coiled-coil alpha-helical domains.

OFDI is a centrosomal protein localized at the basal bodies at the origin of primary cilia and to the nucleus. *OFDI* protein is able to self-associate, and this interaction is mediated by its coiled-coil-rich region. *OFDI* may be part of a multiprotein complex and could play different biological functions in the centrosome-primary cilium organelles as well as in the nuclear compartment. *OFDI* has a crucial role in the biology of primary cilia [351].

Animal Models/Main Features

Knockout animals lacking *OfdI* have the main features of the disease, albeit with increased severity, possibly owing to differences of X inactivation patterns between human and mouse. Failure of left–right axis specification in mutant male embryos, and a lack of cilia in the embryonic node were found. Formation of cilia was defective in cystic kidneys from heterozygous females, implicating ciliogenesis as a mechanism underlying cyst development. Additionally, impaired patterning of the neural tube and altered expression of the 5′ Hoxa and Hoxd genes in the limb buds of mice lacking *OfdI* were observed, suggesting that *OfdI* could have a role beyond primary cilium organization and assembly [352].

Clinical Description

Main features

- Short stature
- Brachydactyly of second through fifth fingers, with various degrees of syndactyly and clinodactyly
- Unilateral polysyndactyly, short big toe that can be deviated in a fibular direction

- A mild learning disability in 45% of patients
- Bilateral polycystic kidneys present in 50% of patients

Cranio-oro-dental features

Typical faces characterized by:

- Frontal bossing
- Lateral displacement of the inner canthi of the eyes (dystopia canthorum)
- Thinning of the nose with hypoplasia of the alar cartilages
- Pseudocleft in the midline of the upper lip produced by a thick frenulum extending from the lip mucosa to the vestibular mucosa with partial erasure of the vestibular sulcus
- Small mandible

Intraorally:

- Thick frenula can be seen at other levels of the mucobuccal fold.
- Seventy-five per cent of patients with *OFDI* syndrome present several thick fibrous bands traversing the mandibular mucobuccal fold.
- The hard palate is divided by deep lateral grooves into three segments.
- The anterior segment contains the incisors and the canines, the rest of the dentition is contained into two lateral palatal segments.
- Asymmetrical cleft soft palate is seen in around 80% of patients.
- The most striking intraoral finding is a bi- or tri-lobulated tongue.
- A small white harmatomatous node can be found in 70% of patients between the abnormal tongue lobes. Histologically this node is composed of fibrous connective tissue containing salivary glands, fat, smooth muscle fibres and even cartilage.
- Ankyloglossia can be found in 30% of patients.

Teeth:

- Hypodontia, missing upper lateral incisors
- Supernumerary teeth
- Enamel hypoplasia
- Teeth malposition as well as supernumerary canines and hypoplastic lateral maxillary incisors found in around 20% of patients and thought to be associated to the abnormal frenula

References

[351,353–362]

Pictures Illustrating the Oro-Dental Features

Figure 2.16

2.3.10 Williams (Beuren) Syndrome (WBS)

Definition

Williams syndrome (sometimes called Williams-Beuren syndrome) is characterized by distinctive craniofacial features (elfin faces), cardiovascular abnormalities and behavioural characteristics.

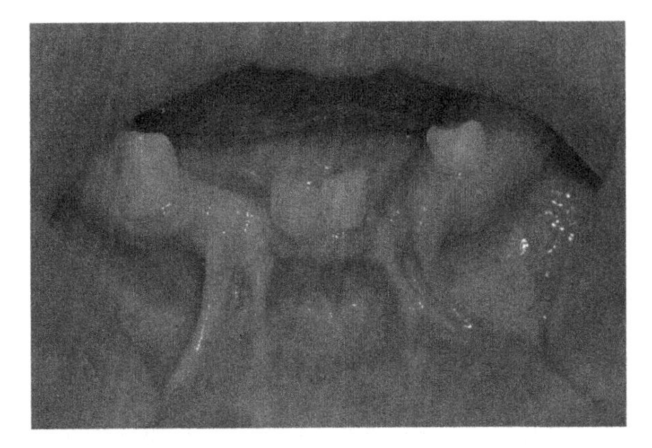

Figure 2.16 OFDI. Thick, fibrous bands traversing the mandibular mucobuccal fold. Reference Centre for Orodental Manifestations of Rare Diseases, Pôle de Médecine et Chirurgie Bucco-dentaires, Hôpitaux Universitaires de Strasbourg, Université de Strasbourg, M-C Maniere.

OMIM Number

#194050

Prevalence

1 in 20,000 live births.

Inheritance

Autosomal dominant.

Aetiology – Molecular Basis – Gene

This contiguous gene deletion syndrome results from a deletion of the 7q11.23 region of chromosome 7, encompassing the elastin gene (*ELN*) and the neighbouring other genes such as *RFC2* (replication factor C, subunit 2), *LIMK1* (*LIM* kinase-1), *GTF2IRD1/GTF2I* (member of the *TFII-I* family of transcription factors) and *FKBP6* (FK506-binding protein). Three genes are involved in craniofacial development: craniofacial development protein 1 (Cfdp1), Sec23 homolog A (Sec23a) and nuclear receptor-binding *SET* domain protein 1 (Nsd1) and are direct *TFII-I* (*GTF2I* and *GTF2IRD1*) targets [363].

Animal Models/Main Features

Homozygous loss of either *Gtf2ird1* or *Gtf2i* function results in multiple phenotypic manifestations, including embryonic lethality; brain haemorrhage and vasculogenic, craniofacial and neural tube defects in mice. *Gtf2ird1* and *Gtf2i* heterozygotes display microcephaly, retarded growth and skeletal and craniofacial defects as observed in WBS. *Gtf2ird1* pattern of expression is found predominantly in musculoskeletal tissues, the pituitary gland, craniofacial tissues, the eyes and tooth buds. *Gtf2i* and *Gtf2ird2* are also expressed during tooth development [364–366].

Clinical Description

Main features

- Unusual ("elfin") face (100%).
- Cardiovascular disease (most commonly supravalvar aortic stenosis 80%).
- Coarctation of the aorta, renal artery stenosis, and systemic hypertension are complications that when present may worsen over time. Because the elastin protein is an important component of elastic fibres in the arterial wall, any artery may become narrowed.
- Learning disability (75%), but remarkable musical and verbal ability.
- Characteristic cognitive profile (90%).
- Idiopathic hypocalcaemia (15%).
- Attention deficit disorder.
- Renal dysfunction.
- Developmental delay
- Short stature.

Cranio-oro-dental features

Ocular manifestations:

- Strabismus
- Esotropia
- Blue irides
- Brown irides
- Green irides
- Retinal vascular tortuosity

Face:

- Round face and full cheeks
- Thick lips and large mouth that is usually held open
- Broad nasal bridge with nostrils that flare forward (anteverted nares)
- Unusually short eyelid folds and flare eyebrows
- Small mandible
- Prominent ears

Oro-dental abnormalities:

- Tooth agenesis and hypodontia
- Microdontia; small permanent teeth (mesio-distal and labio-lingual)
- Tapered teeth
- Possibility of high caries due to enamel hypoplasia/hypomineralization
- Cleft palate occasionally reported

References

[363,367–381]

Pictures Illustrating the Oro-Dental Features

Figure 2.17

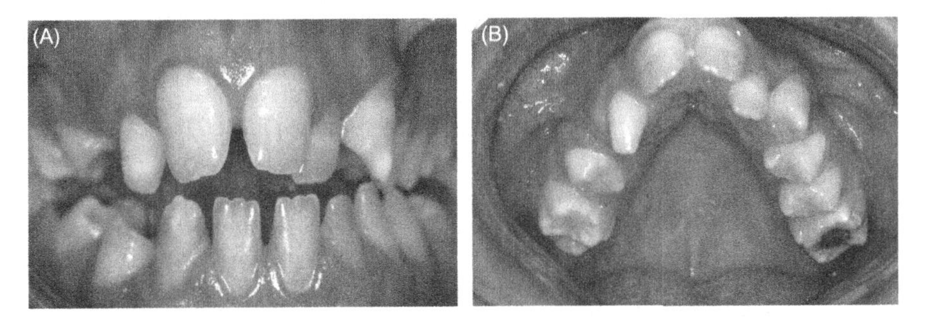

Figure 2.17 Williams. (A) Dental abnormalities encountered in Williams syndrome: hypodontia, microdontia and tapered teeth (maxillary incisors). (B) Note the high arched palate; the maxilla is relatively short. Enamel hypoplasia type defects are visible.

2.4 TGFbeta Superfamily

Syndromes	Associated Features	Gene	Molecules	Inheritance	Locus
TGFbeta superfamily					
Angel-shaped phalango epiphyseal dysplasia (ASPED) #105835	Peripheral dysostosis Hyperextensibility of interphalangeal joints	*CDMP-1 (GDF5)*	TGFbeta family member Growth factor	AD	20q11.2
Oligodontia [382]	Short stature Increased bone density in the spine and base of skull	*LTBP3*	TGFbeta-binding protein 3 and extracellular matrix protein required for osteoclast function	AR	11q13

The transforming growth factor-beta (TGF-beta) superfamily comprises nearly 30 growth and differentiation factors that include TGF-betas, activins, inhibins and bone morphogenetic proteins (BMPs). Multiple members of the TGF-beta superfamily serve key roles in stem-cell fate commitment. The various members of the family can exhibit disparate roles in regulating the biology of embryonic stem (ES) cells and tumour suppression [383]. During tooth development, epithelio-mesenchymal interactions are mediated by signal molecules belonging mostly to four conserved families: transforming growth factor (TGF) beta, WNT, fibroblast growth factor (FGF) and Hedgehog. The continuous growth of the mouse incisors, as well as their subdivision into the crown and root domains, is dramatically altered by modulating a network of FGF and two TGF beta signals, bone morphogenetic protein (BMP) and Activin. This network is responsible for the regulation of the maintenance, proliferation and differentiation of epithelial stem cells that are responsible for growth and enamel production [384].

2.4.1 Angel-Shaped Phalango Epiphyseal Dysplasia (ASPED)

Definition

Angel-shaped phalangoepiphyseal dysplasia (*ASPED*) is a specific bone dysplasia characterized by the association of brachydactyly with particular radiological features, abnormal dentition (hypodontia), delayed capital femoral ossification and early onset degenerative hip arthrosis.

OMIM Number

#105835

Prevalence

ASPED is rare with 15 cases or families reported in the literature [259].

Inheritance

Autosomal dominant.

Aetiology – Molecular Basis – Gene

ASPED is caused by mutations in the *GDF5* (growth/differentiation factor 5) gene, also known as *CDMP-1* (cartilage-derived morphogenetic protein 1) gene (chromosome 20q11.2).

CDMP-1 is a secreted signalling molecule that participates in skeletal morphogenesis.

Du Pan syndrome (fibular hypoplasia and complex brachydactyly (**#228900**; AR), Grebe syndrome (**#200700**; AR), brachydactyly type C (**#113100**; AD), Hunter–Thompson (**#201250**; AR) are allelic diseases caused by mutations in *GDF5*.

Animal Models/Main Features

The transgenic mice present developmental defects of the limbs/digits/tail and skeleton [385,386].

Clinical Description

Main features

- Short to normal stature
- Delayed ossification

- Osteoarthritis of the hip
- Hyperextensible interphalangeal joints of the fingers
- Brachydactyly type C
- Angel-shaped middle phalanges (on radiographs)

Cranio-oro-dental features
Teeth:

- Hypodontia
- Abnormal enamel
- Delay in tooth eruption
- Persistence of primary teeth
- Premature loss of teeth
- Malocclusion of teeth

Very little information exists on orodental aspects of these diseases. Hattab et al. [387], however, describe oral manifestations of severe short-limb dwarfism resembling Grebe chondrodysplasia. Delayed development and eruption of teeth, severe oligodontia of permanent dentition, microdontia, supplemental incisor, enamel hypoplasia of primary teeth, doubled and abnormal frenal attachments, bifid uvula, small maxilla and malocclusion are reported.

References

[385,386,388–393]

Pictures Illustrating the Oro-Dental Features

Figure 2.18

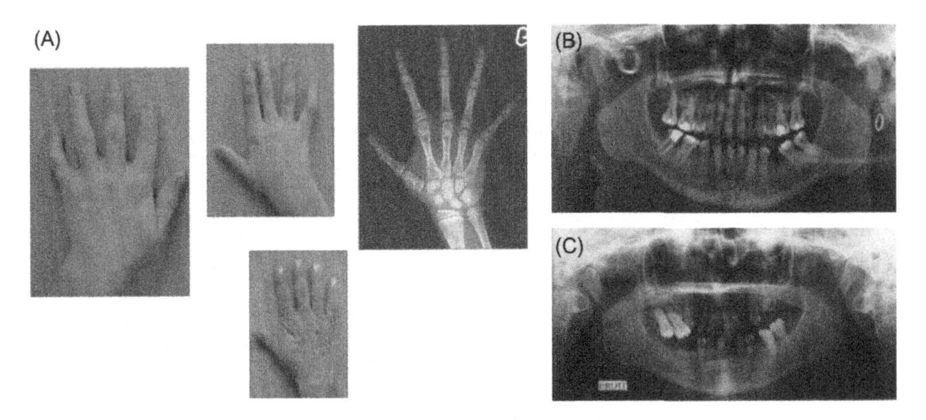

Figure 2.18 ASPED. (A) Brachydactyly type C and angel-shaped middle phalanges (on radiographs). Service de Génétique Médicale, Hôpital Jeanne de Flandre, Lille, France, M. Holder-Espinasse (B–C) Oral phenotype with missing teeth in individuals of the same family presenting with ASPED (panoramic radiograph), Nicolas Chassaing, Toulouse, France.

2.5 SHH Signalling Pathway

Syndromes	Associated Features	Gene	Molecules	Inheritance	Locus
SHH signalling pathway					
Solitary median maxillary central incisor (*SMMCI*) #147250	Short stature, GH deficiency Choanal atresia, midnasal stenosis, ED Holoprosencephaly	*SHH*	Signalling protein	AD	7q36

In the hedgehog signalling network (Figure 2.19), mutations result in various phenotypes, including, among others, holoprosencephaly, nevoid basal cell carcinoma syndrome, Pallister–Hall syndrome, Greig cephalopolysyndactyly, Rubinstein–Taybi syndrome, isolated basal cell carcinoma and medulloblastoma [394–399].

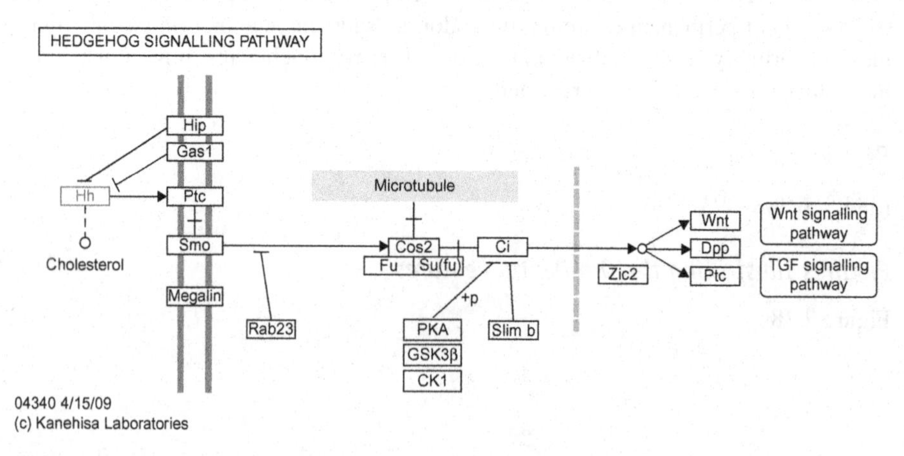

Figure 2.19 Shh signalling KEGG pathway database.

SHH: Holoprosencephaly 3	Solitary Median Maxillary Central Incisor, Microphtalmia Isolated #147250
DHCR7 sterol delta-7-reductase	Smith–Lemli–Opitz syndrome #270400
PATCH1	Gorlin syndrome basal cell naevus syndrome #109400, basal cell carcinoma, holoprosencephaly 7
PATCH2	Basal cell carcinoma, medulloblastoma
SMO smoothened	Basal cell carcinoma, medulloblastoma
GLI1	Basal cell carcinoma, glioblastoma, rhabdomyosarcoma, osteosarcoma, predicts sarcoma grade
GLI3	Greig syndrome #109400, Pallister–Hall syndrome #109400, postaxial polydactyly A, IV, A/B
CREBBP or CBP	Rubinstein–Taybi syndrome #180849
TWIST	Saethre–Chotzen syndrome #101400

In mammals, Hh signals are triggered by the interaction of three distinct proteins (Shh, Sonic Hh; Ihh, Indian Hh and Dhh, Desert Hh) with a common receptor (Ptc1 and Ptc2, Patched 1 and 2) plus additional co-receptors. This interaction relieves the inhibitory activity of Ptc (Patch) upon the transducer Smo (Smoothened). Smoothened is another transmembrane protein, which acts downstream of Ptc and is an essential positive mediator of the Hh signal activating the downstream transcription factors belonging to the Gli family (Gli1, Gli2 and Gli3, homologs of the drosophila bifunctional transcription factor Cubitus interruptus [Ci]). These transcription factors enhance the expression of a number of target genes, including Gli1 itself, which amplifies the signalling response. Gli3 and to a lesser extent Gli2 are also processed into a transcription-repressor form, whereas Gli1 behaves exclusively as a transcription activator.

Megalin (also known as gp330) is a multiligand receptor belonging to the low-density lipoprotein (LDL) receptor family.

Growth Arrest Specific 1 (GAS1) is a 45 kDa glycophosphatydlinositol (GPI)-linked protein originally identified for arresting the cell cycle when overexpressed.

Additional molecules influence in a positive and negative way the signalling pathway. Fused Fu is a serine/threonine kinase. For instance, suppressor of fused (Sufu), binds Gli proteins, thereby inhibiting nuclear accumulation and interfering with their transcriptional activity. Sufu plays a crucial role in negative regulation of the pathway. PKA (protein kinase A) has been shown to function as a repressor.

The cAMP-regulated enhancer (CRE) binds many transcription factors, including CRE-binding (CREB) protein (CPB), which is activated as a result of phosphorylation by PKA. CPB is a 2441 amino acid 265 kDa nuclear protein which is a co-activator for a number of transcription factors, including Ci.

2.5.1 Solitary Median Maxillary Central Incisor (SMMCI)

Definition

Solitary median maxillary central incisor is a midline developmental defect and represents one minor clinical manifestation of holoprosencephaly.

It is encountered in many different syndromes or associations affecting the forebrain and frontonasal process derivatives (*VACTERL, CHARGE* syndrome, XXX, del(18p) or r(18), del(7q36ter)).

The incisor is unique and 'neutral' – neither right nor left.

Recently an association between *SMMCI* and oromandibular-limb hypogenesis syndrome type 1 was reported [400].

OMIM Number

#147250

Prevalence

1 in 50,000 live births.

Inheritance

Variable and autosomal dominant for *SHH* dependent holoprosencephaly (37% of AD holoprosencephaly cases).

Aetiology – Molecular Basis – Gene

The gene involved is *SHH* (7q36).

Different genes involved in holoprosencephaly have been shown to be associated with single maxillary central incisor *SHH* (7q36), *SIX3* (2p21), *ZIC2* (13q32), *TGIF* (18p11.3), *PTCH* (9q22.3) and *GLI2* (2q14). This helps explaining the association with deletions at 18p and 7q36.1 seen over the years with the solitary median maxillary central incisor syndrome. A mutation in *SALL4* (20q13.13–q13.2) was associated with cranial midline defects and single central incisor. Holoprosencephaly occurs in 25% of 13 trisomy syndrome cases, 15% of 18p-syndrome, 10% of 13q cases, and often in triploidy.

CHARGE (coloboma, heart defects, choenal atresia, retarded growth and development, genital hypoplasia, ear anomalies and deafness) syndrome is due to mutations in *CHD7* gene encoding chromodomain helicase DNA-binding protein 7. No evidence yet links *CHD7* to single central incisor phenotype.

Animal Models/Main Features

Growth arrest specific 1 (Gas1) encodes a membrane glycoprotein previously identified as a Shh antagonist in the somite. Gas1(−/−) mice exhibited microform HPE, including midfacial hypoplasia, premaxillary incisor fusion, and cleft palate, in addition to severe ear defects; however, the forebrain remained intact. These defects were associated with partial loss of Shh signalling in cells at a distance from the source of transcription, suggesting that Gas1 can potentiate hedgehog signalling in the early face [401]. Noggin(−/−) mice exhibited a solitary median maxillary incisor that developed from a single dental placode, early midfacial narrowing as well as abnormalities in the developing hyoid bone, pituitary gland and vomeronasal organ. In Noggin(−/−) mice, the expression domains of Shh, as well as the Shh target genes *Ptch1* and *Gli1*, were reduced in the frontonasal region at key stages of early facial development [402].

Clinical Description

Main features

- Short stature
- Growth hormone deficiency
- Abnormal pituitary gland
- Learning disorders

Cranio-oro-dental features

- Microcephaly
- Hypotelorism

- Choanal atresia, midnasal stenosis, congenital nasal pyriform aperture stenosis
- V-shaped palate
- Prominent midpalatal ridge
- Absent maxillary frenum
- Single median maxillary incisor

Management/Oral Health

Multidisciplinary management involves paediatric dentistry, orthodontics, periodontology, implantology and prosthodontics.

When a solitary median maxillary incisor tooth is present, a paediatrician and a geneticist should carefully examine the patient for other craniofacial malformations, especially midline systemic problems.

References

[400–425]

Pictures Illustrating the Oro-Dental Features

Figure 2.20

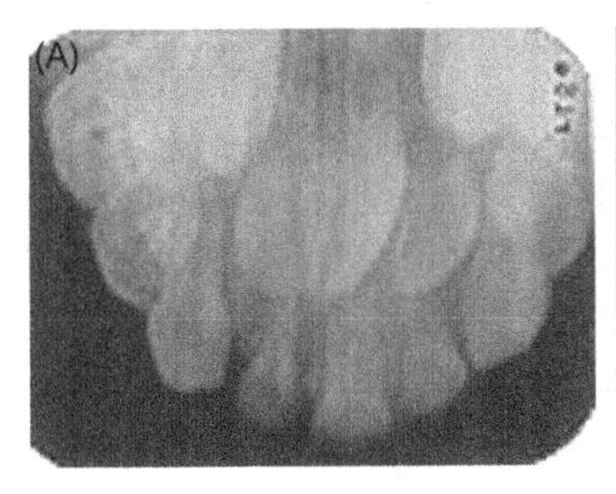

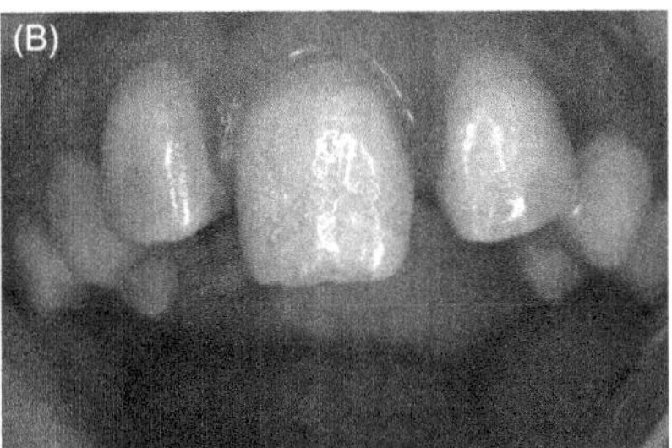

Figure 2.20 SMMCI. (A) The single maxillary median central incisor is visible both in the primary and permanent dentition on the occlusal radiographs (6-year-old child). (B) A SMMCI encountered in CHARGE association.

2.6 Fibroblast Growth Factors (FGF) Pathway

FGF pathway				
Lacrimoauriculodentodigital syndrome #149730	Hypodontia, absent parotid gland, absent Stensen duct, small lacrimal glands and ducts, renal agenesis	*FGFR2* *FGFR3* *FGF10*	AD	10q26 4p16.3 5p13–p12

(*Continued*)

(*Continued*)

Kallmann syndrome 2 KAL2 #147950	Central hypogonadism, lack of sense of smell, in some cases renal aplasia, deafness, syndactyly, cleft lip/palate and teeth agenesis	*FGFR1* KAL1, located on the X, KAL3 PROKR2, KAL 4 PROK2, KAL5 CHD7, KAL 6 FGF8	Fibroblast growth factor receptor-1	AD	8p11.2–p11.1 Xp22.31 (KAL1)
Isolated tooth agenesis [336]	Preferential premolars agenesis	*FGFR1*	Fibroblast growth factor receptor-1	AD	8p11.2–p11.1

The fibroblast growth factors (FGFs) represent a family of extracellular signalling peptides involved in key developmental events like the establishment of the mesoderm, neural patterning, morphogenesis, myogenesis, limb development and the establishment of right–left asymmetry.

FGF ligands (22 in mammals) are small polypeptides with a partially conserved core of 120–130 amino acids that bind heparin with high affinity.

Four genes code for FGF receptors (FGFRs), which are receptor tyrosine kinases (RTKs) and are activated by ligand binding. FGFRs have three immunoglobulinlike domains Ig1, Ig2, and Ig3 that bind ligand and heparan sulphate glycosaminoglycan (HSGAG), and are important in receptor dimerization.

FGF	**LIGANDS SUBFAMILY**
FGF1	*FGF1, FGF2*
FGF4	*FGF4, FGF5, FGF6*
FGF7	*FGF3, FGF7, FGF10, FGF22*
FGF8	*FGF8, FGF17, FGF18*
FGF9	*FGF9, FGF16, FGF20*
FGF19	*FGF19, FGF21, FGF23*
FGF11	*FGF11, FGF12, FGF13, FGF14*

Receptor preference

FGF1 activates all FGFRs; FGF2 prefers FGFR1c and FGFR2c

FGF4	*FGFR1c, FGFR2c*
FGF7	*FGFR2b, FGFR1b*
FGF8	*FGFR3c, FGFR4, FGFR1c*
FGF9	*FGFR3c, FGFR2c*
FGF19	Hormone class, very weak activation of *FGFR1c, FGFR2c*
FGF11	No activation of *FGFRs*

Formation of the FGF:FGFR:HS signalling complex causes the activation of the intracellular kinase domains and the crossphosphorylation of tyrosines on the FGFRs. The FGF pathway is negatively regulated by Sulf (sulfatases and sulfotransferases), Spry (Sprouty) and MAPK phosphatases.

Sef (similar expression to FGF) is a transmembrane protein originally identified in zebrafish and negatively regulates FGF signalling in development and cell culture.

The three FLRT (fibronectionleucinerichtransmembrane) genes originally identified in human adult skeletal muscle act as a positive regulator of FGF signalling and are required for normal responses to FGF signalling in embryo tissues.

The FGF signalling pathway interplays with the canonical WNT signalling pathway.

The FGF pathway is involved in a wide range of pathologies, including skeletal dysplasias, neurodegenerative disease, metabolic disorders and cancer [426].

Craniosynostosis, the premature fusion of the cranial sutures, is due to mutations in the *FGFR2*, *FGFR3* (gain-of-function mutations in *FGFR*), *TWIST1* and *EFNB1* genes [19,427,428].

2.6.1 Lacrimoauriculodentodigital Syndrome (LADD)

Definition

LADD syndrome (Levy–Hollister syndrome) is determined by the presence of cup-shaped ears, anomalies of the teeth and lacrimal ducts, mixed hearing loss and digital malformations.

OMIM Number

 # 149730

Prevalence

Over 20 family cases published [259].

Inheritance

Autosomal dominant.

Aetiology – Molecular Basis – Gene

LADD syndrome is due to mutations in genes involved in the *FGF* signalling pathway: mutations in *FGFR2*, *FGFR3* and *FGF10*.

FGFR2	10q26
FGFR3	4p16.3
FGF10	5p13–p12

Mutation in *FGF10* gene also results in aplasia of the lacrimal and salivary glands (ALSG), an allelic disorder (#180920).

Animal Models/Main Features

Extra teeth anterior to the first molar are a characteristic for knockouts of Sprouty genes that are intracellular inhibitors of FGF signalling. These mice also have extra incisors.

FGF has a crucial role as stimulator of tooth formation, and Sprouty genes/proteins are important endogenous inhibitors modulating FGF activity in tooth formation [429,430].

The germs of *FGF10*-null mice proceed to cap stage normally. However, at a later stage, the cervical loop is not forming, suggesting that *FGF10* maintains the stem-cell compartment in the developing incisor tooth germ [431].

Hypoplastic submandibular glands (SMG) are seen in *Fgf10* and *Fgfr2b* heterozygote mice, whereas SMG aplasia is seen in *Fgf10* and *Fgfr2b* null embryos [432].

Clinical Description

Main features

Eye:

- Dry conjunctival mucosae
- Small-absent lacrimal glands
- Nasolacrimal duct obstruction
- Absent lacrimal gland puncta
- Chronic dacryocystitis
- Recurrent kerato-conjunctivitis
- Nasolacrimal duct fistulas
- Absent tearing
- Reduced visual acuity

Ear:

- Cup-shaped ears unilateral, bilateral or asymmetrical
- Mixed conductive-sensorineural hearing loss, unilateral or bilateral

Kidney:

- Renal agenesis, sclerosis

Hypospadias

Limb:

- Preaxial digital anomalies
- Polydactyly
- Syndactyly
- Clinodactyly of the fifth finger
- Bifid thumb
- Digitalized thumb
- Triphalangeal thumb
- Broad halluces
- Shortening of radius and ulna
- Radioulnar synostosis
- Absent radius

Cranio-oro-dental features

- Xerostomia
  - Salivary gland aplasia/hypoplasia
  - Absent Stensen duct
  - Absent parotid gland
  - Salivary gland anomalies
- Small epiglottis
- Underdeveloped condyle
- Short ramus
- Enlarged mandibular foramen
- Cleft lip and palate reported

Teeth:

- Hypodontia
- Peg-shaped incisors
- Malformed molars
- Microdontia
- Thin rooted teeth
- Taurodontism
- Enamel hypoplasia
- Hypomineralized enamel
- Enamel thinning and wear
- Tooth discolouration
- Delayed eruption of primary teeth
- Severe dental caries

References

[259,278,429–456]

Pictures Illustrating the Oro-Dental Features

Figure 2.21

2.6.2 Kallmann Syndrome (KS; KAL2)

Definition

Kallmann syndrome (KS) is a rare genetic disorder characterized by central hypogonadism with a decreased ability to smell and in some cases renal anomalies, deafness, syndactyly, cleft lip/palate and teeth agenesis.

OMIM Number

#147950

Prevalence

7.7 in 100,000 live births.

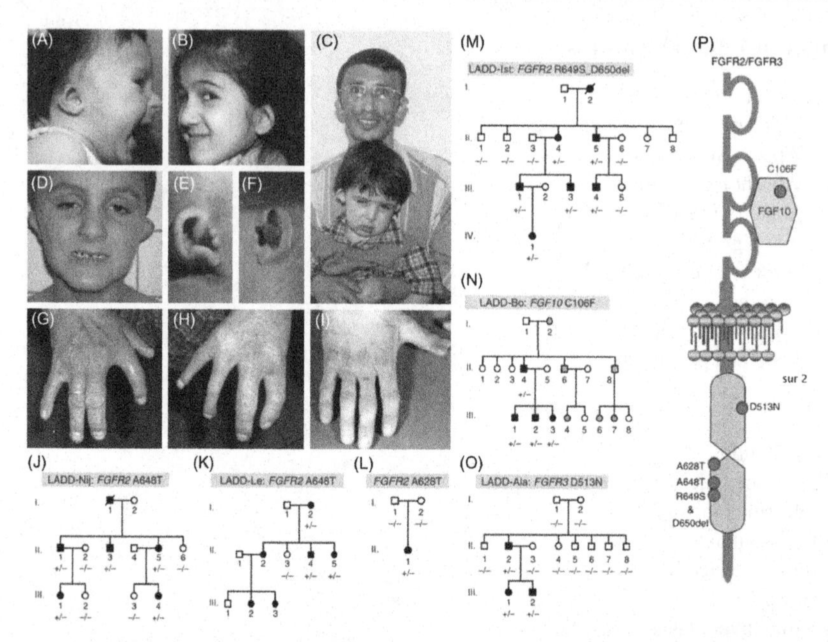

Figure 2.21 LADD. Clinical findings, pedigrees and mutations in *FGFR2, FGFR3* and *FGF10* in LADD syndrome. (A–I) Phenotypic characteristics of LADD patients from different families: (A, D) LADD-Ala, III-2; (B) LADD-Ala, III-1; (C) LADD-Ist, II-5 and III-4; (E) LADD-Be, II-1; (F) LADD-Nij, III-4; (G, H) LADD-Nij, II-3; (I) LADD-Bo, II-4. (A–F) Photographs show facial appearance and typical ear anomalies in LADD patients. Digital anomalies included hypoplastic (G), absent (H) and bifid (I) thumbs. (J–O) Pedigrees of LADD families. Family name, gene involved and identified mutation are given in the grey box on top of each pedigree. Symbols: +, mutation present; –, mutation absent. Filled black symbols indicate affected individuals; filled grey symbols in the LADD-Bo family represent individuals who were probably affected but for whom a detailed clinical description is lacking. (P) Schematic model of *FGFR2* and *FGFR3*. The locations of different mutations are marked by red dots on the receptor or ligand (*FGF10*). The intracellular tyrosine kinase domains of *FGFR2* and *FGFR3* are shown in orange. Mutations in different components of *FGF* signalling in LADD syndrome.
Source: From Ref. [441] Reprinted by permission from Macmillan Publishers Ltd, copyright (2006).

Inheritance

Autosomal dominant.

Aetiology – Molecular Basis – Gene

To date, five genes for KS have been identified—*KAL1*, located on the X chromosome; *FGFR1*, *PROKR2*, *PROK2*, *CHD7* and *FGF8*; and recently, *WDR11*—which are involved in autosomally transmitted forms.

Mutations in fibroblast growth factor receptor-1 cause both Kallmann syndrome and normosmic idiopathic hypogonadotropic hypogonadism.

It has been hypothesized that KS represents a milder allelic variant of CHARGE syndrome, which has been supported by the identification of heterozygous *CHD7* mutations in KS [457].

Animal Models/Main Features

Mice homozygous for a hypomorphic allele of *Fgfr1* have craniofacial defects, some of which appeared to result from a failure in the early development of the second branchial arch [458].

In K14-Cre; *Fgfr1*(fl/fl) mice (tissue-specific inactivation of *Fgfr1* in mouse), severe enamel defects mimic amelogenesis imperfecta (AI), with a rough, irregular enamel surface [459].

Clinical Description

Main features

- Intellectual disability
- Isolated GnRH deficiency
- Anosmia
- Short stature
- Hypogonadotropic hypogonadism
- Congenital heart defect
- Cryptorchidism
- Micropenis
- Absence of secondary sexual features
- Adult females have little or no breast development and primary amenorrhoea
- Delayed skeletal maturation
- Synkinesia of the digits
- Unilateral renal agenesis
- Neurosensory hearing loss

Cranio-oro-dental features

- Choanal atresia
- Cleft lip and palate
- Extreme retrognathism of both maxilla and mandible

Teeth:

- Teeth agenesis, hypodontia, oligodontia in both dentition
- Most frequently missing: lateral mandibular incisors; then second premolars of upper and lower jaws, and lateral maxillary incisors
- Microdontia
- Screwdriver-shaped mandibular incisors
- Thin molar roots

References

[457–465]

Pictures Illustrating the Oro-Dental Features

Figure 2.22

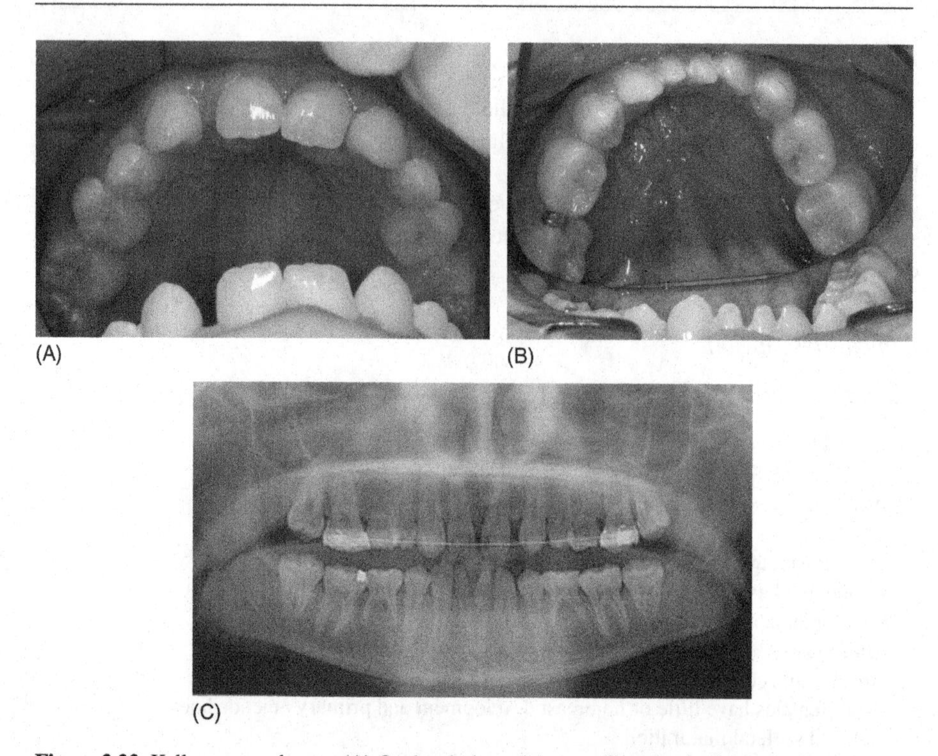

Figure 2.22 Kallmann syndrome. (A) Occlusal view of the maxilla of an individual with Kallmann syndrome. (B) Occlusal view of the mandible of an individual with Kallmann syndrome. (C) Orthopantomogram of an individual with Kallmann syndrome. I. Bailleul-Forestier, Paediatric Dentistry, UPS, Toulouse, France See also [460].

2.7 Other Pathways

Syndromes	Associated Features	Gene	Molecules	Inheritance	Locus
Other pathways					
Bloom syndrome #210900	Pre- and postnatal growth deficiency; sun-sensitivity, telangiectatic, hypo- and hyperpigmented skin; predisposition to malignancy and chromosomal instability	*RECQL3*	DNA helicase RecQ protein-like-3 DNA-dependent ATPase, DNA helicase, and 3-prime-to-5-prime single-stranded DNA translocation activities	AR	15q26.1

(*Continued*)

(*Continued*)

Syndromes	Associated Features	Gene	Molecules	Inheritance	Locus
Rothmund–Thomson syndrome #268400	Genodermatosis poikiloderma, small stature, skeletal and dental abnormalities, cataract and an increased risk of cancer Microdontia, delayed eruption, supernumerary teeth Missing teeth, multiple crown shape defects	*RECQL4*	DNA helicase gene RECQL4	AR	8q24
Diastrophic dysplasia #222600	Osteochondrody splasia	*DTDST*	Sulphate transporter	AR	5q32–q33.1
Johanson–Blizzard syndrome #243800	Pancreatic insufficiency, microcephaly Intellectual disability, short stature, hypo/oligodontia, permanent dentition	*UBR1*	Ubiquitin-protein ligase E3-alpha	AR	15q15.2
Kabuki syndrome #147920	Intellectual disability, growth deficiency, and peculiar face Missing upper lateral incisors and inferior central incisors	*MLL2*	Myeloid/lymphoid or mixed lineage leukaemia 2 Histone methyltransferase	AD	12q12–q14
Karvajal/ Naxos syndrome [466]	Wooly hair, palmoplantar keratoderma, and biventricular dilated cardiomyopathy, hypo/oligodontia	*DSP*	Desmoplakin Desmosomal protein	AR	6p24

2.7.1 Bloom Syndrome (BS, BLS)

Definition

Bloom syndrome, first recognized in 1954, is an autosomal recessive disorder characterized by proportionate pre- and postnatal growth deficiency; telangiectatic erythematous skin lesions which are sunlight sensitive, hypo- and hyperpigmented skin lesions; immune deficiency; hypogonadism and infertility in males; reduced fertility in females; predisposition to malignancy; and chromosomal instability with a tendency to chromosomal breakage with a high frequency of sister chromatid exchanges. Most patients are of Jewish origin, but cases with other ethnic backgrounds have been reported.

OMIM Number

#210900

Prevalence

- There is a 1 in 231 prevalence of heterozygotes among the Ashkenazi Jewish population.
- Carrier frequency for common allele in Ashkenazi Jewish population is ~1%.
- Parental consanguinity is not increased in affected Jewish families, but parental consanguinity in non-Jewish families is very high.

Inheritance

Autosomal recessive.

Aetiology – Molecular Basis – Gene

The responsible gene (*RECQL3=RECQ2=BLM*) encoding DNA helicase RecQ protein-like-3 has been mapped to 15q26.1.

DNA helicases are ubiquitous enzymes that unwind DNA in an ATP-dependent and directionally specific manner essential for DNA repair, recombination, transcription and replication. Five human DNA helicases sharing sequence similarity with the *E. coli* RecQ helicase have been identified. Three of the human RecQ helicases are implicated in cancer-prone genetic diseases (Bloom syndrome, Werner syndrome and Rothmund–Thomson syndrome).

Animal Models/Main Features

Mouse embryos homozygous for a targeted mutation in the murine Bloom syndrome gene (*Blm*) are developmentally delayed and die by embryonic day 13.5. Other viable genetically modified mice for *Blm* have elevated rates of both cancer and chromosomal aberrations [467,468].

Clinical Description

Main features

- Short stature
- Mild learning disabilities
- Respiratory diseases: bronchiectasis, chronic lung disease
- Syndactyly, polydactyly, fifth finger clinodactyly
- Facial telangiectasia
- Skin photosensitivity and hypo/hyperpigmentations
- Immune deficit
- Predisposition to malignancy

Cranio-oro-dental features

- Dolichocephaly/scaphocephaly, microcephaly
- Narrow, long face, malar hypoplasia
- Prominent ears
- Prominent, large, small/short nose
- Conjunctivitis, conjunctival telangiectasia, eyelid lesions and eyelashes may be lost.
- Small mandible/micrognathia
- Teeth: hypodontia/oligodontia, absent upper lateral incisors

Management/Oral Health

Multidisciplinary management.

References

[467–476]

Pictures Illustrating the Oro-Dental Features

Figure 2.23

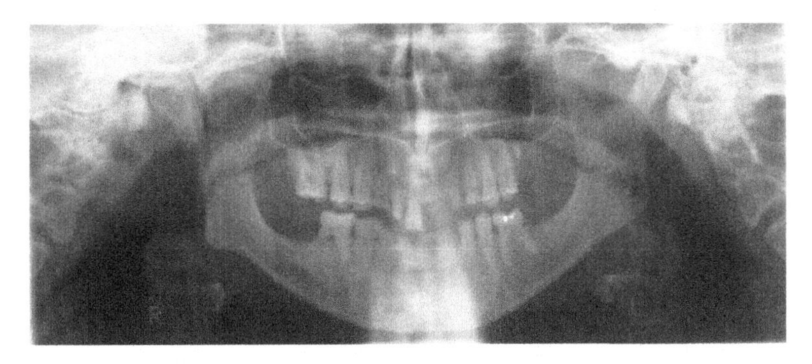

Figure 2.23 Bloom syndrome. Panoramic radiograph showing oligodontia and anomalies of tooth shape and size in Bloom syndrome.

2.7.2 Rothmund–Thomson Syndrome

Definition

Rothmund–Thomson syndrome is characterized by variable expression of typical cutaneous changes (poikilodermal rash), juvenile cataracts, sensitivity to sunlight, skeletal anomalies, short stature, abnormal hair growth and defective nails and teeth, intellectual disability, hypogonadism and a typical facial appearance. It is a cancer-prone genetic disease.

Patients are at increased risk for the development of tumours like osteosarcoma.

OMIM Number

#268400

Prevalence

300 cases have been published [259].

Inheritance

Autosomal recessive.

Aetiology – Molecular Basis – Gene

The syndrome is due to mutation in the DNA helicase gene *RECQL4* (8q24.3) involved in chromosomal stability. There is genetic heterogeneity in this syndrome.

Baller–Gerold syndrome and Rapadilino syndrome have overlapping phenotypes and are due also to mutation in *RECQL4*.

Animal Models/Main Features

A viable *Recql4*-mutant mouse model exhibits a distinctive skin abnormality, birth defects of the skeletal system, genomic instability and increased cancer susceptibility [477].

Clinical Description

Main features

- Short stature
- Congenital skin rash
- Erythematous skin lesions in infancy
- Poikiloderma
- Skeleton defects
- Osteoporosis
- Small hands and feet
- Small, thin nails
- Sunlight intolerance/sensitivity
- Premature aging
- Increased risk of cancer
- Osteosarcoma
- Basal cell carcinoma
- Squamous cell carcinoma
- Lymphoma
- Lymphatic leukaemia
- Intellectual disability (5–13% of cases)
- Gastrointestinal problems

Cranio-oro-dental features

- Redness of cheeks between 3 and 6 months of age
- Sparse hair, eyebrows and eyelashes; alopecia
- Frontal bossing
- Cataract
- Microphtalmia
- Strabismus

- Glaucoma
- Small nose with concave ridge
- Prognathism

Teeth:

- Missing teeth
- Supernumerary teeth
- Crown malformations
- Short roots
- Microdontia
- Delayed eruption
- Caries
- Loose teeth
- Periodontitis

References

[259,477–492]

Pictures Illustrating the Oro-Dental Features

Figure 2.24

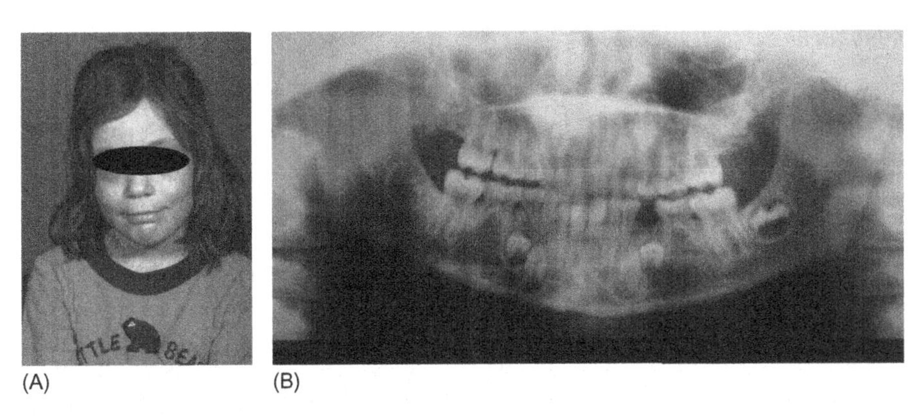

Figure 2.24 Rothmund–Thomson. (A) Redness of cheeks, or poikiloderma, meaning areas of altered pigmentation, prominent blood vessels, and thinning of the skin. (B) Hypodontia and missing permanent premolars (14, 15, 24, 25, 35, 45) visible on this panoramic radiograph (G. Winter, EDI, UCL).

2.7.3 Diastrophic Dysplasia Syndrome

Definition

The term *diastrophic* is derived from the Greek *diastroph*, meaning 'distorted'.

The syndrome consists of chondrodysplasia with dwarfism, with numerous and severe skeletal abnormalities from cervical spine to the feet.

OMIM Number

#222600

Prevalence

In Finland, where it is most commonly found, there is a prevalence of 1 in 30,000.

Inheritance

Autosomal recessive.

Aetiology – Molecular Basis – Gene (5q32–q33.1)

The solute-linked carrier (SLC) 26 family of transporter proteins includes SAT-1, diastrophic dysplasia sulphate transporter (DTDST), DRA/CLD, pendrin, prestin, PAT-1/CFEX and Tat-1, which are structurally related and transport one or more substrates (sulphate, chloride, bicarbonate, iodide, oxalate, formate, hydroxyl or fructose). Mutations in *DTDST* gene (*SLC26A2*), which codes for a sulphate transporter protein, lead to diastrophic dysplasia. DTDST is necessary and sufficient to induce fibronectin matrix assembly [493].

Animal Models/Main Features

Homozygous-mutant mice were characterized by growth retardation, skeletal dysplasia and joint contractures [494,495].

Clinical Description

Main features

Although the development and growth of cartilaginous structures are disturbed, the intramembranous ossification and appositional growth pattern are not primarily affected. Physeal, epiphyseal, and articular cartilages are defective, leading to the characteristic findings: patients with diastrophic dysplasia have epiphyseal involvement and are at risk for degenerative joint diseases. There is wide variability in the phenotypic expression.

- Short stature, short limbs
- Cervical kyphosis (spinal compression due to abnormalities in cervical vertebrae and to dorsiflexion necessitated by intubation and velopharyngeal surgery may be a dangerous operative complication)
- Persistent extension limitation in elbow and knee joints
- Club feet
- Ulnar deviation of hands
- Shortened phalanges, and, in particular, abduction of thumbs ('hitchhiker thumbs') and big toes
- Thin hair/underdeveloped glands

Cranio-oro-dental features

Typical facial features:

- Prominent cheek and circumoral fullness
- Narrow nasal bridge without flattening or depression
- Broad mid-nose and flared
- Long, full appearance of the face with a high, broad forehead and square jaw
- Midline haemangiomas are sometimes present.

Intraorally:

- Cleft palate, either complete or partial, is seen in approximately 50% of patients. This cleft palate may contribute to aspiration pneumonia. Forty-one of 95 Finnish patients (43%) with diastrophic dysplasia had open cleft palate (CP)
- Highly arched palate
- Bifid uvula
- Submucous clefts or its microforms observed in an additional 30 patients (32%) of Finnish patients
- The cartilage of the larynx and trachea is abnormally soft and may contribute to narrowing of respiratory passages. The abnormal palate may be responsible for the production of a somewhat characteristic voice.

Teeth:

- Hypodontia affects 31% of patients.

References

[493–503]

2.7.4 Johanson–Blizzard Syndrome

Definition

Johanson–Blizzard syndrome (JBS) is an autosomal recessive inherited disorder that is characterized by pancreatic insufficiency, short stature, a distinct appearance with marked underdevelopment of the alae nasi, intellectual disability, hearing loss and dental anomalies.

OMIM Number

#243800

Prevalence

Over 23 cases published [259].

Inheritance

Autosomal recessive.

Aetiology – Molecular Basis – Gene

Johanson–Blizzard syndrome is caused by a defect in the *UBR1* gene, ubiquitin-protein ligase E3 component N-recognin 1 gene (15q15.2). Protein quality control and subsequent elimination of terminally misfolded proteins occur via the ubiquitin-proteasome system. Misfolded proteins are tagged with ubiquitin involving an ubiquitin activating enzyme (E1), ubiquitin conjugating enzymes (E2) and ubiquitin ligases (E3).

Animal Models/Main Features

The knockout mice with absent *Ubr1* expression suffer from exocrine pancreatic insufficiency but present teeth [504].

Clinical Description

Main features

- Exocrine pancreatic insufficiency (selective defect of acinar tissue, whereas the islets of Langerhans and ducts are preserved)
- Diabetes mellitus in older children
- Short stature
- Low birth weight, failure to thrive
- Delayed bone age
- Hypothyroidism
- Malabsorption
- Intellectual disability (moderate to severe, but normal intelligence can occur)
- Hypotonia
- Anorectal abnormalities (anal imperforation)
- Malrotation of small intestines
- Congenital heart defects
- Malformation of the uterus and vagina in female infants
- Genitourinary defects
- Skin defects (café-au-lait spots, etc.)
- Small breast and nipples

Cranio-oro-dental features

- Microcephaly
- Oddly patterned hair growth, midline ectodermal scalp defects
- Prominent forehead
- Congenital clefting of bones surrounding the optical orbit
- Absent eyebrows and eyelashes
- Strabismus
- Absent lacrima puncta
- Flat ears
- Congenital deafness, sensorineural hearing loss
- Often marked underdeveloped nasal alae
- Small maxilla and mandible

Teeth:

- Oligodontia
- Lacking most of the permanent teeth except the first permanent molars and the upper permanent central incisors
- Hypoplastic primary teeth

References

[259,504–554]

Pictures Illustrating the Oro-Dental Features

Figure 2.25

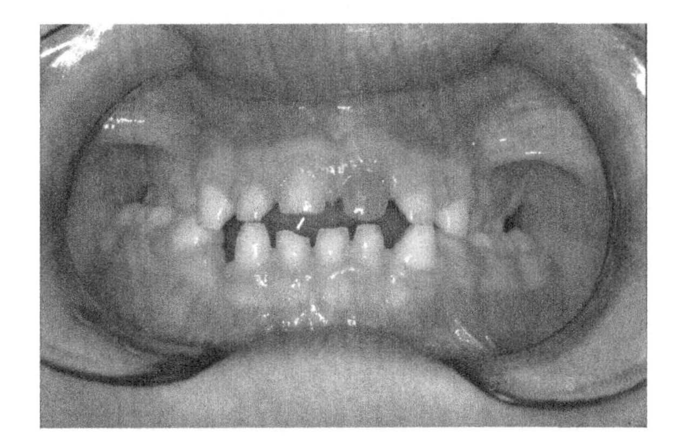

Figure 2.25 Johansson–Blizzard. This patient with Johansson–Blizzard syndrome only presented with the first permanent molars in his permanent dentition. All the other permanent teeth were missing. Note the missing lower primary incisor – the shape and size anomalies as well as the hypodevelopment of the maxilla with bilateral crossbite (S. Dewhurst, EDI, UCL).

2.7.5 Kabuki Syndrome

Definition

This syndrome is defined by a typical facial appearance in association with postnatal short stature, hypotonia, joint laxity, developmental delay, persistent foetal fingertip pads and additional congenital abnormalities, including cardiac anomalies and ectodermal manifestations. The facial appearance is like a Kabuki actor's mask in the traditional Japanese play.

OMIM Number

#147920

Prevalence

1.16 in 100,000; 1 in 32,000 Japanese individuals.

Inheritance

Autosomal dominant. Simplex cases make up the vast majority of the reported cases with Kabuki syndrome (*de novo* mutations).

Aetiology – Molecular Basis – Gene

Kabuki syndrome-1 (KABUK1) is caused by mutations in *MLL2* (12q13.12), a gene that encodes a Trithorax-group histone methyltransferase, a protein important in the epigenetic control of active chromatin states. Nonsense or frameshift mutations lead to haploinsufficiency.

Kabuki syndrome-2 (KABUK2) is caused by mutation in the *KDM6A* gene (Xp11.3)

Animal Models/Main Features

Loss of *Mll2* function in mouse slowed growth, increased apoptosis and retarded development, leading to embryonic failure before E11.5 [555].

Clinical Description

Main features

- Short stature
- Intellectual disability
- Epilepsy
- Hypotonia
- Cardiac anomalies, vascular anomalies
- Skeletal anomalies
- Renal, urogenital anomalies
- Gastrointestinal anomalies
- Immunological defects (increased susceptibility to infection)
- Dermatoglyphic anomalies
- Persistent fingerpads
- Café-au-lait spots
- Hirsutism

Cranio-oro-dental features

- Craniosynostosis reported
- Microcephaly
- Wide forehead
- Long palpebral fissures
- Arched eyebrows
- Eversion of the lateral third of the lower eyelids

- Thick eyelashes
- Blue sclerae
- Prominent ears
- Preauricular pits
- Broad ears
- Low set ears
- Broad and depressed nasal tip
- Trapezoid philtrum
- Everted lower lip
- Microforms of lower lip fistula
- Paramedian elevation of the lower lip
- Cleft lip/palate
- Highly arched palate
- Bifid tongue and uvula

Teeth:

- Missing teeth (permanent upper lateral incisors and lower central incisors being the most commonly absent, upper molars)
- Missing upper lateral incisor (14.3% of cases)
- Agenesis of permanent upper canine reported
- Hypodontia found in *MLL2* identified patients but also in patients with no *MLL2* mutations.
- Supernumerary teeth
- Microdontia
- Abnormal tooth shape
- Conical teeth
- Screwdriver-shaped upper incisor
- Fusion/gemination of teeth
- Taurodontic molars
- Large pulp chambers
- Intrapulpal calcification
- Division of the lower third of the root in normally single rooted teeth
- Short roots
- Overretention of primary teeth
- Delayed tooth eruption
- Ectopic upper molars
- Widely spaced teeth
- Malocclusion
- Micrognathia
- Unilateral posterior crossbite
- Retrognathia of the upper jaw – midface hypoplasia
- Anterior open bite

References

[555–573]

Pictures Illustrating the Oro-Dental Features

Figure 2.26

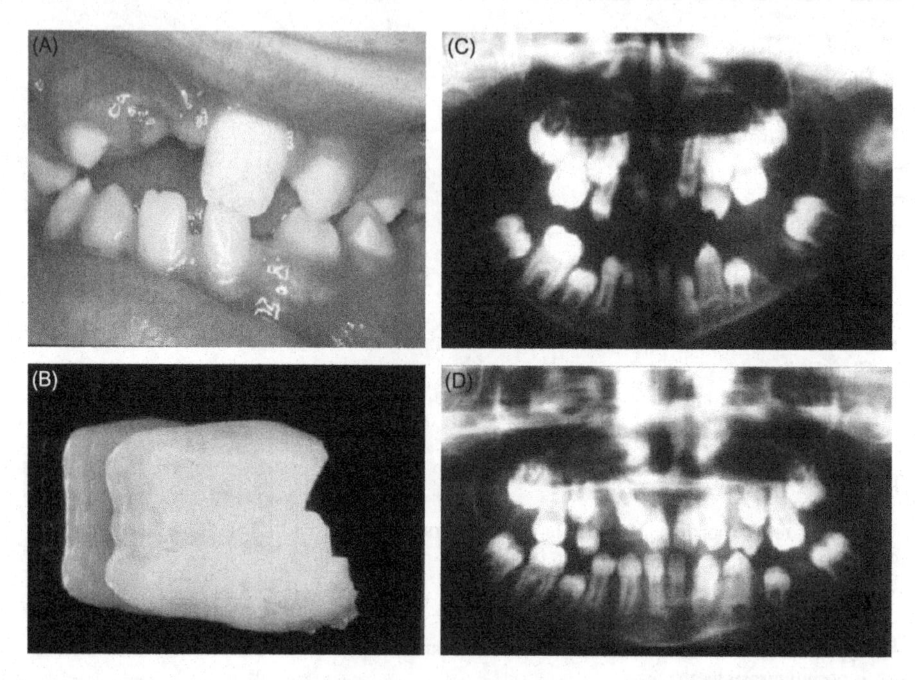

Figure 2.26 Kabuki syndrome. (A) The oral cavity of patient 1 and (B) the tooth fragment at 11 years of age: the pictured tooth was taken in a mirror. The screwdriver shape of the crown is clearly visible. (C) Dental tomograms of patient 1 at (top) 9 and (bottom) 11 years of age with non-attachment of 12, 22, 31, 41 and large pulp lumen of the remaining molars with denticle deposits. Of note is the division of the lower third of the roots of teeth 45, 34, 35, the root resorption of 11 and 21 and the shortened roots of 32 and 42. The Kabuki syndrome: four patients with oral abnormalities.
Source: From Ref. [559] by permission of Oxford University Press.

3 Supernumerary Teeth

Syndrome	Associated Factors	Gene	Molecules	Inheritance	Locus
Cleidocranial dysplasia #119600	Bone dysplasia	*RUNX2 (CBFA1)*	Transcription factor	AD	6p21
Familial adenomatous polyposis (FAP) #175100	Colonic polyposis, odontomes, supernumerary teeth, unerupted teeth	*APC*	Negative regulator of Wnt	AD	5q21–q22
Robinow syndrome #180700	Mesomelic limb shortening facial and genital anomalies	*WNT5A*		AD	3p
#268310 [574] [575] [576]		*ROR2*		AR	9q22.31
	Triangular mouth with exposed incisors and upper gums Hypodontia Supernumerary teeth Root malformation Gum hypertrophy, ankyloglossia and micrognathia Wide retromolar ridge, alveolar ridge deformation and malocclusion				
Nance-Horan syndrome #302350	Cataracts, unusual face Mesiodens, cone-shaped incisors	*NHS*	Nuclear protein	X	Xp22.13
Tricho-Rhino-Phalangeal Syndromes TRPS1 #190350 TRPS2 #150230 TRPS3 #190351	Hair, face, selected bones, fine and slow-growing scalp hair, laterally sparse eyebrows, sparse eyelashes, broad nasal tip, long and flat filtrum, thin upper lip, large and protruding ears, cone-shaped epiphyses of phalangeal bones and hip malformations	*TRPS1* *TRPS1* and *EXT1* *TRPS1*	Nuclear protein with nine predicted zinc finger (Zfn) domains, including one GATA-type Znf and a carboxy-terminal IKAROS-like double Zfn	AD	8q23.3 8q24.11–q24.13 8q23.3

(*Continued*)

Dento/Oro/Craniofacial Anomalies and Genetics. DOI: 10.1016/B978-0-12-416038-5.00003-2

(*Continued*)

Syndrome	Associated Factors	Gene	Molecules	Inheritance	Locus
Opitz G/BBB syndrome #300000	Eye anomalies, laryngotracheoesophageal cleft, congenital heart disease, genitourinary anomalies and gastrointestinal disorders Supernumerary teeth in mandibular anterior region, ankyloglossia, CLP [577,578]	*MID1*	Ubiquitin-specific regulation of the microtubule-associated catalytic subunit of protein phosphatase 2Ac	X-linked recessive	Xp22
Ellis-van Creveld #225500	Neonatal teeth, hypodontia, eruption, skeletal dysplasia, short limbs, short ribs, postaxial polydactyly and thin malformed nails and teeth [305] Congenital cardiac defects	*EVC1*, *EVC2*	Cilia transmembrane proteins positive regulators of shh signalling	AR	4p16
Craniosynostosis and dental anomalies Kreiborg-Pakistani syndrome #614188 [579]	Craniosynostosis, delayed and ectopic tooth eruption, supernumerary teeth Maxillary hypoplasia Malocclusion Classe III Digit anomalies	*IL11RA*	Interleukin11 receptor alpha cytokine	AR	9p13.3

Teeth numbers are restricted. Remnants of teeth in the shape of evolving and then involuting dental lamina could be seen in the mouse in the upper anterior and in the diastema regions, demonstrating that an extended dental formula could be achieved provided appropriate timing and signalling [580]. In the mouse incisor and diastema regions, dental placodes are transiently distinct, being morphologically similar to the early tooth primordia in reptiles. Two large vestigial buds emerge in front of the prospective first molar and presumably correspond to the premolars eliminated during mouse evolution. These data could explain the frequency of supernumerary teeth found in the upper maxilla region, and suggest that in syndromic conditions expression and formation of supernumerary teeth is indeed possible.

Various pathways with their associated signalling molecules and transcription factors are indeed involved in supernumerary teeth formation and syndromes.

3.1 Cleidocranial Dysplasia (CCD)

Definition

Previously known as cleidocranial dysostosis, this syndrome is characterized mainly by cranial, dental and clavicular anomalies.

OMIM Number

#119600

Prevalence

Over 1000 cases reported.

Inheritance

Autosomal dominant. Fifty percent of cases do not have a family history. This can be explained either by *de novo* mutations (one-third of patients) or low penetrance, or even germline mosaicism in one of the parents.

Aetiology – Molecular Basis – Gene

- The gene is *RUNX2 (CBFA1)*, which maps to the short arm of chromosome 6 (6p21).
- Runt-related transcription factor 2 is the master regulator of osteoblast differentiation and bone formation. Its mutation in human leads to haploinsufficiency of the *RUNX2/CBFA1* protein.
- Chromosomal translocations, deletions, insertions, nonsense and splice-site mutations, as well as missense mutations of the *RUNX2* gene, have been described in CCD patients.
- Individuals with identical gene mutations showed a wide variation in supernumerary tooth formation.

Animal Models/Main Features

In the mouse, one allele for *Runx2/Cbfa1* (*Runx2/Cbfa1* +/− mice) is sufficient for an undisturbed [60] and an apparently normal formation of the periodontium [581]. Mice lacking the *Runx2* gene (*Runx2*−/−) die at birth and lack bone and tooth development [60]. Developing teeth fail to advance beyond the bud stage [60,581–585].

Clinical Description

Clinical variability is seen with classic CCD, mild CCD and even isolated primary dental anomalies.

Main features

- Bilateral or unilateral, total or partial absence of clavicles
- Dropping shoulders
- Short neck
- Spinal and pelvic anomalies
- Coxa vara or coxa valga
- Scoliosis
- Moderately short stature

Cranio-oro-dental features

Cranio-facial features:

- The skull is brachycephalic with marked frontal, parietal and occipital bossing.
- Fontanelles and sutures remain open, often for life.

- Wormian bones are formed through secondary centres of ossification in the suture lines.
- The paranasal sinuses are often underdeveloped or absent.
- Development of the premaxilla is poor, resulting in a false or relative prognathism.
- The face appears small.
- The nasal bridge is depressed.
- The nose is broad at the base.
- There is hearing loss.

Oro-dental findings:

- Submucous or complete cleft palate has been reported in some patients.
- Delayed union occurs at the mandibular symphysis.
- Multiple supernumerary teeth: The number of supernumerary teeth at times simulates a third dentition. Mostly upper permanent incisors and lower premolars are involved.
- Pseudo-anodontia occurs due to delayed eruption and impaction of deciduous, permanent and supernumerary teeth.
- Dentigerous cyst formation around the impacted teeth also has been described.
- The impacted teeth appear to lack a layer of cellular cementum.
- Enamel hypoplasia has been found.
- Malocclusion is present.

References

[428,581–600]

Pictures Illustrating the Oro-Dental Features

Figure 3.1

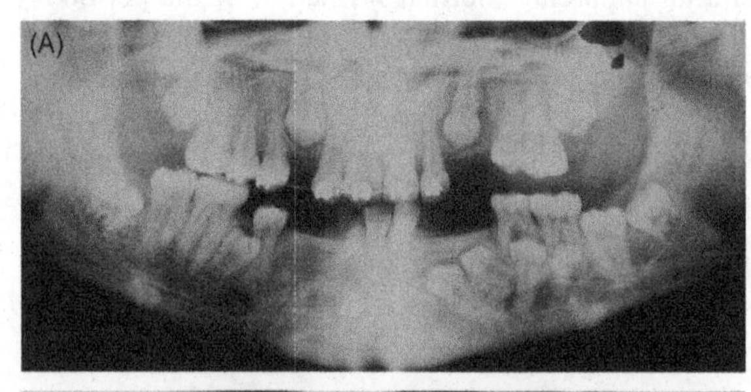

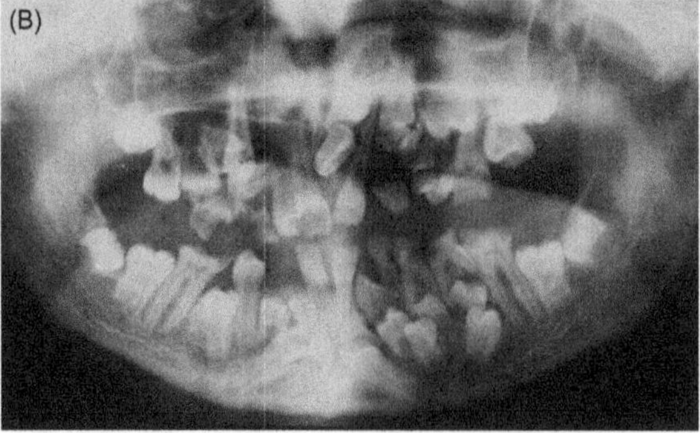

Figure 3.1 Cleidocranial dysplasia. (A) Female 14 years old with cleidocranial dysplasia. Numerous supernumerary teeth have been extracted, and the X-ray shows still several extra teeth present within bone. (B) Multiple supernumerary teeth in a cleidocranial dysplasia patient.

3.2 Familial Adenomatous Polyposis (FAP)

Definition

FAP is an inherited genetic disorder characterized by multiple adenomatous intestinal polyps, osteomas, fibromas and epidermal cysts.

Synonyms

Gardner syndrome, familial polyposis coli.

OMIM Number

#175100

Prevalence

5.25 in 100,000 [259]. In the European Union, prevalence has been estimated at 1 in 11,300 to 37,600 [601]. It accounts for less than 1% of colorectal cancer (CRC) cases.

Inheritance

Autosomal dominant, complete penetrance variable expressivity.

Aetiology – Molecular Basis – Gene

5q21–q22

- Due to the tumour-suppressor gene *APC* (adenomatous polyposis coli), strong repressor of the beta catenin–Wnt pathway.
- Partial and whole gene deletions represent a large proportion (4–33%) of the *APC* mutations found in polyposis patients.
- Mosaicism occurs in a significant number of *APC* mutations, and it is estimated that one-fifth of the *de novo* cases of FAP are mosaic.
- Tumours develop as a result of Wnt pathway activation.
- Another gene *MUTYH* (*MYH*) (1p34.3–p32.1) has been involved in an attenuated form of FAP, AFAP (#608456) transmitted in an autosomal recessive way.

Animal Models/Main Features

Loss of function of the tumour-suppressor gene *Apc* has been associated with the development of murine neoplasia, principally of the intestinal epithelium. Supernumerary teeth can form from multiple regions of the jaw [602–606].

Clinical Description

Main features

- Adenomatous polyps of the colon and rectum.
- Colorectal cancer developing by the fourth decade of life in an untreated patient.

- Polyps of the colon and sometimes of the stomach and small intestine are associated with osseous and soft tissue tumours.
- Multiple other tumours are found (adrenal carcinoma, thyroid papillary carcinoma, astrocytoma, glioma, medulloblastoma, desmoid tumours).
- Skin lesions (keloid, pigmentation, fibromas, lipomas, epidermoid cysts) are present.
- Mammary fibrosis is present.

Cranio-oro-dental features

- Congenital hypertrophy of the retinal pigment epithelium
- Jaw osteomas
- Dental anomalies:
  - Missing teeth
  - Supernumerary teeth
  - Impacted teeth
  - Abnormal roots (long, pointed in molars)
  - Odontomas
  - Dentigerous cysts

Management/Oral Health

Early diagnosis of FAP is crucial and may be life saving. As oral signs usually precede gastrointestinal symptoms, the dentist may play an important role in the diagnosis of FAP.

References

[601–618]

Pictures Illustrating the Oro-Dental Features

Figure 3.2

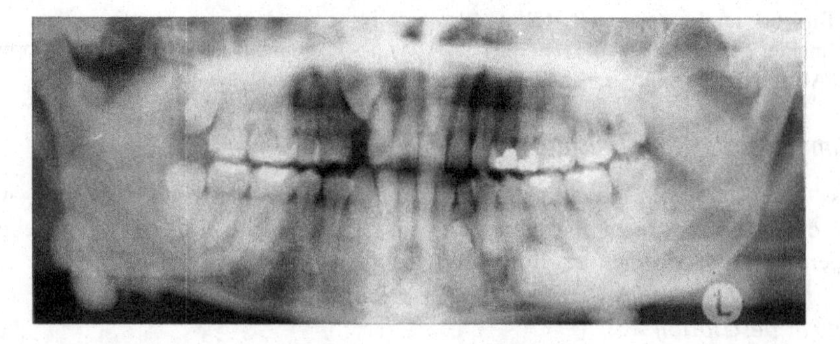

Figure 3.2 Gardner syndrome. DPT showing several well-defined opacities around the periphery of the mandible.
Source: From Ref. [619] Reprinted by permission from Macmillan Publishers Ltd, copyright (2002).

3.3 Nance-Horan Syndrome (NHS)

Definition

Nance-Horan syndrome is characterized by the association in male patients of dense congenital cataracts with microcornea, microphthalmia, dental anomalies and facial dysmorphism.

Synonyms

X-linked cataract with dental anomalies, cataract-dental syndrome.

OMIM Number

#302350

Prevalence

Around 40 families have been reported in the literature.

Inheritance

X-linked semi-dominant transmission, with high penetrance in heterozygotes.

Aetiology – Molecular Basis – Gene

NHS (Xp22.13).

The function of the *NHS* gene and encoded corresponding protein is important for coordinating actin remodelling and maintaining cell morphology. The regulation of the *NHS* gene is complex as it transcribes several isoforms as for example *NHS-A* and *NHS-1A*. The endogenous and exogenous *NHS-A* isoform proteins localizes to the cell membrane of mammalian cells in a cell-type-dependent manner, and it co-localizes with the tight junction (TJ) protein ZO-1 in the apical aspect of cell membrane in epithelial cells. *NHS-1A* protein isoform is a cytoplasmic protein.

NHS is allelic to X-linked cataract [620].

Animal Models/Main Features

There is an animal model in the *Mus musculus* mouse, the mutant Xcat (*Nhs1* gene). Animals have abnormal lens development [621,622].

Clinical Description

Clinical manifestations in heterozygous females are identical to those of affected males, but they are attenuated and are often limited to infraclinical (subclinical) findings. There is no intellectual impairment in heterozygotes.

Main features

Eye:

- Congenital cataract (in 100% of cases, bilateral, usually severe, dense and most often total)
- Microcornea (96% of cases)
- Microphthalmia
- Severe visual impairment (93% of cases) evidenced by nystagmus (93%), sometimes associated with strabismus (43%), and surgery is generally required (89% of cases)
- Glaucoma occur

Unusual face:

- A long, sometimes narrow, often rectangular face
- Marked, long sometimes vertical chin, and prognathism in all cases
- A large nose, with a high, narrow nasal bridge
- Large, often protruding, ears

Intellectual impairment is observed in about 30% of cases. It is usually (80%) mild or moderate, homogeneous and without motor delay, but profound retardation is possible (20%) and is associated with autistic features.

Oro-dental features

Dental abnormalities are present in 100% of cases, concern the primary and permanent dentition and are of high diagnostic value.

- Diastema between the maxillary incisors
- Screwdriver-shaped or conical maxillary central incisors
- Notched incisors
- Existence of cingular cusps (talon cusps) in incisors is characteristic
- Canines often enlarged, globular
- Rounded, globular and sometimes small premolars and molars, with a mulberry or lotus flower appearance due to supernumerary cusps
- Supernumerary teeth (incisors or posterior teeth)
- Impacted teeth
- Missing teeth
- Delayed eruption
- Late persistence of deciduous teeth
- Pulp chambers anomalies (taurodontism, wide pulp chambers, abnormally calcified pulps, pulp stones)
- Teeth malposition: germ translation, ectopic position

Management/Oral Health

The recognition of the specific oro-dental features could lead to early diagnosis of NHS and easier recognition of female carrier status. The management is multidisciplinary.

References

[620–637]

Pictures Illustrating the Oro-Dental Features

Figure 3.3

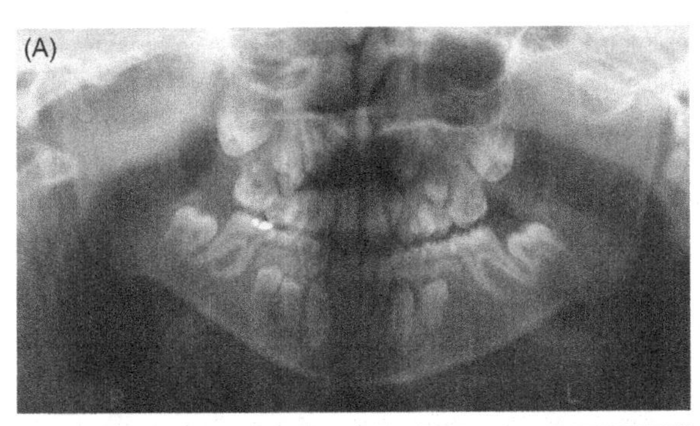

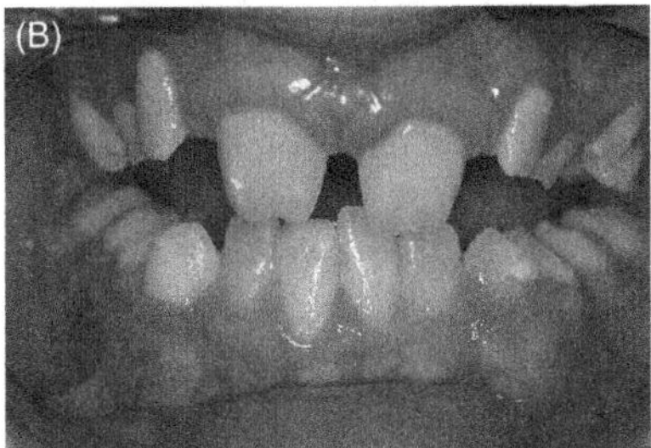

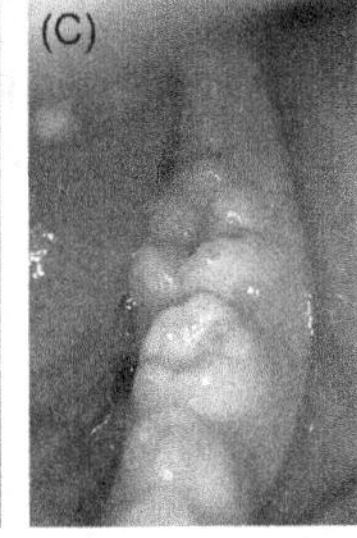

Figure 3.3 Nance-Horan syndrome. (A) Panoramic radiograph of a 10-year-old boy showing different features encountered in Nance-Horan syndrome: agenesis of 35, screwdriver maxillary central incisors, rounded premolars and molars, anomalies of root shape (46, 36) delayed eruption and late persistence of primary teeth. (B) Screwdriver-shaped permanent maxillary incisors. (C) Rounded aspect of molars.

3.4 Tricho-Rhino-Phalangeal Syndromes

Definition

Tricho-rhino-phalangeal syndromes are a group of developmental disorders (type I, II, III) characterized by abnormalities of the hair, face and selected bones.

Synonyms

Langer-Giedion syndrome (type II), Sugio-Kajii syndrome (type III).

OMIM Number

TRPS1 #190350
TRPS2 #150230
TRPS3 #190351

Prevalence

not known

Inheritance

Autosomal dominant.

Aetiology – Molecular Basis – Gene

Tricho-rhino-phalangeal syndrome types I and III are caused by haploinsufficiency of a specific gene coding for a zinc finger protein, *TRPS1*, that is a putative transcription factor. The *TRPS1* gene encodes a GATA-type transcriptional repressor.

Type II is a microdeletion syndrome involving deletions of both *TRPS1* and *EXT1* genes.

The expression of *TRPS1* modulates mineralized bone matrix formation in differentiating osteoblast cells through the regulation (repressor) of osteocalcin transcription.

TRPS1 can bind the promoter of *RUNX2* and inhibit the activity of the *RUNX2* promoter *in vitro*.

Animal Models/Main Features

Trps1 largely influences the development of hair follicles and is important in epithelio-mensenchymal interactions. The mouse embryos homozygous mutant for *Trps1* do not present any obvious dental anomalies but can display a cleft palate.

Clinical Description

Main features

- Short stature
- Scoliosis, lordosis
- Hip dysplasia at young age
- Cone-shaped epiphyses of phalangeal bones (hands)
- Short metacarpal (hands)
- Short metatarsal (feet)
- Thin nails, koilonychia, leukonychia

Cranio-oro-dental features

- Sparse, fine and slow-growing scalp hair
- Large and protruding ears
- Laterally sparse eyebrows
- Sparse eyelashes
- Broad tip of the nose
- Long and flat philtrum
- A thin upper vermillion
- Micrognathia
- Mandibular prognatism
- Highly arched palate
- Supernumerary teeth
- Supernumerary teeth in lateral sectors appear to have the morphology of premolar
- Small teeth
- Microdontia
- Delayed eruption
- Malocclusion

References

[638–657]

Pictures Illustrating the Oro-Dental Features

Figure 3.4

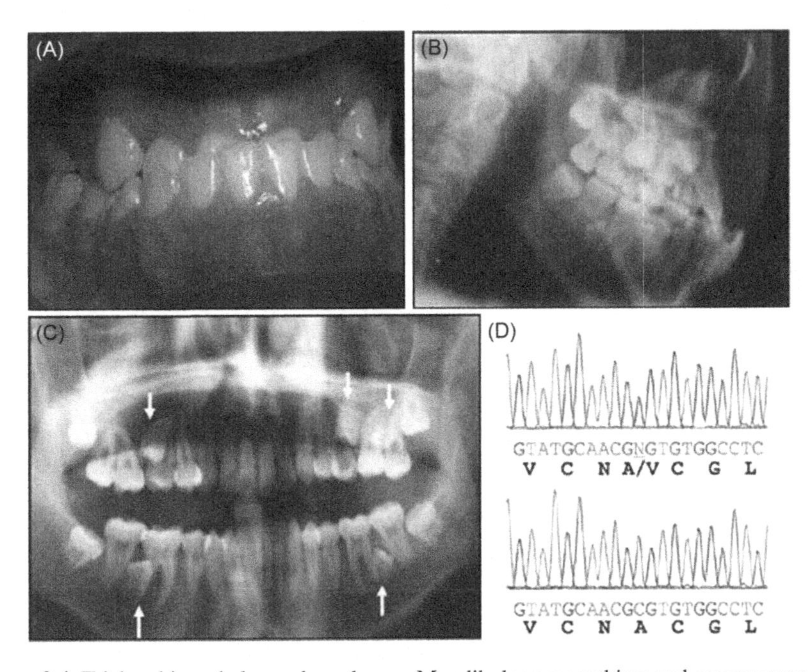

Figure 3.4 Tricho-rhino-phalangeal syndrome. Mandibular prognathism and supernumerary teeth observed in a 17-year-old Thai male with TRPS. (A) Mandibular prognathism observed at 14 years old. (B) X-ray at age 17, showing a very prognathic mandible. (C) Panoramic X-ray at age 17 shows five non-erupted supernumerary teeth exhibiting premolar-like shape (arrows). (D) Electrophoretograms of the sequencing of PCR products. A heterozygous C > T mutation was found at nucleotide position 2756 (*TRPS1* gene).
Source: From Ref. [638] copyright © 2008 by (Kantaputra) Reprinted by Permission of SAGE Publications.

4 Abnormalities of Tooth Shape and Size

Syndrome	Associated Factors	Gene	Molecules	Inheritance	Locus
MOPD II microcephalic osteodysplastic primordial dwarfism, type II #210720 [658,659]	Extreme short stature, microcephaly Extreme microdontia Opalescent and abnormally shaped teeth Rootless molars Hypoplastic alveolar bone	*PCNT2*	Pericentrin 2, a key centrosomal protein	AR	21q22.3
Microdontia/oligodontia (Figure 4.1)	Microdontia/oligodontia	*SMOC2*	BM40 family secreted modular calcium-binding 2 protein	AR	6q27
Deafness with LAMM #610706 [660–665]	Deafness, complete labyrinthine aplasia, microtia and microdontia Conical teeth	*FGF3*	Growth factor	AR	11q13.3
Schimke immuno-osseus dysplasia #242900 [666]	Spondyloepiphyseal dysplasia (SED), resulting in disproportionate short stature, nephropathy and T-cell deficiency Microdontia	*SMARCAL1*	Swi/snf-related, matrix-associated, actin-dependent regulator of chromatin, subfamily a-like 1 (SMARCAL1)	AR	2q35
Turner syndrome [131] (Figure 4.2)	Short stature, ovarian failure, visceral anomalies, especially cardiac Small teeth, short roots, thin enamel, retrognathia				X0

(*Continued*)

Dento/Oro/Craniofacial Anomalies and Genetics. DOI: 10.1016/B978-0-12-416038-5.00004-4

(Continued)

Syndrome	Associated Factors	Gene	Molecules	Inheritance	Locus
Klinefelter syndrome [667] (Figure 4.3)	Tall stature, learning deficit, behavioural disturbances, genital anomalies, larger teeth, taurodontism underdeveloped maxillae and mandibular prognathism				XXY
Rubinstein-Taybi syndrome #180849	Talon cusps, microcephaly, learning disability, Dysmorphic features, typical face	*CREBBP*	CREB-binding protein Transcriptional coactivator	AD	16p13.3
	Broad thumbs and toes	*EP300*	Shh pathway		22q13
Otodental syndrome #166750	Globodontia (large bulbous molars), sensorineural hearing loss, coloboma	*FGF3*	Growth factor	AD	11q13.3
OFCD syndrome #300166	Typical face, cataract, heart anomalies Radiculomegaly, oligodontia, fused teeth, persistent primary teeth, delayed dentition	*BCL-6 co-repressor (BCOR)*	Interacting transcriptional co-repressor BCOR	X	Xp11.4
KBG syndrome #148050	Short stature, characteristic facies (telecanthus, wide eyebrows, brachycephaly), macrodontia (large upper incisor), learning disability and skeletal anomalies	*ANKRD11*	a family member of ankyrin repeat-containing cofactors interacting with p160 nuclear receptor coactivators and inhibiting ligand-dependent transcriptional activation	AD	16q24.3
Ekman-Westborg and Julin [668–671] (Figure 4.4)	Macrodontia of incisors Multituberculism Gigantic third molars Single conical molar roots Prognathism Crossbite	?	?	?	?

The shape and size of a tooth is dependent on the number of post-mitotic cells (odontoblasts and ameloblasts), their spatiotemporal localization and their functional activities (matrices secretion and mineralization) involving specific epigenesis and cell kinetics events [81,82,672–674] .

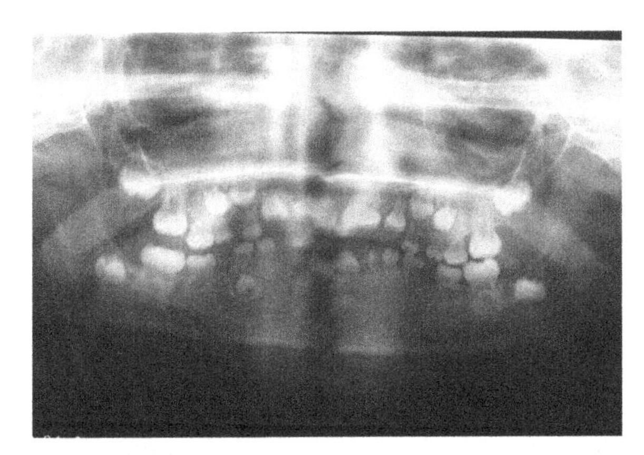

Figure 4.1 Extreme microdontia, oligodontia and short roots associated with *SMOC2* mutation.

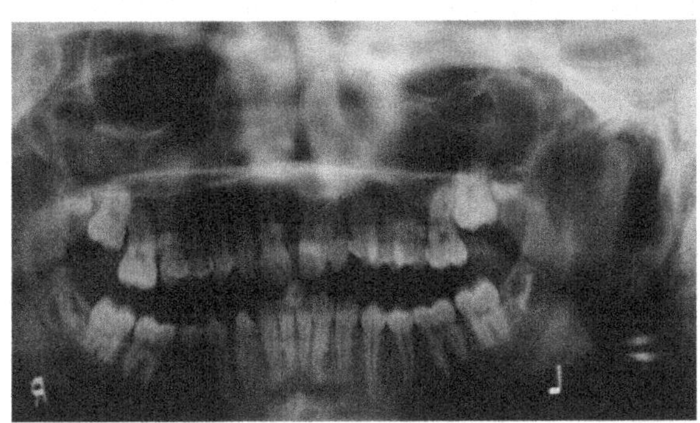

Figure 4.2 Turner syndrome. Small teeth, short roots and thin enamel visible on this panoramic radiograph.

4.1 Rubinstein-Taybi Syndrome (RTS)

Definition

Rubinstein-Taybi syndrome is a disorder characterized by mental and growth retardation, broad thumbs and great toes, and unusual facial characteristics.

OMIM Number

#180849

Prevalence

1 in 100,000 [298].

Inheritance

Autosomal dominant.

Aetiology – Molecular Basis – Gene

It is caused by either a microdeletion at 16p13.3 mutations in the CREB-binding protein gene (*CREBBP* or *CBP*) or *EP300* gene (at 22q13).

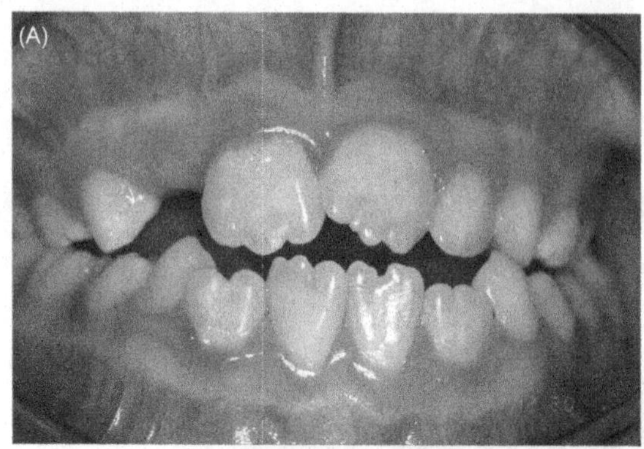

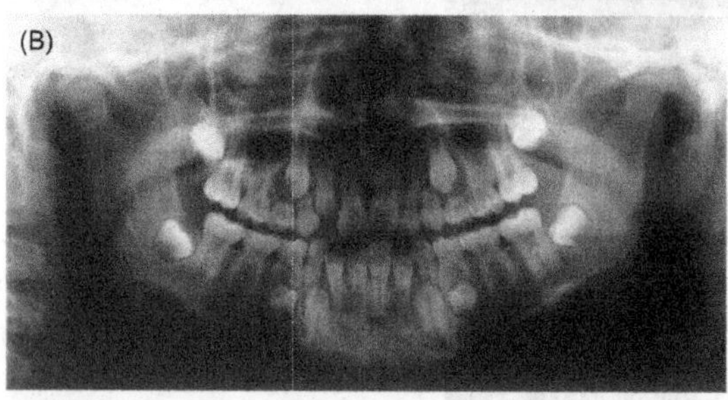

Figure 4.3 Klinefelter syndrome. (A) Large incisors visible in a boy with Klinefelter syndrome. (B) Note the taurodontic molars and missing premolars on this panoramic radiograph.

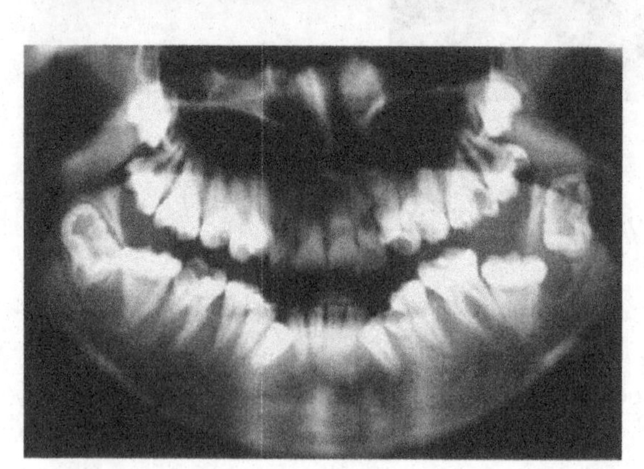

Figure 4.4 Ekman-Westborg and Julin. Radiograph showing huge third molars, radiolucent area mesial to third molar, and second molars with single conical roots *Source: From Ref. [668]. Reproduced by permission of John Wiley and Sons, copyright (2003).*

These two genes show strong homology and encode histone acetyltransferases (HATs), which are transcriptional co-activators (required by many transcription factors for transactivation) involved in many signalling pathways. Mutations in *CBP* and *EP300* (rare) are only found in around 60% of cases.

Animal Models/Main Features

Homozygous mouse embryos die between embryonic days 9.5 (E9.5) and E10.5 and have a defect in neural tube closure. Heterozygous mice have skeletal, cardiac and haematopoietic defects, retarded growth and haematologic tumours [675–679].

Clinical Description

Main features

- Mental, motor, language and social retardation
- Short stature
- Skin: hirsutism, keloid formation, supernumerary nipples, café-au-lait spots
- Heart anomalies (30% of patients)
- Eye microphthalmia, glaucoma, cataract, nystagmus, coloboma and lacrimal duct obstruction
- Hearing loss

Limb and Skeletal anomalies:

- Delayed maturation
- Vertebral anomalies
- Limb anomalies
- Hands: broad thumbs, single transverse palmar crease, clinodactyly, interdigital webbing, persistent foetal fingertip pads
- Feet: broad halluces, duplicated halluces, sandal gap, deep plantar crease
- Sternal anomalies

Genitourinary anomalies, especially in males

Cranio-facio-oro-dental features

- Microcephaly
- Delayed closure of fontanelles

Face:

- Prominent forehead
- Low anterior hairline
- Highly arched heavy eyebrows
- Downward slanting palpebral fissures
- Ptosis
- Convex nasal ridge, broad nasal bridge, deviated nasal septum
- Anomalies of size, shape position and rotation of ears
- Facial asymmetry
- Haemangiomas
- 'Grimacing' smile
- Small mouth
- Thin upper vermillion
- Micrognathia, retrognathia
- Relative prognathism
- Poor tongue mobility, marked median grove of the tongue
- Narrow V-shaped maxilla with highly arched and narrow palate
- Cleft lip and/or palate; or bifid or malformed uvula
- Wide alveolar ridges
- Recurrent subluxation of temporo-mandibular-joint (TMJ)

Teeth:

The dental anomalies affect both primary and permanent dentitions.

- Anomalies of tooth number:
  - Hypodontia
  - Supernumerary (i.e. mesiodens)
  - Double teeth
- Anomalies of tooth size and shape:
  - Talon cusps – the most striking feature, affecting upper incisors in around 70% of cases in the permanent dentition
  - Screwdriver incisors, present in one-third of patients
  - Elongation of the upper central incisors
  - Extra cusps on the primary molars
- Anomalies of tooth structure:
  - Demarcated enamel opacities and hypoplasia
- Anomalies of tooth eruption/exfoliation:
  - Normal to delayed eruption of the primary teeth
  - Natal teeth
  - Enamel wear associated with bruxism, oral habits (gagging, vomiting) and gastro-oesophageal reflux
- Occlusion:
  - Crowding
  - Crossbite

Individuals with *EP300* mutations may have a slightly different phenotype compared to individuals with *CREBBP* mutations, with milder cognitive impairment, more pronounced microcephaly, absent or mild downslanting of palpebral fissures, distinct arched eyebrows and greater degree of retrognathia [680].

Management/Oral Health

Prevention is an essential part of the dental management of patients with RTS because some of them have increased levels of caries attributed to the following factors: brushing difficulties associated with reduced mouth opening, decreased manual dexterity and gag reflex; crowding and malposition of teeth; presence of talon cusps and deep grooves that predispose to plaque accumulation; and enamel defects and difficulties in cooperation. Dietary advice, oral hygiene and health instructions; appropriate use of fluoride and regular dental check-ups are highly recommended. Behaviour management can be difficult due to communication problems, increased anxiety and short attention span, stubbornness, lack of persistence, a need for continuous attention from carers and sudden mood changes. Non-pharmacological techniques include acclimatization, empathy, 'tell-show-do', positive reenforcement and others. Inhalation sedation with nitrous oxide can be very helpful for anxious patients. General anaesthesia (GA) can be used for very uncooperative patients and when extensive dental treatment is required.

The problems associated with talon cusps include interference with occlusion, displacement of tooth due to premature contact, speech problems, irritation of the tongue and caries susceptibility.

The treatment options are as follows:

- Fissure sealants or composite resin restorations are used in the deep grooves to prevent caries formation.

- Selective grinding is used to promote formation of secondary dentine plus fluoride as a desensitizing agent.

The removal of the talon cusp may require pulp treatment. Joint orthodontic, endodontic and prosthodontic treatments might be necessary. In extreme cases, the extraction of the tooth might be discussed.

Early consultation with an orthodontist is advisable in order to keep the orthodontic requirements as simple as possible.

In general terms, the dental care/treatment of patients with RTS requires a multidisciplinary approach aimed at prevention. The collaboration of a paediatric dentist, an orthodontist, the cleft palate team and the treating general practitioner and medical team is of the outmost importance.

References

[675–694]

Pictures Illustrating the Oro-Dental Features

Figure 4.5

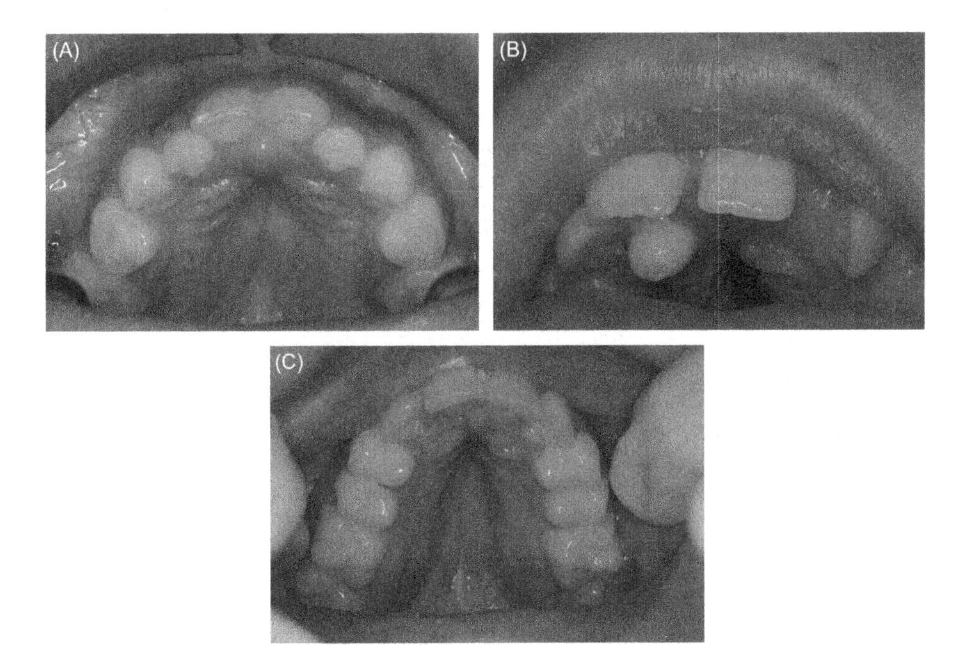

Figure 4.5 Rubinstein-Taybi syndrome. (A) Talon cusps or hyperdeveloped palatal cingulum of maxillary primary lateral incisors. (B) Extra teeth or mesiodens in the upper incisor region. (C) Wide alveolar ridge, narrow V-shaped maxilla with highly arched and narrow palate and grinded talon cusp on palatal face of 12 and 22.

4.2 Otodental Syndrome

Definition

This syndrome is characterized by enlarged bulbous teeth (globodontia) affecting both primary and permanent dentitions, and sensorineural hearing loss.

OMIM Number

#166750

Prevalence

Not yet determined. At least nine families have been described in the literature.

Inheritance

Autosomal dominant with variable expressivity and penetrance.

Aetiology – Molecular Basis – Gene

The *FGF3* gene, when mutated, is the gene involved in this syndrome [695].

Animal Models/Main Features

None reported in the literature. Transgenic animals display anomalies of hearing/vestibular/ear, nervous system, the limbs, digits and tails. No information exists about dental anomalies in these mutants [696–698].

Clinical Description

Main features

- Progressive bilateral sensorineural hearing loss above 1000 Hz which can start as early as in childhood

Cranio-oro-dental features

- Primary dentition more severely affected
- Large bulbous crowns of canines, premolars and molars
- Possible fusion of molar-premolar teeth: macrodontic teeth seemingly formed by fusion of smaller malformed teeth
- Partial or total absence of conoid premolars
- Incisors all having normal morphology
- Conoid supernumerary teeth
- Yellow hypoplastic spots in the enamel of canines and molars
- Vertical fissures in enamel
- Septated pulp chambers in molars
- Short roots
- Pulp stones
- Odontomas
- Delayed eruption

Management/Oral Health

Multidisciplinary management

- A preventive programme is mandatory in order to maintain proper oral hygiene.
- Due to the abnormal crown morphology, there is a great propensity to caries formation.
- There is a high rate of endodontic-periodontic lesions due to the aberrant coronal and pulpal morphology.
- Endodontic therapy can be difficult because of duplicated pulp canals in the affected posterior teeth.
- Multiple extractions may be needed and fixed, or removable prostheses should be constructed.
- Implants are also a potential consideration.

References

[695–708]

Pictures Illustrating the Oro-Dental Features

Figure 4.6

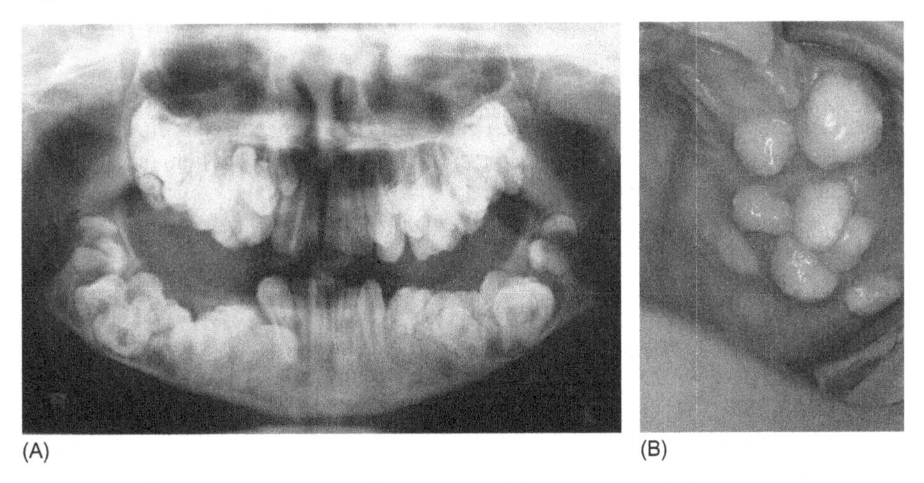

Figure 4.6 Otodental syndrome. (A) Panoramic radiograph showing multiple supernumerary teeth eventually fused to generate giant globodont teeth. The root formation is impaired. The anterior incisor sector seems normal. (B) Intraoral view of some of these multiple teeth elements encountered in the canine and posterior areas in otodental syndrome.

4.3 Oculo-Facio-Cardio-Dental Syndrome (OFCD)

Definition

OFCD is a very rare syndrome characterized by dental radiculomegaly, congenital cataract, facial dysmorphism and congenital heart disease.

Synonyms

Microphthalmia syndromic 2, MCOPS2.

OMIM Number

#300166

Prevalence

Around 20 affected families have been reported in the literature.

Inheritance

X-linked dominant (lethal in male). A skewed pattern of X inactivation has been demonstrated.

Aetiology – Molecular Basis – Gene

BCL6 co-repressor gene (*BCOR*) Xp11.4 is causative of the syndrome.

BCOR can mediate transcriptional repression by the oncoprotein BCL6 and has the ability to reduce transcriptional activation by AF9, a known mixed-lineage leukaemia (MLL) fusion partner.

BCOR mutation increases the osteo-dentinogenic potential of mesenchymal stem cells (MSCs) providing a molecular explanation for abnormal root growth [709].

Animal Models/Main Features

Some *Bcor* transgenic animals have been generated but no phenotype is currently published in the literature [710]. Bcor is expressed in both dental epithelium and the mesenchyme at mouse E11.5. Silencing of *Bcor* expression in dental mesenchymal cells at E14.5 causes dentinogenesis defects and retardation of tooth root development [711].

Clinical Description

Main features

Eye:

- Bilateral congenital cataracts
- Microphthalmia
- Visual impairment
- Secondary glaucoma
- Ptosis
- Exotropia, commonly

Heart:

- Cardiac anomalies: ventricular septal defect, atrial septal defect, mild cardiomegaly, ventricular and atrial hypertrophy, benign peripheral pulmonary stenosis, mitral valve prolapse

Face:

- Long faces
- Thick and curved eyebrows
- Thin nasal ridge with bifid tip
- Long philtrum

Oro-dental features

- Radiculomegaly of canines and sometimes premolars
- Oligodontia
- Fused teeth
- Supernumerary teeth
- Delayed tooth eruption
- Persistence of primary teeth
- Cleft palate or submucous cleft palate, bifid uvula in 50% of cases.

Management/Oral Health

Early diagnosis, recognition of this unique diagnostic feature that is the radiculomegaly and multidisciplinary management of oculo-facio-cardio-dental syndrome are essential.

References

[709–727]

Pictures Illustrating the Oro-Dental Features

Figure 4.7

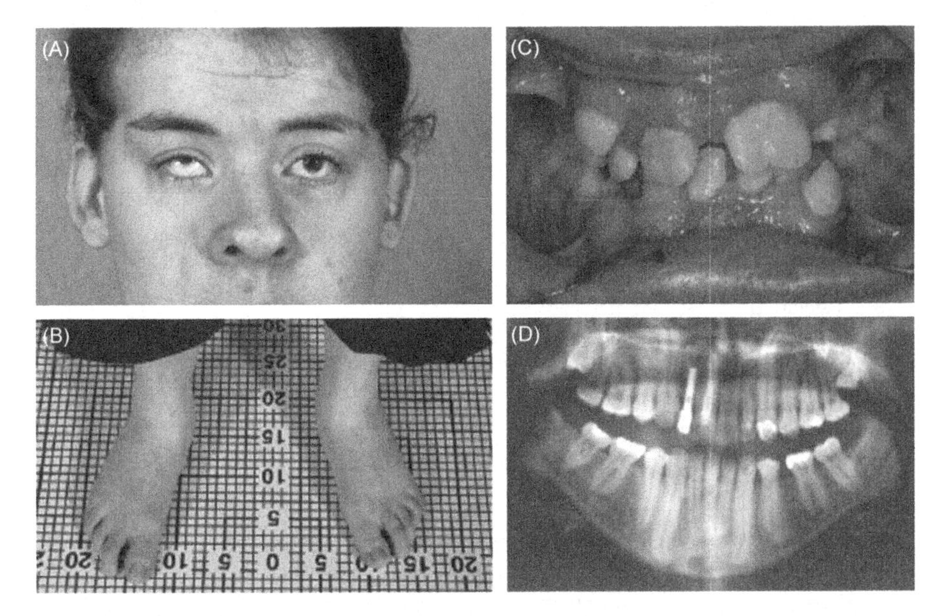

Figure 4.7 Oculo-facio-cardio-dental syndrome. Clinical features of OFCD. (A−D) Features include mild microphthalmia and dysmorphic appearance with (A) simple ears, elongated face and broad nasal tip, (B) hammer-toe deformity, (C) dental abnormalities with bilateral fused permanent upper incisors and (D) radiculomegaly of the canines and eventually premolars shown on orthopantomogram.
Source: From Ref. [724] Reprinted by permission from Macmillan Publishers Ltd, copyright (2004).

4.4 KBG Syndrome (Herrmann-Pallister-Opitz Syndrome)

Definition

The joint description by Herrmann, Pallister, Tiddy and Opitz was based on two families, whose initials made up the original term: KBG. The syndrome consists of typical facial dysmorphism associated with short stature, learning disability, skeletal anomalies and macrodontia of upper permanent central incisors.

OMIM Number

#148050

Prevalence

Around 50 patients have been described in the literature.

Inheritance

Autosomal dominant. A male-to-female ratio of 21 to 8 is reported.

Aetiology – Molecular Basis – Gene

Mutations in *ANKRD11* encoding ankyrin repeat domain 11, also known as ankyrin repeat-containing cofactor 1, cause KBG syndrome.

Animal Models/Main Features

None reported in the literature.

Clinical Description

Main features

- Short stature (less than 10th percentile)
- Learning disability (moderate to severe)
- Skeletal developmental anomalies
  - Vertebrae
  - Ribs (cervical ribs)
  - Hands (short hand tubular bones, brachyclinodactyly of fifth finger)
  - Delayed bone age
  - Scoliosis, kyphosis
  - Anomalies of the sternum (pectus excavatum)
  - Congenital hip dislocation, hip dysplasia
  - Dysplasia of the femoral heads and necks
- Abnormal EEG

Cranio-facio-oro-dental features

Face:

- Microbrachycephaly
- Round face

- Coarse hair, low frontal hairline
- Broad eyebrows
- Mild synophrys
- Telecanthus, hypertelorism
- Elevated nasal bridge
- Anteverted nostrils
- Long philtrum
- Thin upper vermillion
- Prominent ears
- Facial asymmetry

Teeth:

- Macrodontia (typically large upper central permanent incisors but may be generalized)
- Multiple mammelons on the incisal edge
- Shovel-shaped incisors
- Teeth agenesis (lateral upper permanent incisor)
- Fusion of teeth (central and lateral permanent incisors)
- Enamel hypoplasia
- Enamel pits
- Premature loss of teeth

4.4.8 Management/Oral Health

Multidisciplinary management

4.4.9 References

[728–745]

4.4.10 Pictures Illustrating the Oro-Dental Features

Figure 4.8

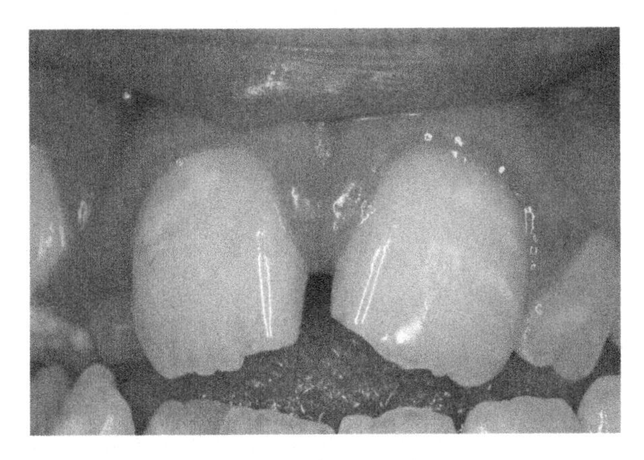

Figure 4.8 KBG syndrome. Macrodont permanent upper central incisors in KBG syndrome (notice the multiple incisor edge mammelons and the feeling of tooth fusion). Microdont, peg-shaped notched permanent maxillary lateral incisors – 12, 22. Enamel hypoplasia is present.

5 Anomalies in Structure of Teeth – Dentine

Syndrome	Associated Features	Gene	Molecules	Inheritance	Locus
Dentinogenesis imperfecta II DGI-II DGI-1 #125490	Defective dentine progressive hearing loss No other defects	*DSPP*	Three matrix proteins: DSP, DGP, DPP	AD	4q21.3
Dentinogenesis imperfecta III DGI-III #125500	Defective dentine	*DSPP*	Three matrix proteins: DSP, DGP, DPP	AD	4q21.3
Dentine dysplasia type II #125420	Coronal dentine dysplasia Opalescent deciduous teeth Pulp stones, short roots	*DSPP*	Three matrix proteins: DSP, DGP, DPP	AD	4q21.3
Dentine dysplasia type I #125400	Rootless teeth Short roots, pointed ends Radicular dentine dysplasia Aberrant dentine formation Premature loss of teeth Both dentition normal tooth morphology and colour	SMOC2?		AD	
Osteogenesis imperfecta (OI) with or without dentinogenesis imperfecta type I	Defective dentine and bone	*COL1A1*	Matrix protein (type I collagen)	AD, AR	17q21.31–q22
		COL1A2			7q22.1
		CRTAP	Cartilage-associated protein	AR	3p22
DGI-I		*LEPRE1*	Leprecan		
		Or P3H1		AR	1p34
		SERPINF1		AR	17p13.3
		SERPINH1		AR	11q13.5
		SP7, OSX		AR	12q13.13

(*Continued*)

Dento/Oro/Craniofacial Anomalies and Genetics. DOI: 10.1016/B978-0-12-416038-5.00005-6

(*Continued*)

Syndrome	Associated Features	Gene	Molecules	Inheritance	Locus
Ehlers–Danlos syndrome type I #130000 [746]	Skin hyperextensibility, articular hypermobility and tissue fragility Clinically normal-appearing dentition Abnormal appearance of the dentine in TEM	*COL1A1*	Matrix protein (type I collagen)		17q21.33
	Narrow maxilla	*COL5A1*	Collagen alpha-1(V)		9q34.3
		COL5A2	Collagen alpha-2(V)		2q31
Ehlers–Danlos syndrome type VIIc #225410 [746]	Severe joint hyperextensibility and mild stretchability and bruisability of the skin Hypodontia Deciduous teeth shape anomalies Dysplastic dentine defects Dysplastic roots Enamel attrition Gingival hyperplasia Open bite Micrognathia Recurrent mandibular subluxations	*ADAMTS2*	Procollagen protease	AR	5q35.3
Spondylometaphyseal dysplasia with dentinogenesis imperfecta (Goldblatt syndrome) #184260 [747,748]	Spondylometa-physeal dysplasia Joint laxity Dentinogenesis imperfecta			AR?	
Elsahy–Waters Branchio-skeleto-genital syndrome 211380 [749]	Intellectual disability, hypospadias and characteristic craniofacial morphology Radicular dentine dysplasia Dental cysts Prognathism Maxillary hypoplasia Bifid uvula, cleft palate	?	?	AR	

(*Continued*)

(*Continued*)

Syndrome	Associated Features	Gene	Molecules	Inheritance	Locus
Odontodysplasia	Regional, ghost teeth Affects primary and permanent teeth Defective enamel and dentine More frequent in the maxilla Might be associated with vascular malformation	?	?	?	?

Hereditary dentine disorders are a group of autosomal dominant inherited conditions affecting the primary and/or the permanent dentition and are due to mutations in the genes encoding the major protein constituents of dentine, that are, collagens and phosphoproteins.

These phosphoproteins have recently been grouped in the family of the SIBLINGs (small integrin-binding ligand, N-linked glycoproteins), containing osteopontin, bone sialoprotein, dentine matrix protein (DMP1), dentine sialo/phosphoproteins (DSPP), matrix extracellular phosphoglycoprotein (MEPE) and enamelin.

5.1 Dentinogenesis Imperfecta Type II and DGI-II or Hereditary Opalescent Dentine

Definition

Clinically, the abnormal dentine appears dull and bluish-brown, amber or opalescent, the teeth rapidly becoming worn from occlusal or biting stresses. Radiographically, the crowns appear bulbous, with a marked cervical constriction and pulp chambers; root canals are narrow or totally obliterated, thus absent, and roots are short. Both dentitions are involved.

OI is not a feature.

Synonyms

Hereditary opalescent dentine, DGI type II, DGI-II, opalescent dentine without OI, DGI Shields' type II, dentinogenesis imperfecta 1, DGI-1, Capdepont teeth.

OMIM Number

#125490

Prevalence

1 in 6000 to 1 in 8000.

Inheritance

Autosomal dominant with almost complete penetrance.

Aetiology – Molecular Basis – Gene (4q21.3)

DSPP gene on chromosome 4q21.3 encodes three major non-collagenous dentine matrix proteins: dentine sialoprotein (DSP), dentine glycoprotein (DGP) and dentine phosphoprotein (DPP). Defects in this gene cause both dentinogenesis imperfecta and dentine dysplasias that are allelic conditions. The dentine disease phenotype is dependent on mutation location.

Animal Models/Main Features

DSPP knockout mouse teeth display a widened predentine zone and develop defective dentine mineralization similar to human dentinogenesis imperfecta type III.

In a conditional *DPP* knockout, a significant recovery in the dentine volume, but not in the dentine mineral density, is observed, suggesting that DSP might regulate initiation of dentine mineralization and that DPP might be involved in the maturation of mineralized dentine [750,751].

Clinical Description

Main features

- Progressive sensorineural hearing loss is a rare feature (OMIM #605594) [751].
- There are no other systemic defects.

Cranio-oro-dental features

- The primary (usually more severely) and permanent dentitions are affected.
- The occlusal surface of the dental enamel is generally abraded.
- Hypoplastic enamel might be associated.
- Cervical crown constriction and bulbous crowns are present.
- Defective dentine: The dentine is heavily worn and uniformly shaded brown.
- The dental pulp chambers over time become partially or completely obliterated.
- Multiple pulp exposures exist.
- Multiple abscesses are present.
- Roots are short and constricted.

Management/Oral Health

A medical history should aim to establish whether the dental condition is a 'syndromic' form of DGI as this is a variable feature of a number of heritable conditions. It is important to ask patients with DGI about histories of bone fracture with minimal trauma, joint hyperextensibility, short stature, hearing loss and scleral hue. Genetic assessment and counselling is important.

In the primary dentition, stainless steel crowns on the molars may be used to prevent tooth wear and maintain the occlusal vertical dimension. The aesthetics may be improved using composite facings or composite strip crowns. Over-dentures might be recommended. General anaesthesia may facilitate treatment.

As the permanent dentition erupts, it should be closely monitored in relation to the rate of tooth wear with intervention only if necessary. Cast occlusal onlays on the first permanent molars, and eventually the premolars, help to minimize tooth wear and maintain the occlusal vertical dimension. The emphasis should be on minimal tooth preparation until the child reaches adulthood. At this point, if clinically indicated, a full mouth rehabilitation may be considered. Obliteration of the pulp chambers and root canals in teeth that develop abscesses makes endodontic therapy difficult if not impossible.

The replacement of premature tooth loss is with dentures. Dental implants may be considered after growth is complete at about 18 years of age. Maxillo-mandibular atrophy is a consequence of no or rudimentary root development and early tooth loss. Ridge augmentation prior to implants is often required.

Exposed dentine is more susceptible to tooth decay than enamel. For all patients, regular dental checkups and prevention of tooth decay in the form of oral hygiene instruction, dietary advice and appropriate use of fluoride is essential. Early diagnosis and regular dental care, however, cannot prevent premature tooth loss due to short or absent roots and spontaneous abscess formation that often occurs [753,754].

References

[72,750–765]

Pictures Illustrating the Oro-Dental Features

Figure 5.1

5.2 Dentinogenesis Imperfecta Type III (DGI-III)

Definition

This is a form of dentinogenesis imperfecta found in a population called the Brandywine (an inbred tri-racial population of Caucasians, African Americans and Native American Indians) isolate from Maryland and Washington, DC.

Synonyms

DGI type III, DGI-III, DGI Shields' type III, DGI-III Brandywine- type dentinogenesis imperfecta.

OMIM Number

#125500

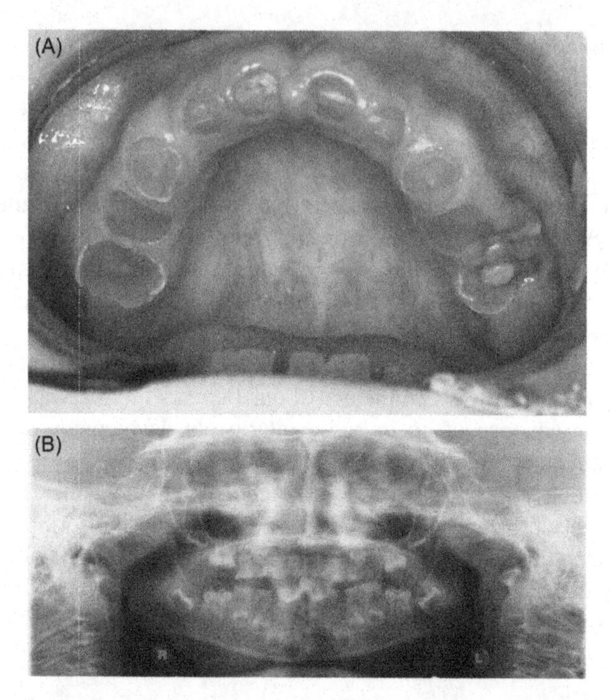

Figure 5.1 Dentinogenesis imperfecta type II. (A) Typical clinical aspects of the oral cavity in dentinogenesis imperfecta. The teeth appear amber and opalescent. The enamel shed off, leading to the rapid wear of the dentine. (B) Panoramic radiograph of the same patient affected with DI type II: cervical crown constriction, short roots and obliteration of pulp spaces are visible.

Prevalence

Within the isolate, the prevalence is estimated at 1 in 15 [766].

Inheritance

Autosomal dominant with almost complete penetrance.

Aetiology – Molecular Basis – Gene (4q21.3)

DSPP gene on chromosome 4q21.3 encodes three major non-collagenous dentine matrix proteins: DSP, DGP and DPP. Defects in this gene cause dentinogenesis imperfecta type III. DGI-II and DGI-III are not separate diseases, but rather the phenotypic variation of a single disease: they are allelic conditions [72,767,768].

Animal Models/Main Features

DSPP knockout mouse teeth display a widened predentine zone and develop defective dentine mineralization similar to human dentinogenesis imperfecta type III [750].

Clinical Description

Main features

Affected persons do not have stigmas of OI.

Cranio-oro-dental features

- Deciduous and permanent teeth are opalescent.
- There is marked attrition.
- Enamel pitting is present on some permanent teeth.
- The primary teeth show multiple pulp exposures.
- Radiographically,
  - The primary teeth often look like 'shell' teeth, indicating dentine hypotrophy [769].
  - Pulps of developing teeth are larger than normal during early development but rapidly become obliterated.
  - Large pulp chambers and root canals exist.
  - Radiolucencies of the apices are present.
  - Increased constriction at the cementoenamel junctions [770] occurs.
- Anterior open bites might be found in persons with complete permanent dentitions.

Management/Oral Health Multidisciplinary Management

See previous section on DGI-II. Differential diagnosis, genetic assessment and counselling and early intervention are essential.

References

[72,750,753,755,758,766–771]

5.3 Dentine Dysplasia

Definition

Dentine dysplasia consists of radicular dentine dysplasia characterized by short-rooted teeth with sharp conical apical constrictions and aberrant growth of dentine in the pulp chamber, leading to reduced pulp space in permanent teeth and total pulpal obliteration in the primary dentition.

Synonyms

Dentine dysplasia type I, dentine dysplasia type II, DD-I, DD-II, rootless teeth, radicular dentine dysplasia.

OMIM Number

#125420 dentine dysplasia type II
#125400 type I

Prevalence

1 in 100,000 (type I).

Inheritance

Autosomal dominant with complete penetrance.

Aetiology – Molecular Basis – Gene (4q21)

The only identified mutations causative of DD (type II) are located within the *DSPP* gene [772–774].

Animal Models/Main Features

None reported in the literature. Reduced expression of DSPP is associated with dysplastic dentine in mice overexpressing TGFβ-1 in teeth [774].

Clinical Description

Main features

Cranio-oro-dental features

There are two varieties, type I (radicular) and type II (coronal); only type I is associated with premature loss of teeth. Four distinct forms of dentine dysplasia type I and one form of dentine dysplasia type II are identified. The enamel seems normal but wears off. Type I, or radicular dentine, dysplasia:

- Type I affects both dentitions.
- Morphology and colour of the crown of the teeth are normal.
- Hypermobility of teeth: The roots are seen on radiographs as short and with pointed ends and conical apical constrictions.
- Aberrant growth of dentine leads to reduced pulp space in permanent teeth and total pulpal obliteration in the primary dentition.
- Teeth are lost generally due to trauma, which will easily induce exfoliation because of the roots shape and short length.
- There is delayed eruption.
- Opacity of the incisional margins is present.
- Periapical radiolucencies are often seen in non-carious teeth.

Type II:

- Primary teeth are amber and translucent, resembling DGI-II.
- Pulp chambers are obliterated by abnormal dentine.
- The permanent teeth have a normal coronal morphologic character and colour and seem either unaffected or show mild radiographic abnormalities.
- Permanent teeth have 'thistle tube'–shaped pulp chambers and multiple pulpal calcifications.

The clinical features characteristic of various forms of DGI and DD can be seen in different individuals of the same kindred.

Management/Oral Health

See previous sections on DGI type II. Patients with DD-I have mobile teeth due to very short roots and as a result tend to lose teeth early in the primary and permanent dentition. Until growth is complete, the treatment of choice for the replacement of missing teeth is dentures; thereafter, implants should be considered.

References

[109,753,755,756,758,759,772–786]

Pictures Illustrating the Oro-Dental Features

Figure 5.2

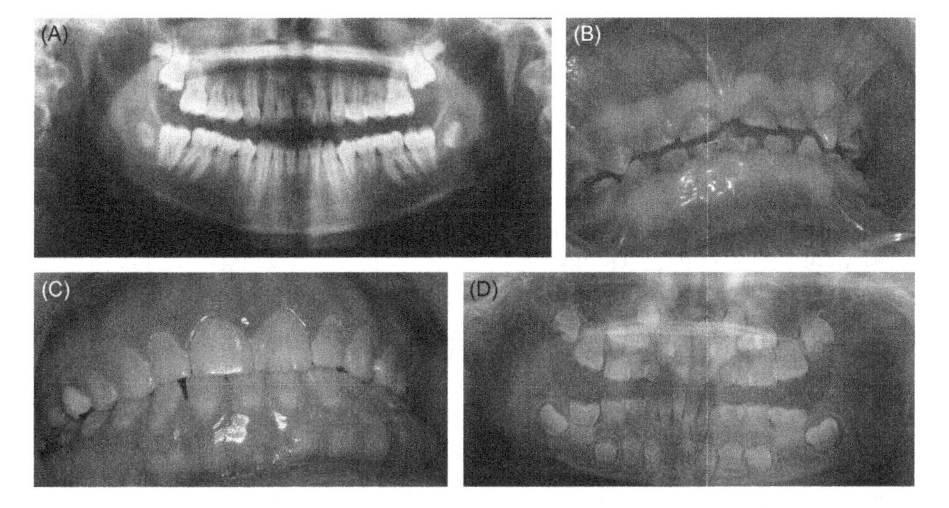

Figure 5.2 Dentine dysplasia. (A) DD type I with normal appearance of the teeth and short roots visible on X-ray. (B) DD type II in a boy. The dental phenotype in the primary dentition is similar to dentinogenesis imperfecta type II. (C) DD type II is the mother of the patient in (B). The clinical appearance of teeth is subnormal. (D) Another appearance of dentine dysplasia (Faculté de Chirurgie Dentaire de Nancy, Université Henri Poincaré, D Droz).

5.4 Osteogenesis Imperfecta (OI)

Definition

OI represents a group of conditions characterized by the formation of an abnormal bone with low bone mass and tendency to fractures. Some of those disorders are occasionally associated with abnormal dentine formation. The classic OI syndromes have been divided in four types: I, II, III and IV. Recently, types V, VI and

VII (AR) have been added to the list – but these types are not the result of mutations in the type I collagen gene. Additionally, several other varieties of OI have been recognized.

- Type I OI has been subdivided into IA, IB and IC.
- Type II OI has been subdivided into IIA, IIB and IIC.
- Type IV has been subdivided into IVA and IVB.

Recently, the discovery of new genes involved in OI, especially the recessive forms, has led to the reconsideration of the traditional classification in the light of the genetic and molecular data.

The correspondence and linking between OI and dentinogenesis imperfecta is still to be worked out.

Abnormal dentine formation is also a variable feature of other syndromes [787]: Ehlers–Danlos syndrome [746], Goldblatt syndrome [788], Schimke immuno-osseous dysplasia [789], branchio-skeleto-genital syndrome [790,791], osteodysplastic and primordial short stature with severe microdontia and opalescent teeth and rootless molars [792].

OMIM Number

#166200	OI type I	AD	
#166240	OI type IA with DI (10%)	AD	
#166210	OI type II, OI type IIA	AD	
#610854	OI type IIB perinatal lethal	AR	*CRTAP*
#259420	OI type III with DI (50%)	AD, AR	
#166220	OI type IVA without DI, IVB with DI (80%)	AD	
%610967	OI type V	AD	
#610968	OI type VI	AR	*SERPINF1*
#610682	OI type VII	AR	*CRTAP*
#610915	OI type VIII without DI	AR	*LEPRE1*
#259440	OI type IX with DI	AR	*PPIB*
#613848	OI type X with DI	AR	*SERPINH1*
#613849	OI type XI without DI but with delayed tooth eruption	AR	*SP7, OSX*

Prevalence

OI occurs in about 1 in 10,000–20,000 births.

Inheritance

- Type I and II are inherited as autosomal dominant AD. Individuals affected with type II will die during or shortly after birth.
- Type III: Around three-quarters of the cases are due to a new autosomal dominant mutation, and the remaining quarter of cases is inherited as autosomal recessive AR.
- Types IVA and IVB are inherited as autosomal dominant.

Aetiology – Molecular Basis – Gene

In approximately 90% of individuals with OI, mutations in either of the genes encoding the pro-alpha-1 or pro-alpha-2 chains of collagen type I *COL1A1* (17q21.31–q22.05) or *COL1A2* (7q22.1) can be identified.

Of those without collagen mutations, a number of them will have mutations involving the enzyme complex responsible for post-translational hydroxylation of the position 3 proline residue of COL1A1. Two of the genes encoding proteins involved in that enzyme complex – leprecan *LEPRE1*, also named prolyl 3-hydroxylase P3H1 (OI type VIII), and cartilage-associated protein (*CRATP*) (type IIB, VII) – when mutated have been shown to cause autosomal recessive OI, which has a moderate to severe clinical phenotype, often indistinguishable from OI types II or III.

Recessive forms of OI result from mutations in collagen-modifying enzymes and chaperones like *CRTAP*, *LEPRE1*, *PPIB* and *FKBP10*, and in *SERPINH1*, which encodes the collagen chaperone-like protein HSP47 [793].

Animal Models/Main Features

Mutations in the *Col1a1* locus cause variable phenotypes, from embryonic lethal to viable/fertile with altered fibrillogenesis. Homozygotes can show impaired bone formation and fragility, osteoporosis, dermal fibrosis, impaired uterine postpartum involution and aortic dissection. In *Col1a2* mutation, dentinogenesis imperfecta is present in both homozygote and heterozygote mice. A deletion in the gene encoding sphingomyelin phosphodiesterase 3 (*Smpd3*) results in osteogenesis and dentinogenesis imperfecta in the mouse.

The various animal models duplicate the phenotype and biochemistry of the human disease [794–798].

Clinical Description

Main features

Type I:

- Mild to moderately severe bone fragility
- Blue sclera
- Hearing loss

Type II (all subtypes):

- Infants are stillborn or die shortly after birth.
- Severe bone fragility.
- All bones can be affected.
- Bowing of legs.
- Curvature of the spine (kyphosis and scoliosis).
- Shortening of arms and legs.
- Blue sclerae.
- Newborns present with many fractured bones at birth.

Type III:

- Moderately severe to severe bone fragility
- Blue sclera in infancy which becomes normal with age
- Not lethal in infancy, but death common during the first and second decades of life

Type IV (all subtypes):

- Mild to moderately severe bone fragility.
- Sclerae, which may be pale blue during childhood, are normal.
- Hearing loss.

Cranio-oro-dental features

- **Type IA:** Teeth are normal.
- **Type IB:** Teeth are opalescent.
- **Type IC:** Teeth present alterations similar to those seen in dentine dysplasia type II (coronal dentine dysplasia).
- Crowns and roots are generally smaller than normal.
- Pulp chambers are also smaller.
- Teeth are opalescent or translucent at eruption time, but they darken with age.
- The enamel is normal but generally lost because the abnormal dentine cannot provide adequate support.

Type II:

- Deformity of the skull
- Globular dentine
- No predentine
- Abnormal pulp–dentine junction
- Grossly dilated capillaries in the coronal pulp

Type III:

- Some patients present dental changes.
- The dental abnormalities are similar to those described for the previous types of OI.

Type IVA:

- Teeth are normal.

Type IVB:

- Teeth are opalescent and similar to those seen in OI type IB.
- Dental changes are more common in OI type IV than in OI type I.

Management/Oral Health

- Oral hygiene and health protocol
- Reconstruction of anterior teeth especially maxillary ones with direct composite veneers
- Stainless steel crowns
- Extractions
- Orthodontics
- Use of osteodistraction and implants

- Biphosphonates alendronate and pamidronate could be used in the systemic treatment of patients with OI

References

[794–797,799–823]

Pictures Illustrating the Oro-Dental Features

Figure 5.3

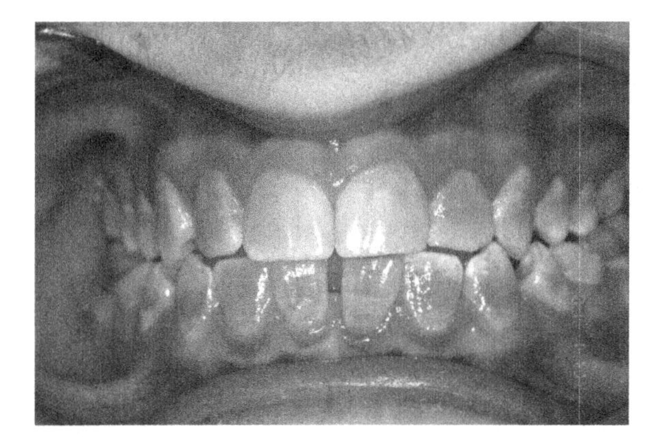

Figure 5.3 OI with dentinogenesis imperfecta.

6 Anomalies Structure of Teeth – Enamel

The amelogenesis imperfecta(s) (AI) represent a group of inherited alterations in the composition of tooth enamel that are not associated with a metabolic disorder or a syndrome, but some of them are seen in conjunction with other anomalies like taurodontism, dentine defects [825], eruption/resorption anomalies or even with skeletal anterior open bite [786,826]. The prevalence of all forms of AI varies with different countries, and it ranges from 1 in 4000 in Sweden to 1 in 15,000 in the United States.

The AIs have been classified following their clinical manifestations and the sequence observed in normal tooth development. Thus, if the alteration took place during enamel matrix formation, it results in hypoplasia, a rather quantitative defect; if it took place during the first step of apposition (e.g. calcification and mineralization), it results in hypocalcification; and if it took place in the second step of apposition (maturation), it results in hypomaturation, better described as qualitative defects. A combination of hypoplasia and hypomineralization, hypomaturation is also recognized.

The majority of AIs (63%) are transmitted according to an autosomal dominant pattern, 12% are autosomal recessive, 6% X-linked and in 19% the modalities of transmission were impossible to define. There have been a plethora of publications dealing with the molecular genetic aspects of AI [108,130,827–834].

Several genes have been identified as the responsible genes for some of the AI varieties, but not for all. Among the identified genes are *AMELX* [828], *ENAM* [835,836], *ENAMELYSIN* or *MMP20* [837], *KALLIKREIN 4* [833], *DLX3* [838], *FAM83H* [839–841] and *WDR72* [842]. Other genes involved in enamel formation have not yet been identified in association with amelogenesis imperfecta *AMELY* [843], *AMELOBLASTIN* [844], *TUFTELIN* [845], *AMELOTIN* [80] and odontogenic ameloblast-associated protein (*ODAM*) [846].

Clinically, it is not always easy to properly differentiate among the various types.

Further genetic typing is necessary in order to properly match clinical appearance with genetic make-up and consequently lead to adequate genetic counselling and treatment.

Dento/Oro/Craniofacial Anomalies and Genetics. DOI: 10.1016/B978-0-12-416038-5.00006-8

Syndrome	Associated Features	Gene	Molecules	Inheritance	Locus
Amelogenesis imperfecta					
Amelogenesis imperfecta AIH1 #301200	Hypoplasia, hypomineralization, hypomaturation	*AMELX*	Matrix protein (amelogenin)	X	Xp22.3–p22.1
Amelogenesis imperfecta Type IB; AI1B = AIH2 #104500 Type IC; AI1C #204650	Hypoplastic form • Smooth thin • Local hypoplastic • Pits • Anterior open bite	*ENAM*	Matrix protein (enamelin)	AD AR	4q21
Amelogenesis imperfecta, type III, AI3 #130900	Hypomineralization	*FAM83H*	family with sequence similarity 83, member H protein of unknown function	AD	8q24.3
Amelogenesis imperfecta and gingival hyperplasia #614253 [824]	Gingival hyperplasia	*FAM20A*	Secreted glycoprotein	AR	17q24.2
Amelogenesis imperfecta #204700 IIA1; AI2A1	Hypomaturation pigmented	*KLK4*	Enamel matrix serine proteinase 1 emsp1 protease, serine, 17	AR	19q13.3–q13.4
Amelogenesis imperfecta #612529 IIA2, AI2A2	Hypomaturation	*MMP20*	Enamelysin matrix metalloproteinase 20	AR	11q22.2
Amelogenesis imperfecta #613211 IIA3, AI2A3	Hypomaturation	*WDR72*	Beta propeller scaffold for protein–protein interactions	AR	15q21.3
Amelogenesis imperfecta AIH3 %301201	Hypoplasia, hypomaturation	?		X	Xq22–q28
Amelogenesis imperfecta #104510	Hypoplasia Hypomaturation	*DLX3*	Transcription factor	AD	17q21.3–q22

TYPE IV; AI4	Taurodontism No associated osseous, hair, non-dental features				
Tricho-dento-osseous syndrome #190320	Enamel hypoplasia Taurodontism Hair and bone defects	*DLX3*	Transcription factor	AD	17q21.3
Tricho-onycho-dental (TOD) syndrome	Hypoplastic enamel, taurodontism, dentine dysplasia type II bones in TOD are normal	?		AD	
Amelogenesis imperfecta and nephrocalcinosis 204690	Amelogenesis imperfecta hypoplastic type, nephrocalcinosis, enuresis and polyuria	?	?	AR	?
Amelogenesis imperfecta and cone–rod dystrophy #217080		*CNNM4*	Putative metal transporter	AR	2q11
Amelogenesis imperfecta and platispondyly #601216	Short stature Platispondyly Absent enamel Multiple abscesses	?	?	AR	?
Kohlschutter-(Tonz) syndrome #226750 (Schossig...Zschocke et al, AJHG in press)	Amelogenesis imperfecta, epilepsy and learning impairment	ROGDI	?	AR	16p13.3
Amelo-onycho-hypohidrotic syndrome %104570	Enamel, fingernails and toenails and sweat abnormalities	?	?	AD	?
Amelogenesis imperfecta, Heimler syndrome 234580	Sensorineural hearing loss, enamel hypoplasia (thin enamel) and nail abnormalities	?	?	AR	?

(*Continued*)

(Continued)

Syndrome	Associated Features	Gene	Molecules	Inheritance	Locus
Syndromes with enamel defects					
Epidermolysis bullosa simplex	Oral blisters	*Keratin 5 (KRT5)*		AD	12q13
		Keratin 14 (KRT14)			
#131760 (EBS Dowlin-Meara type)	Teeth not affected				17q12–q21
#131800 (EBS Weber-Cockayne type)					
#131900 (EBS Koebner type)					
Epidermolysis bullosa junctional EBJ #226700 and non-Herlitz type (#226650)	Pitted enamel, skin and oral blisters	*LAMA3*	Matrix protein (laminin-5)	AR	18q11.2
		LAMB3			1q32
		LAMC2			1q25–q31
		COL17A1			10q24.3
Epidermolysis Bullosa Dystrophic EBD #226600		*COL7A1*		AR	3p21.3
		MMP1, gene modifier			11q22–q23
Tuberous sclerosis # 191100	Hamartoma formation in the skin, nervous system, heart, kidney Pitted enamel	*TSC1*	Hamartin	AD	9q34
		TSC2	Tuberin Tumour growth suppressors		16p13.3

APECED #240300	Enamel hypoplasia, polyendocrinopathy candidosis	*AIRE*	Transcription factor	AR	21q22.3
Albright hereditary osteodystrophy	Enamel hypoplasia, short stature, obesity, round faces	*GNAS1*	Signal transducer	AD	20q13.2
#103580	subcutaneous ossifications, brachydactyly and so on	*Guanine nucleotide-binding proteins alpha-stimulating 1*			
Oculodentodigital dysplasia ODDD #164200	Enamel hypoplasia, microdontia	*GJA1*	Gap junction protein connexin 43	AD	6q21–q23.2
Pycnodysostosis # 265800	Short stature, increased bone density, fragile bones Enamel hypoplasia, oligodontia, delayed tooth eruption, malocclusion	*CTSK*	Cathepsin K	AR	1q21
Vitamin D-dependent rickets #264700 (type I), #277440 (type II)	Growth failure, hypotonia, weakness, rachitic rosary, convulsions, tetany and pathologic fractures Hypoplastic enamel similar to hypocalcified type of amelogenesis imperfecta	*CYTOCHROME P450, CYP27B1*	Renal 1-alpha-hydroxylase	AR	12q13.3

6.1 Amelogenesis Imperfecta

6.1.1 Amelogenesis Imperfecta 1, Hypoplastic/Hypomaturation Type AIH1

Definition

The enamel phenotype consists of decreased enamel matrix formation (hypoplasia). Enamel is thinner than normal in focal or generalized areas. Radiodensity of enamel is greater than that of dentine, or the enamel can be hypomineralized or hypomature (radiodensity of enamel is decreased).

The teeth may present a thin layer of enamel of normal colour and translucency, or the enamel may be of normal thickness but poorly mineralized with loss of translucency and/or a yellowish-brown discolouration. In some families, the phenotype appears to be both hypoplasia and abnormal mineralization occurring together. When hypoplasia is the exclusive or predominant phenotype, there may be marked sensitivity of the teeth to thermal and osmotic stimuli.

Synonyms

Amelogenesis imperfecta, X-linked 1; AIH1 enamel hypoplasia, X-linked amelogenesis imperfecta, hypomaturation type, with snow-capped teeth.

OMIM Number

#301200

Prevalence

Reports vary widely with values from 1 in 14,000 to 1 in 700, all types of AI included.

Inheritance

X-linked dominant trait.

Aetiology – Molecular Basis – Gene

AMELX codes for the amelogenin (Xp22.3–p22.1).

AIH1 shows the typical pattern of X-linked inheritance. Heterozygous females can pass on the mutant gene to children of either sex, with the risk of this being 50%. The condition affects males and females in strikingly different ways. Males show the trait fully. By contrast, females who inherit the mutant gene have vertical markings of the enamel as a result of X chromosome inactivation or Lyonization. Thus, these heterozygous females may have teeth with vertical ridges and grooves as a result of hypoplasia of the enamel or have vertical bands of alternating normal and discoloured enamel.

Animal Models/Main Features

The amelogenin null phenotype shows an amelogenesis imperfecta phenotype and reveals that the amelogenins are not required for initiation of mineral crystal formation but rather for the organisation of crystal pattern and regulation of enamel thickness.

Enamel from affected mice with a deletion at the *Arhgap 6/Amel X* locus appears chalky white, and molars show excessive wear. The enamel layer is hypoplastic and non-prismatic, whereas other dental tissues are normal [847–849].

Clinical Description

Main features

Cranio-oro-dental features

- All teeth are affected.
- The enamel defects can be hypoplasia: smaller and pitted teeth, reduced enamel thickness; hypomineralization or hypomaturation, implying variation in colour from white opaque to yellow to brown.
- The pulps and dentine are normal.
- A skeletal anterior open bite is seen in approximately 50% of patients with X-linked AI.

Management/Oral Health

Multidisciplinary management:

Appropriate diagnosis, genetic assessment and counselling are essential. Prevention, oral health advice and appropriate use of fluoride are mandatory.

In infancy, the primary dentition is protected by the use of preformed metal crowns on posterior teeth. Either polycarbonate crowns or composite restorations are used on anterior teeth. Treatment might require the use of sedation (e.g. nitrous oxide and midazolam) or general anaesthesia.

The eruption of the permanent dentition presents a particularly difficult period. Some of the forms of AI present with hypersensitive teeth or with teeth that crumble, and both presentations provide a very real disincentive to good oral hygiene and are also very difficult to restore. Those cases with enamel which is reasonably hard (i.e. less hypomineralized) and thin (i.e. more hypoplastic) lend themselves fairly readily to the use of preformed metal crowns on posterior teeth as they erupt and composite restorations on anterior teeth. These latter may need to be added to as more of the cervical part of the tooth is revealed. Restorative treatment requires local analgesia.

It is important that both a restorative dentist and an orthodontist be involved with the paediatric dentist in the care plan from the child's early age. The anterior open bite seen in some cases of AI requires consideration of surgical as well as restorative management.

The longer-term care still revolves around either crowns or, more frequently these days, adhesive, plastic restorations and implants [834].

References

[129,783,828–830,834,843,850–860]

Pictures Illustrating the Oro-Dental Features

Figure 6.1

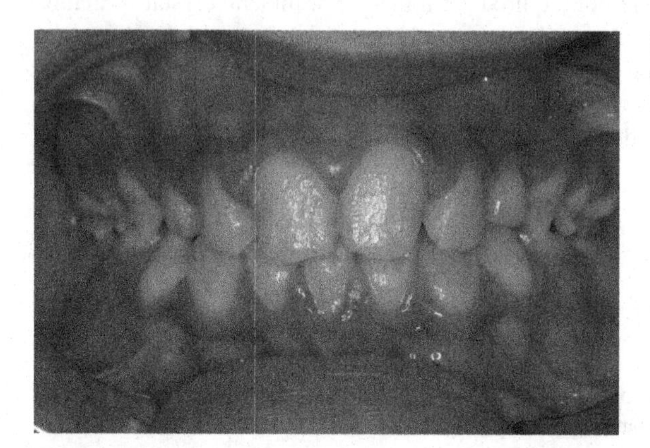

Figure 6.1 Hypoplastic form of amelogenesis imperfecta.

6.1.2 Amelogenesis Imperfecta (AI) Hypoplastic Type IB, IC

Definition

The enamel is thinner than normal, and on radiographs it contrasts normally with dentine. Enamel matrix formation is diminished and abnormal.

OMIM Number

#104500 type IB
#204650 type IC

Prevalence

For all varieties, up to 1 in 15,000.

Inheritance

Autosomal dominant (type IB) (with variation in expressivity as per the number of teeth affected and the severity of the lesion) or autosomal recessive (Type IC).

Aetiology – Molecular Basis – Gene

The *ENAM* (enamelin) gene has been mapped to 4q21. It encodes an enamel matrix protein.

Animal Models/Main Features

The phenotypes due to enamelin inactivation in mice, identified through a dominant ethylnitrosourea screen (large-scale production of mouse mutants using the alkylating agent ethyl-nitrosourea (ENU), a highly potent mutagen known to induce point mutations; it has been used in forward (phenotype-based) genetic screens with which one can identify and study a phenotype of interest and resembles hypoplastic AI from a hypomaturation type to complete loss of enamel on incisors and molars. Homozygotic mutants showed total enamel aplasia with exposed dentinal tubules, whereas heterozygotic mutants showed a significant reduction in enamel thickness. Dentine defects were also observed [861–863].

Clinical Description

Cranio-oro-dental features

- AI may affect only primary teeth or both dentitions.
- All or only some teeth may show the defect.
- Some teeth may appear normal in both dentitions.
- Random pits occur on the enamel, pinpoint to pinhead in size.
- Pits are often arranged in rows and columns.
- AI is present mostly in labial and buccal surfaces, mainly in anterior teeth.
- The lingual surface is affected to a minor degree.
- The enamel can also be rough and thin.
- Or it can be local hypoplastic enamel.
- Horizontal pits, grooves or linear depressions occur in the middle third of the buccal surface of the teeth.
- Skeletal anterior open bite is present in 50% of cases.

References

[832,835,836,850,863–871]

Management Oral/Health

See 6.1.1.

Pictures Illustrating the Oro-Dental Features

Figure 6.2

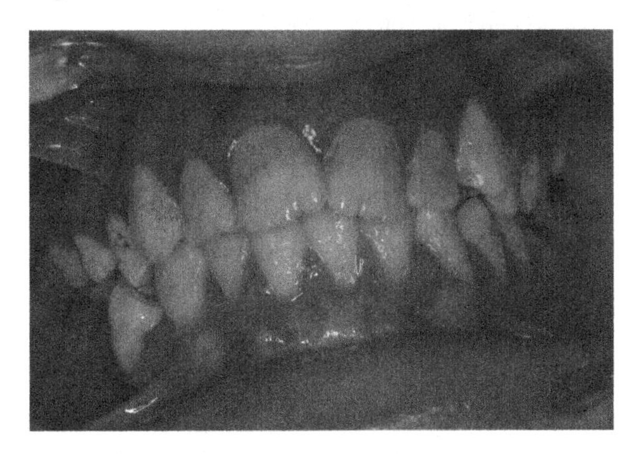

Figure 6.2 Hypoplastic form of amelogenesis imperfecta with enamel pitting.

6.1.3 Amelogenesis Imperfecta, Type III (AI3)

Definition

The enamel has normal thickness and a markedly decreased mineral content.

OMIM Number

#130900

Prevalence

For all varieties AI, up to 1:15,000.

Inheritance

Autosomal dominant.

Aetiology – Molecular Basis – Gene

The *FAM83H* gene has been mapped to 8q24.3. The phenotypes may demonstrate variations according to the mutations. Fam83h localizes in the intracellular environment, is associated with vesicles and plays an important role in dental enamel formation. *FAM83H* is the first gene involved in the aetiology of amelogenesis imperfecta that does not encode a secreted protein [872].

Animal Models/Main Features

The mouse genome informatics website (http://www.informatics.jax.org/) mentions some possible mouse models, but no phenotypic data are available [710].

Clinical Description

Cranio-oro-dental features

- Hypocalcified amelogenesis imperfecta.
- Soft enamel.
- In some families, the phenotype may affect primarily the cervical enamel.
- Normal hardness of dentine.

Management/Oral Health

See 6.1.1.

References

[840,841,872–875]

Pictures Illustrating the Oro-Dental Features

Figure 6.3

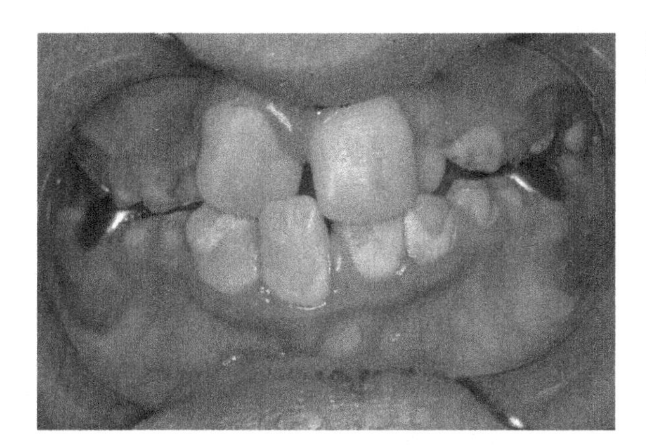

Figure 6.3 Hypomineralized amelogenesis imperfecta.

6.1.4 AI Pigmented Hypomaturation Type IIA1, AI2A1; IIA2, AI2A2; IIA3, AI2A3

Definition

Enamel is of normal thickness, but it is mottled and softer than normal, chipping from the crown. The tooth enamel contains large amounts of enamel matrix and can be penetrated by the tip of an explorer. On radiographs, the enamel has about the same radiodensity as dentine.

OMIM Number

#204700, #612529, #613211

Prevalence

For all varieties, up to 1 in 15,000.

Inheritance

Autosomal recessive.

Aetiology – Molecular Basis – Gene

The hypomaturation type of amelogenesis imperfecta is due to a mutation in the *KLK4* (Kallikrein) (type IIA1, AI2A1) gene which maps to 19q13.2 or due to a mutation in the *MMP20* (enamelysin) (type IIA2, AI2A2) gene which maps to 11q22.3–q23. MMP20 and KLK4 are proteinases secreted by ameloblasts, MMP20 being the early one and KLK4, the later one, and both of them are active during the maturation stage of enamel formation. MMP20 regulates crystal elongation, normal architecture of the dentino-enamel junction and the organisation and maintenance of the enamel rods. KLK4 is a degrading enzyme with proteolytic effects on the enamel matrix, clearing enamel proteins from it.

Mutations in the gene beta propeller *WDR72*, which is critical to ameloblast vesicle turnover during enamel maturation, cause autosomal-recessive hypomaturation AI (type IIA3, AI2A3) [842].

Animal Models/Main Features

MMP-20 appears responsible for enamel matrix organisation and for subsequent resorption of enamel matrix proteins. The murine *Mmp20* null mouse exhibits both hypoplastic and hypomineralized defects [853,876].

Clinical Description

The *MMP20* deficient and the *KLK4* deficient AI share many similar clinical features which will be described in common (*vide infra*), the main difference is the colour being a homogeneous dark yellow hue for the KLK4 teeth, while the MMP20 teeth have an irregular greyish-brown discolouration and are somewhat glossier.

The AI related to *WDR72* mutations has near-normal volumes of organic enamel matrix but with weak, creamy-brown opaque enamel that fails prematurely after tooth eruption.

Cranio-oro-dental features

- The enamel crowns are normal in size and shape.
- The enamel surface is rougher, duller and less reflective than normal enamel.
- Enamel appear to be more brittle and show a tendency to fracture or chip, but it is not particularly susceptible to dental caries.
- The radiodensity of the enamel is generally less than that of normal enamel, but on radiographs it can still be distinguished from the underlying dentine.
- As mentioned earlier, the KLK4 teeth have a homogeneous dark yellow hue, and the MMP20 teeth have an irregular greyish brown discolouration and are a little glossier.

References

[79,130,833,837,853,887–890, El-Sayed, 2009 #2899]

Pictures Illustrating the Oro-Dental Features

Figure 6.4

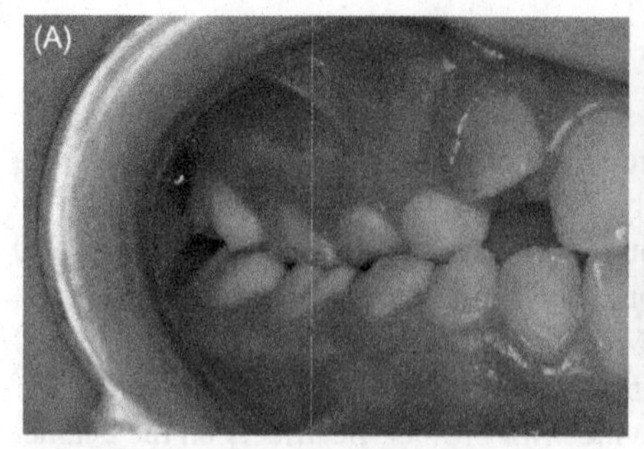

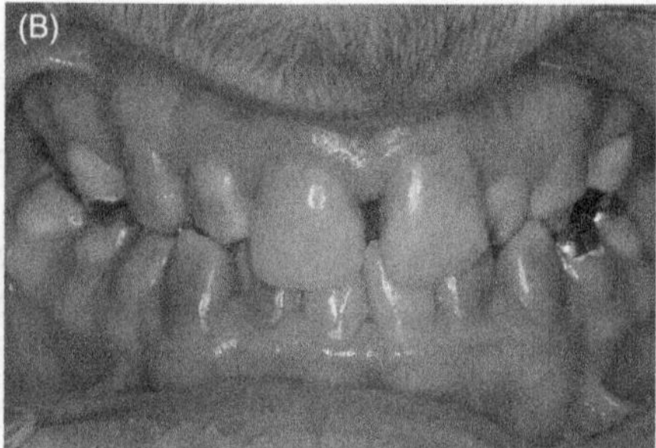

Figure 6.4 (A) Hypomaturation amelogenesis imperfecta with *MMP20* mutation. (B) Hypomaturation amelogenesis imperfecta with *WDR72* mutation.

6.1.5 Amelogenesis Imperfecta AIH3

Definition

Amelogenesis imperfecta hypoplastic type; also named amelogenesis imperfecta, hypoplastic/hypomaturation, X-linked 2.

OMIM Number

%301201

Prevalence

For all varieties AI, up to 1 in 15,000.

Inheritance

X-linked.

Aetiology – Molecular Basis – Gene

A second locus on chromosome X (Xq22–q28) has been associated with AI in one family [891].

Animal Models/Main Features

None reported in the literature.

Clinical Description

Cranio-oro-dental features

- Enamel hypoplasia

References

[891]

6.1.6 Amelogenesis Imperfecta Type IV, AI4

Definition

Amelogenesis imperfecta hypoplastic–hypomaturation with taurodontism.

OMIM Number

#104510

Prevalence

For all varieties AI, up to 1 in 15,000.

Inheritance

Autosomal dominant.

Aetiology – Molecular Basis – Gene

The *DLX3* gene, which encodes a transcription factor, has been mapped to 17q21.3–q22.

This gene is also responsible for the tricho-dento-osseous syndrome: some authors suggest the two conditions may be allelic and overlapping [892].

Animal Models/Main Features

Animal models exist for the inactivation of *Dlx3*, homozygous null mutants die at embryonic day E9.5–10.0 [893].

Selective ablation of *Dlx3* in the epidermis results in complete alopecia – establishing that Dlx3 is essential for hair morphogenesis, differentiation and cycling programmes. No data exist about a possible dental phenotype [894]. This mouse is a model for TDO.

Clinical Description

Cranio-oro-dental features

- Enamel hypoplasia
- Enamel hypomaturation
- Taurodontism
- No associated osseous, hair, non-dental features

References

[838,892,895]

6.1.7 Tricho-Dento-Osseous Syndrome (TDO)

Definition

The syndrome belongs to the ectodermal dysplasias and consists of kinky or curly hair; enamel hypoplasia; and sclerosis of the skull calvarium and/or the long bones. There may be three distinct types of TDO syndrome which may be differentiated mainly by whether the calvaria and/or long bones exhibit sclerosis, thickening and/or density.

OMIM Number

#190320

Prevalence

There are well-documented kindreds in geographical proximity in the United States (Washington County, Virginia; Holston Valley, Tennessee; and Alamance County, North Carolina), and isolated cases around the world.

Inheritance

Autosomal dominant.

Aetiology – Molecular Basis – Gene (17q21.3)

The distal-less (*DLX*) homeobox genes are expressed in both epithelial and mesenchymal craniofacial tissues during embryogenesis. Mutations in the *DLX3* transcription factor are seen in TDO.

Animal Models/Main Features

Animal models exist for the inactivation of *Dlx3*, homozygous null mutants die at embryonic day E9.5–10.0 [893]. The following systems are affected skin/coat/nails, growth/size and endocrine/exocrine.

Selective ablation of *Dlx3* in the epidermis results in complete alopecia – establishing that Dlx3 is essential for hair morphogenesis, differentiation and cycling programs. No data exist about a possible dental phenotype. This mouse is a model for TDO [893].

Enhanced trabecular bone volume and mineral density are observed in these transgenic mice suggesting a novel role for this specific *DLX3* mutation in osteoclast differentiation and bone resorption [896].

Clinical Description

Main features

- Hair: kinky, coarse and/or curly hair is present at birth in 80% of cases.
- Bone: long bones exhibit sclerosis.
- Nails: abnormally thin, brittle.
- The phenotype might be less severe [897].

Cranio-oro-dental features

- Changes in the cranial bones are seen in most individuals with TDO – an increased cranial thickness, obliterated diploë and no visible mastoid pneumatization. Some have craniosynostosis, causing dolicocephaly (head abnormally long and narrow).
- Taurodontism is seen in virtually 100% of dentate individuals with TDO but was not seen in a one family [892].
- Enamel hypoplasia is seen in 100% of the dentate individuals with TDO. Dental abscess formation appears to result from pulp exposures that occur when the thin enamel covering is worn through, allowing the enlarged pulp chambers to be exposed directly to the oral cavity.
- Discoloured teeth were present in 76% of affected individuals. Enamel alterations range from being extremely thin and/or rough and pitted to being of normal colour and only slightly decreased thickness.
- Defective dentine, filling of tooth pulps with amorphous denticle-like material and markedly delayed or advanced dental maturity have also been reported.

References

[892–894,896–917]

6.1.8 Tricho-Onycho-Dental (TOD) Syndrome

Definition

A form of ectodermal dysplasia characterized by scanty, fine, curled hair, thin dysplastic nails, taurodontic molars, hypoplastic–hypomature enamel, dysplastic dentine and hypohidrosis.

OMIM Number

Number still to be assigned

Prevalence

Not yet determined, rare.

Inheritance

Autosomal dominant.

Aetiology – Molecular Basis – Gene

Not yet identified.

Animal Models/Main Features

None reported in the literature.

Clinical Description

Main features

- Scanty, fine, curly scalp hair
- Axillary, pubic and facial hair are also scanty
- Thin malformed nails with longitudinal striations, especially toe nails
- Diminished sweat secretion (hypohidrosis)

Cranio-oro-dental features (detailed)

- Square, taurodontic teeth
- Markedly thin, mottled enamel, brownish-yellow in colour
- Wide spaces between teeth due to loss of enamel with consequent loss of contact points
- Enamel rapidly abraded, exposing the dentine, with consequent early caries formation
- Short teeth roots due to apical migration of the bifurcation
- Radiographs showing obliterated pulp chambers similar to those seen in teeth with dentine dysplasia type I (radicular dentine dysplasia)
- Periapical radiolucencies common, most likely as a consequence to infections subsequent to caries; fistulous tracts also a frequent finding, again, associated with deep caries and pulpar necrosis
- Histologically, enamel thinner than normal, with poorly defined enamel rods
- Pulp chambers partially occupied by abnormal masses of calcified dentine, generally secondary and tertiary dentine

- Dentine traversed by dentinal tubules which are abnormal in size and shape
- Mantle dentine (dentine adjacent to the enamel and the cementum) within normal limits

In TOD, the hair and nail appearance are less marked than in TDO as well as the bone findings. Teeth are also taurodontic, enamel is hypoplastic and the dentine is similar to that see in dentine dysplasia (in TDO only the enamel is affected). The bones in TOD are essentially normal as opposed to in TDO. Both TOD and TDO are inherited as autosomal dominant. They might be allelic diseases.

Management/Oral Health

- Oral hygiene protocol
- Treatment of caries, sinus tracts, periodontal and periapical pathology, if present
- Careful evaluation of the use of prosthetic intraoral appliances, including crowns and/or partial dentures
- Implants, if needed, when the patient reaches the appropriate age
- Dental prosthesis

References

[918]

Pictures Illustrating the Oro-Dental Features

Figure 6.5

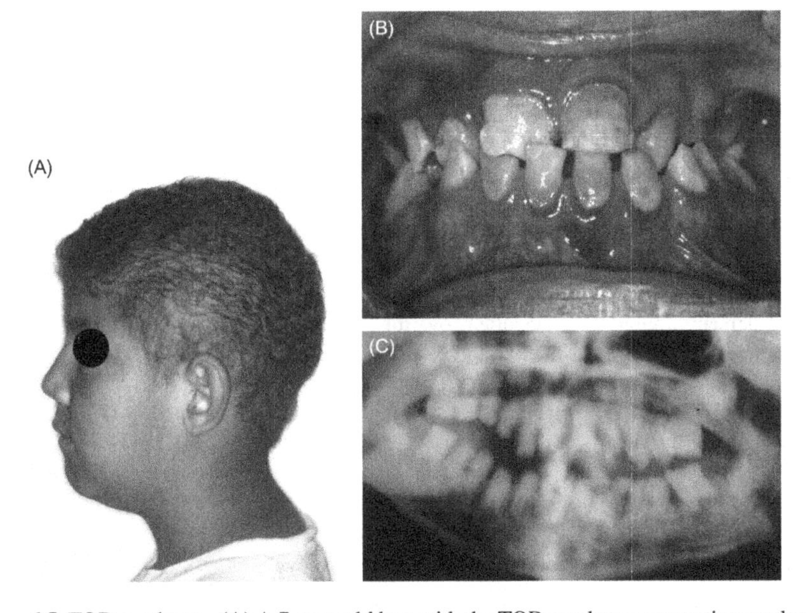

Figure 6.5 TOD syndrome. (A) A 7-year-old boy with the TOD syndrome presenting markedly curly and scanty scalp hair. (B) This intraoral photo of the same patient shows enamel hypoplasia and bilateral crossbite. Several gingival abscesses were also present. (C) This panoramic X-ray shows obliteration of pulp chambers, short roots and almost total absence of enamel. The dentine is similar to that seen in dentine dysplasia type I (radicular dentine dysplasia).

6.1.9 Amelogenesis Imperfecta and Nephrocalcinosis Type IG; AI1G

Definition

Association of amelogenesis imperfecta hypoplastic type, nephrocalcinosis, enuresis and polyuria.

OMIM Number

204690

Prevalence

Unknown.

Inheritance

Autosomal recessive.

Aetiology – Molecular Basis – Gene

Not yet identified.

Animal Models/Main Features

None reported in the literature.

Clinical Description

Main features

- Enuresis and polyuria beginning around 2 years of age.
- Nephrocalcinosis becoming apparent radiographically at around 5 years of age.
- Urinary tract infection and pyelonephritis in late childhood, early adulthood.
- Possible malignant hypertension and uraemia during the third decade of life.
- Reduced urinary excretion of calcium and phosphate, increased serum osteocalcin.

Cranio-oro-dental features (detailed)

- Enlargement of gingival tissue, often covering one-third of the tooth crown.
- Absence of enamel in the primary and permanent dentition.
- A yellower appearance to teeth.
- Several permanent teeth, from canines to molars, remain unerupted, with intra-alveolar resorption of the crowns.
- Enlargement of follicles of unerupted permanent teeth.
- Multiple calcifications of the pulps and the follicles.

Management/Oral Health

The patient should be referred for renal examination.

References

[919–929]

Pictures Illustrating the Oro-Dental Features

Figure 6.6

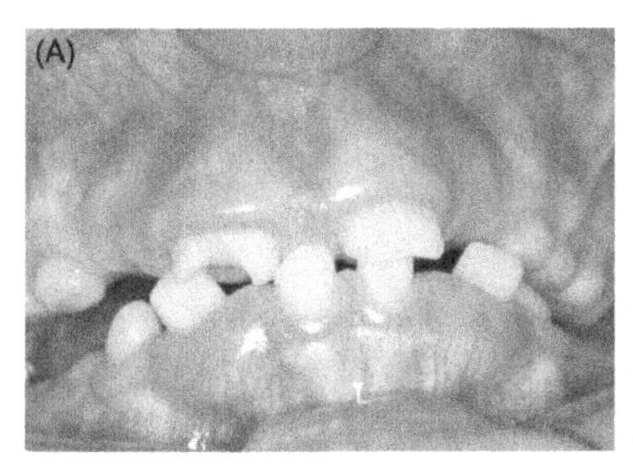

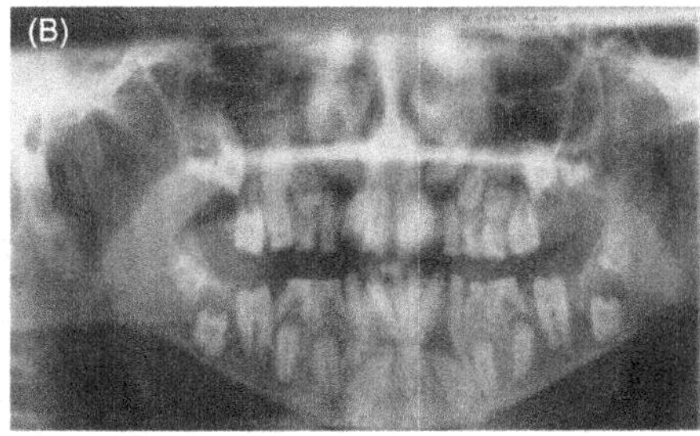

Figure 6.6 Amelogenesis imperfecta and nephrocalcinosis. (A) Clinical photograph of permanent and primary teeth of a patient with AI/NC, showing yellow discolouration and alterations in tooth shape and in enamel thickness. (B) Panoramic radiograph showing erupted and unerupted teeth with large well-defined pericoronal radiolucencies, except for the third molars. Courtesy of Acevedo AC Dental Anomalies Clinic, Dentistry School, Oral Health Faculty, University of Brasilia, Brasilia, Brazil.

6.1.10 Amelogenesis Imperfecta and Cone–Rod Dystrophy

Definition

The recessively inherited association of cone–rod dystrophy (CORD) with amelogenesis imperfecta (AI) (Jalili syndrome) was first reported in a large consanguineous Arabic family from the Gaza Strip [930].

OMIM Number

#217080

Prevalence

There are nine ethnically diverse, non-related families described thus far.

Inheritance

Autosomal recessive.

Aetiology – Molecular Basis – Gene

The causative gene is *CNNM4* (2q11), which encodes a putative metal transporter. The protein has been localized in retina and teeth (specifically mainly in the enamel organ and in ameloblasts). CNNM4 is probably involved in teeth mineralization.

Animal Models/Main Features

None reported in the literature.

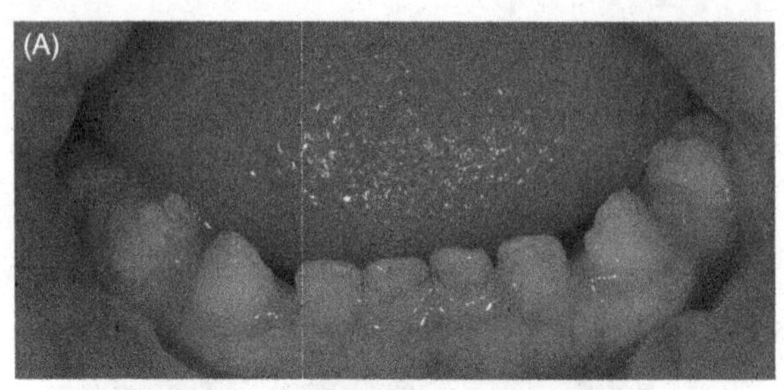

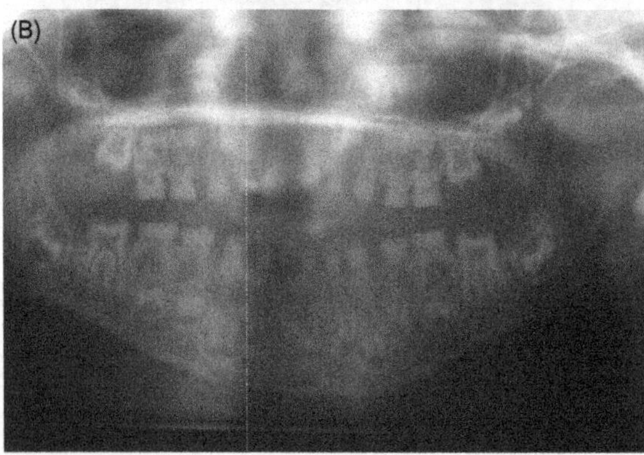

Figure 6.7 Amelogenesis imperfecta and cone–rod dystrophy. (A) Intraoral picture. (B) Panoramic radiograph of a child presenting with amelogenesis imperfecta and cone–rod dystrophy.

Clinical Description

Main features

The retinal dystrophy is of early onset, presenting with photophobia and nystagmus in the first few years of life. Visual acuity is severely reduced in childhood, with complete absence of colour vision.

Cranio-oro-dental features

- The dental phenotype co-segregating with the CORD was that of hypoplastic/hypomineralized AI affecting both the primary and permanent dentitions.
- The teeth look dysplastic and yellowish-brown, with almost no visible enamel – leading to severe teeth erosion/abrasion.
- Radiographic examination reveals absence of the normally distinct contrast between enamel and dentine, and taurodontic first permanent molars and root formation anomalies.
- Dentine formation seems also to be altered, with intrapulpal calcification of primary molars, bulbous appearance of permanent teeth crowns and thin roots.
- Strikingly, the amelogenesis imperfecta appearance is similar in all examined patients and could be considered a diagnostic marker for this rare disease.

Management/Oral Health

See 6.1.1.

References

[930–934]

Pictures Illustrating the Oro-Dental Features

Figure 6.7

6.1.11 Amelogenesis Imperfecta and Platyspondyly

Definition and clinical description

Platyspondyly refers to flatness of the vertebral bodies as a result of poorly formed vertebrae. Platyspondyly affects one, several or all vertebral bodies. Generalized platyspondyly is responsible for brachyolmia (short trunk). Platyspondyly should be differentiated from spondyloepiphyseal dysplasia (SED) and spondyloepimetaphyseal dysplasia (SEMD), where patients also have epiphyseal and epymetaphyseal dysplasia.

Patients with isolated generalized platyspondyly (GP) and some varieties of SED and SEDM have been reported as having various dental anomalies [791,935–938].

The association of amelogenesis imperfecta (AI) and GP without epi- or metaphyseal dysplasia has been reported in three families [935,936,938], inherited as autosomal recessive, but it seems to represent two different syndromes. The case of Houlston et al. [936] and case 3 of Bertola et al. [935] are characterized by GP, AI and taurodontism; the radiologic appearance of the dentition in these two different cases is identical, and it is different from that observed in the case reported by Verloes et al. [938] and cases 1 and 2 which were brothers, of Bertola et al. [935]. These patients only present GP and AI without taurodontism. Additionally, the two brothers reported by Bertola et al. [935] had congenital absence of the second mandibular premolars.

These syndromal associations need to be further explored in order to establish their differences and/or similarities and to ascertain their mode of inheritance as well to determine whether they are allelic variations or independently segregating genetic mutations.

OMIM Number

601216

Prevalence

Three families described so far.

Inheritance

Autosomal recessive (?)

Aetiology – Molecular Basis – Gene

Not yet identified.

Animal Models/Main Features

None reported in the literature.

References

[791,935–938]

Pictures Illustrating the Oro-Dental Features

Figure 6.8

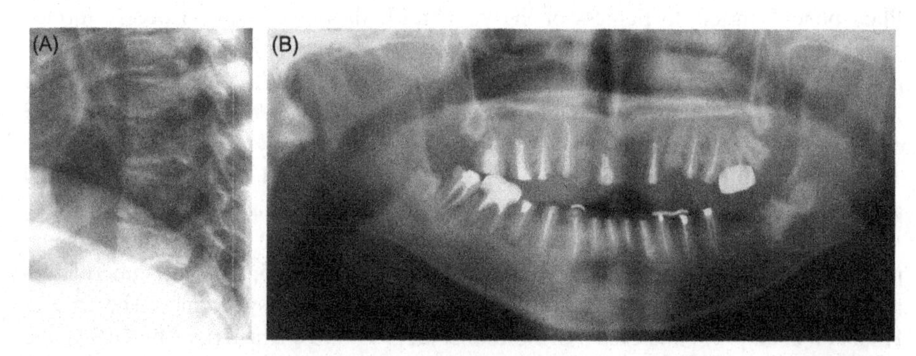

Figure 6.8 Amelogenesis imperfecta and platyspondyly. (A) Flattened vertebral bodies seen in platyspondyly (Courtesy of M Fischbach, Y Alembick, Hôpitaux Universitaires Strasbourg). (B) Panoramic radiograph displaying almost total absence of enamel and impacted teeth.

6.1.12 Kohlschütter(-Tönz) Syndrome

Definition

Kohlschütter et al., in 1974, described a central nervous system (CNS) syndrome characterized by epilepsy, learning disability and amelogenesis imperfecta.

OMIM Number

#226750

Prevalence

Nineteen cases in seven families have been described to date.

Inheritance

Autosomal recessive.

Guazzi et al. [939] reported a family where three siblings, their father and paternal grandfather had amelogenesis imperfecta. In two siblings and the father, the defect was associated with a neurological syndrome. These cases were reported as a possible variation of the Kohlschütter(-Tönz) syndrome, thus suggesting that the syndrome could represent an example of heterogeneity or that the family of Guazzi represents another entity.

Aetiology – Molecular Basis – Gene

ROGDI 16p13.3 (publication in press).

Animal Models/Main Features

None reported in the literature.

Clinical Description

Main features

- Epileptic seizures
- Developmental delay
- Progressive mental and motor deterioration
- Cerebral atrophy
- Hypoplasia of the cerebellar vermis
- Hypohidrosis
- Abnormal EEG
- Broad thumbs

Cranio-oro-dental features

- Amelogenesis imperfecta that has been reported as hypoplastic or hypocalcified enamel
- Yellow teeth

Management/Oral Health

- Oral hygiene and health protocol should be followed.
- Operative dental procedures could be instituted depending on the degree of patient's cooperation and most likely under conscious sedation.

References

[939–947]

Pictures Illustrating the Oro-Dental Features

Figure 6.9

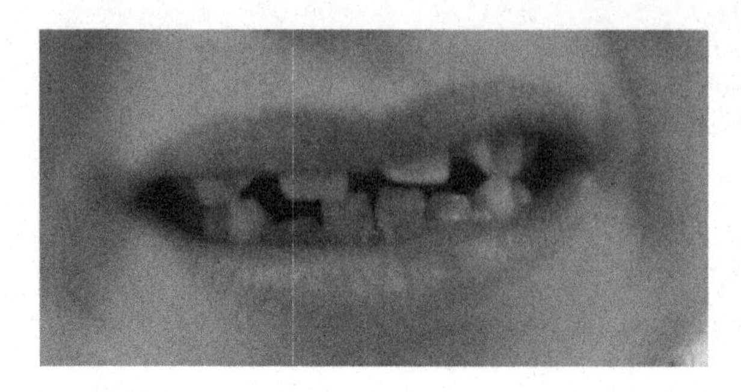

Figure 6.9 Kohlschutter-Tonz syndrome. Courtesy of J Zschocke Human Genetics, Medical University Innsbruck, Austria; N Wolf Dept. of Child Neurology VU University Medical Center Amsterdam, The Netherlands.

6.1.13 Amelo-Onycho-Hypohidrotic Syndrome

Definition

This variety of ectodermal dysplasia is characterized by teeth's enamel, fingernails and toenails and sweat abnormalities (*vide infra*).

Synonyms

Amelogenesis imperfecta and terminal onycholysis (amelo-onycho-hypohidrotic dysplasia).

OMIM Number

%104570

Prevalence

Rare, yet to be determined.

Inheritance

Autosomal dominant.

Aetiology – Molecular Basis – Gene

Not yet identified.

Animal Models/Main Features

None reported in the literature.

Clinical Description

Main features

- Distal onycholysis affecting all fingernails and toenails
- Subungual hyperkeratosis
- Functional hypohidrosis
- Dry skin

Cranio-oro-dental features

- Seborrhoeic dermatitis of the scalp
- Hypoplastic hypocalcified enamel
- Many teeth retained which may undergo resorption before eruption
- Occasionally, absence of the lacrimal punctae

Management/Oral Health

- Oral hygiene protocol
- Aesthetic dental procedures, perhaps including jacket crowns in anterior teeth
- Crowns in posterior teeth
- Orthodontics and endodontics, if indicated

References

[948–951]

Pictures Illustrating the Oro-Dental Features

Figure 6.10

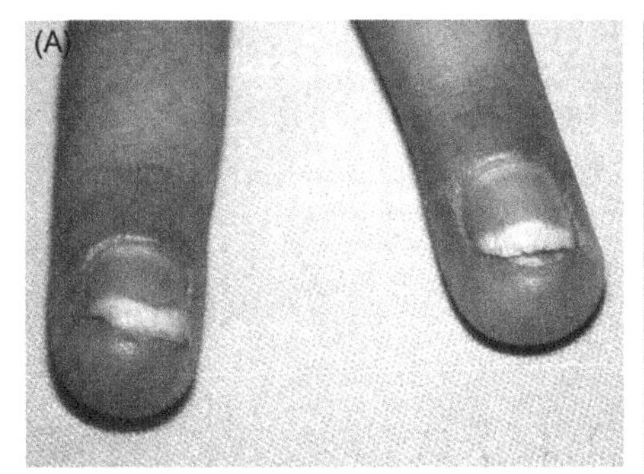

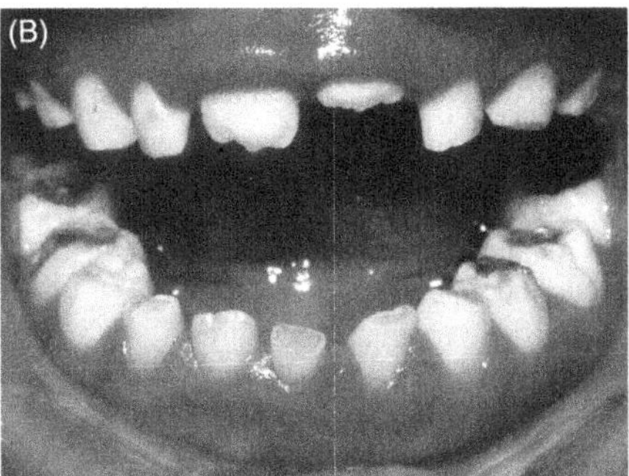

Figure 6.10 Amelo-onycho-hypohidrotic syndrome. (A) Note marked distal onycholisis seen in a patient with the amelo-onycho-hypohidrotic syndrome. (B) The enamel of these teeth is hypoplastic and hypocalcified, with a whitish-grey colour.

6.1.14 Heimler Syndrome

Definition

Heimler et al., in 1991, formally reported a brother and sister with sensorineural hearing loss, enamel hypoplasia (thin enamel) and nail abnormalities.

OMIM Number

234580

Frequency

Six cases in three families have been reported to date.

Inheritance

Autosomal recessive.

Aetiology – Molecular Basis – Gene

Not yet identified.

Clinical Description

Main features

- Bilateral or unilateral sensorineural hearing loss developing between 2 and 6 years of age – Transverse grooves (Beau's lines) on fingernails and toenails
- Punctate leukonychia of nails

Cranio-oro-dental features

- Primary dentition not affected
- Amelogenesis imperfecta
- Overcrowding of teeth
- Enamel discolouration
- Dentine and enamel of normal thickness
- Taurodontic permanent molars
- Delayed eruption of permanent teeth

Management/Oral Health

- Oral hygiene and health protocol
- Orthodontics, if needed
- Restorative dental procedures

Animal Models/Main Features

None reported in the literature.

References

[952–955]

6.2 Syndrome with Enamel Defects

6.2.1 Epidermolysis Bullosa Syndromes (EB)

Definition

EB consists of several hereditary and non-hereditary vesiculo-bullous syndromes affecting skin and oral and other mucosae. Blisters develop as a consequence of even minor trauma, causing painful, open wounds. Heat may also induce blister formation in some types. Additionally, teeth enamel hypoplasia has been described in some types.

There are more than 27 types and a large number of subtypes of EB that have been classified based on the level of epidermis–dermis at which the separation takes place, combined with the resultant clinical signs and the inheritance pattern; additionally, the specific molecular defects are also taken into consideration. The three basic clinical types are EB Simplex, EB Junctional and EB Dystrophica.

Prevalence

EB affects 1 in 17,000 live births (all types of EB). The carrier risk in the United States is 1 in 113 for any type of EB. Here, only those types that present oral manifestations are discussed.

Epidermolysis Bullosa Simplex (EBS)

OMIM Number

#131760 EBS Dowlin-Meara type
#131800 EBS Weber-Cockayne type
#131900 EBS Koebner type

There are several other subtypes that do not present oral manifestations.

Inheritance

Autosomal dominant.

Aetiology – Molecular Basis – Gene

Mutations in the genes that code for keratin 5 (*KRT5*) mapped to 12q13 and keratin 14 (*KRT14*) mapped to 17q12–q21.

Keratin 5 is a human 58-kD type II keratin, which is co-expressed with keratin 14 which is a 50-kD type I keratin.

Clinical Description

These types of EB are characterized by intra-epidermal vesicle formation.

Main features

- Weber–Cockayne: Skin blisters in hand and feet develop from the first to the third decade of life and appear with greater frequency after prolonged walks or during warm weather.
- Koebner: Generalized skin vesicles and bullae that start to develop from birth to around the third year of life.
- In over 24% of patients, the lesions heal with scarring.
- Dowling-Meara: This type is the most severe, with generalized vesicles starting as earlier as the first month of life.
- Occasional haemorrhagic blisters occur.
- It is non-scarring.
- Some patients may have loss of nails due to extensive chronic blisters.

Cranio-oro-dental features

- Weber-Cockayne: Generalized intraoral blisters are present in 9% of patients.
- Koebner: Generalized intraoral blisters are present in 23% of patients.
- Dowling-Meara: Generalized intraoral blisters are present in over 60% of patients.
- The blisters have a herpetiform appearance.
- Teeth are not affected in any of these types of EB.

Epidermolysis Bullosa Junctional (EBJ)

The splitting in EBJ types takes place at the level of the lamina lucida in the skin basal membrane. EBJ is subdivided into the so-called Herlitz type (OMIM #226700) and non-Herlitz type (OMIM #226650). The Herlitz type is characterized by severe clinical manifestations and it is fatal in infancy.

Here only the non-Herlitz type will be presented.

OMIM Number

#226700, #226650

Prevalence

The carrier risk in the United States is at 1 in 350 for the combined junctional types and 1 in 781 for the Herlitz type.

Inheritance

Autosomal recessive.

Aetiology – Molecular Basis – Gene

Mutations in several genes that encode the subunits of laminin – 5:

- *LAMA3* locus in 18q11.2
- *LAMB3* locus in 1q32
- *LAMC2* locus in 1q25–q31

Additionally, mutations in the gene that encodes for collagen XVII alpha 1 *COL17A1* locus in 10q24.3

Animal Models/Main Features

The incisors of the *Col17(–/–)* mice exhibited reduced yellow pigmentation, diminished iron deposition, delayed calcification and markedly irregular enamel prisms, indicating enamel hypoplasia. The molars of *Col17(–/–)* mice demonstrate advanced occlusal wear [956]. These anomalies clearly reproduce the enamel hypoplasia in human patients with junctional epidermolysis bullosa [956–960].

Clinical Description

Main features

- Blisters start soon after birth.
- Hands and extremities are mostly affected.

- Facial skin may also be affected.
- Non-scaring vesicles in the skin are present.
- Skin atrophy occurs.
- Loss of eye lashes occurs.
- Subungual haemorrhage exists.
- Nail dystrophy is present.

Cranio-oro-dental features

- Intraoral blisters can develop at any site.
- Blisters tend to be less frequent and diminish in severity with age.
- Enamel hypoplasia; teeth present a marked yellow colour, and the enamel is pitted.
- Hypodontia is present.

Epidermolysis Bullosa Dystrophic (EBD)

The splitting occurs below the sublamina densa. EBD, also known as Hallopeau-Siemens type, has been subdivided into four clinical subtypes which are difficult to differentiate from each other. EBD may produce significant multi-organ system involvement associated to severe scarring.

OMIM Number

#226600

Prevalence

Prevalence is estimated at 1 in 17,000 live births for all forms of EB.
EBD accounts for 25% of all forms of EB.

Inheritance

Autosomal recessive.

Aetiology – Molecular Basis – Gene

Various mutations in the collagen VII gene (*COL7A1*) 3p21.3

Clinical Description

Main features

- Skin bullae may be present at birth or appear shortly thereafter.
- Bullae develop at sites of trauma or pressure as well as develop spontaneously. The sites most frequently involved are occiput, scapulas, elbows, fingers, buttocks and feet.
- Painful ulcers follow the rupture of the bullae.
- Severe keloid scars are associated with healing, with resultant contraction of the affected areas, and the hands may acquire a mitten or claw-hand appearance as a consequence to the scarring.
- Skin pigmentation or de-pigmentation may accompany the keloid scars.
- Nails are dystrophic or absent.
- Hair may be scanty.

- Eyes may be affected by conjunctivitis, keratitis, corneal opacities, symblepharon and/or blepharitis.
- Bullae of the pharynx or larynx may eventuate in aphonia, dysphagia and stenosis.
- Malnutrition is often a consequence.
- Skin cancers are a later complication.

Cranio-oro-dental features

- Intraoral bullae present in over 90% of patients and perhaps microstomia, ankyloglossia, tongue atrophy with depapillation and obliteration of the vestibular and buccal sulci
- Severe periodontal disease
- Alveolar bone resorption
- Xerostomia
- Oral carcinoma
- Severe enamel hypoplasia with rapid caries development
- Delayed eruption and tooth retention
- Defective root cementum

ALL FORMS OF EB

Management/Oral Health

- Scaling and root planing
- Oral hygiene and health protocol
- Caries prevention
- Routine dental procedures can be performed generally under general anaesthesia

References

[961–1011]

Pictures Illustrating the Oro-Dental Features

Figure 6.11

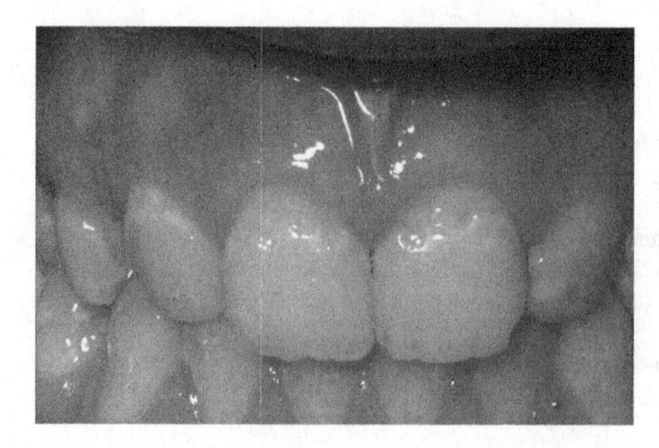

Figure 6.11 Epidermolysis bullosa. Pitting of enamel encountered in Epidermolysis bullosa. Courtesy of Mike Harrison EDH, UCLH.

6.2.2 Tuberous Sclerosis Syndrome (Epiloia; Bourneville-Pringle Syndrome)

Definition

The syndrome consists of hamartomatous lesions in the skin, nervous system, heart, kidney and other internal organs. The name comes from the characteristic *tuber*-like growths in the brain, which calcify and become *sclerotic*.

OMIM Number

#191100

Prevalence

About 1 in 6000.

Inheritance

Autosomal dominant, but many cases result from sporadic mutations.

Aetiology – Molecular Basis – Gene

One responsible gene (*TSC1*) has been mapped to the long arm of chromosome 9 (9q34) and produces a protein called hamartin. The other gene (*TSC2*) is located on the short arm of chromosome 16 (16p13.3) and produces the protein tuberin. Both proteins normally act as tumour growth suppressors.

Inactivation of TSC leads to Notch1 activation, linking TSC pathogenesis to NOTCH pathway [1012].

Animal Models/Main Features

Neuronal loss of *Tsc1* causes dysplastic and ectopic neurones, reduced myelination, seizure activity and limited survival. *Tsc1* null embryos die at mid-gestation from a failure of liver development. *Tsc1* heterozygotes develop kidney cystadenomas and liver haemangiomas at high frequency, but the incidence of kidney tumours is somewhat lower than in *Tsc2* heterozygote mice. Liver haemangiomas were more common, more severe and caused higher mortality in female than in male *Tsc1* heterozygotes [1013–1015].

Clinical Description

Main features

The most frequent presentation is a triad of the following:

- Adenoma sebaceum: actually an angiofibroma with passive involvement of sebaceous glands
- Epilepsy
- Learning disability

- Possibility of hamartomas affecting the brain, eyes, heart, kidneys, lungs or other tissues or organs
- Cyst-like areas within certain skeletal regions, particularly the phalanges
- Sharply defined areas of skin hypopigmentation ('ash-leaf' spots)
- Flat, café-au-lait spots on various areas of the skin
- Fibromas arising around or beneath the nails (subungual fibromas)
- Rough, elevated lesions (shagreen patches) on the lower back

Cranio-oro-dental features

- Gingival angiofibromas specially in the anterior gingiva in around 45% of patients, but they may also be present in other oral mucosae.
- Pitted enamel is present in 50–100% of patients.
- Fibrous intraosseous lesions may be present.

Management/Oral Health

Removal of the intraoral angiofibromas may be indicated for aesthetic reasons.

References

[1012,1015–1036]

Pictures Illustrating the Oro-Dental Features

Figure 6.12

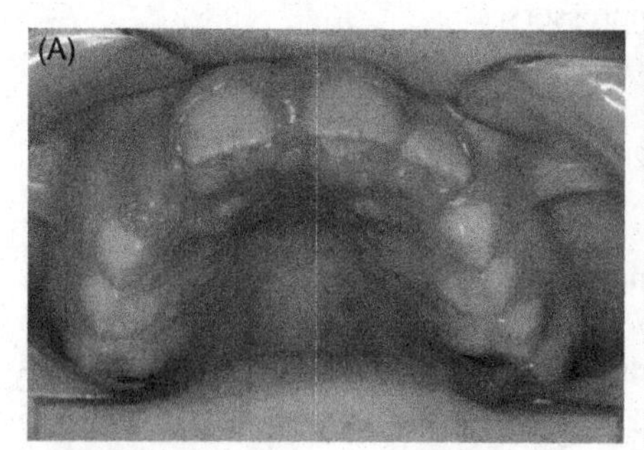

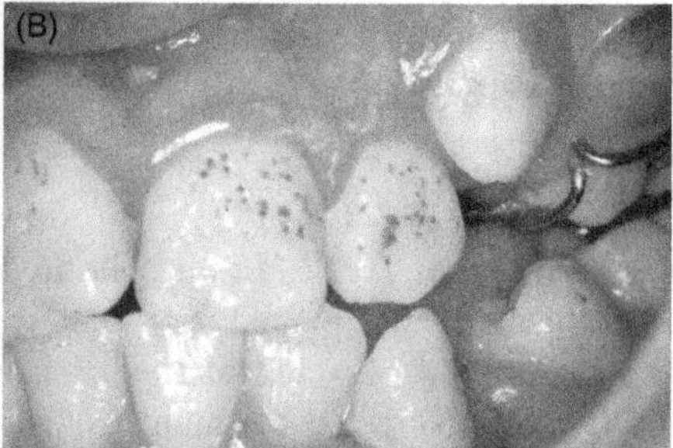

Figure 6.12 Tuberous sclerosis syndrome. (A) Gingival angiofibromas. (B) Enamel pitting.

6.2.3 APECED Syndrome (Autoimmune-Polyendocrinopathy-Candidiasis; Autoimmune Polyglandular Disease Type I)

Definition

The acronym APECED stands for autoimmune polyendocrinopathy candidiasis ectodermal dysplasia. The syndrome is defined as the concurrent subnormal functioning of several endocrine glands (polyglandular autoimmune disease (PGA)).

OMIM Number

#240300

Prevalence

APECED syndrome is a rare disease, most cases having been reported in Finland, where the estimated prevalence is 1 in 25,000.

Inheritance

Autosomal recessive.

Aetiology – Molecular Basis – Gene

The autoimmune regulator gene *AIRE* (21q22.3) codes for a transcription factor. A monogenic mutation of *AIRE* transcription factor leads to an anomaly of normal immunological tolerogenesis which in turn leads to the formation of autoantibodies directed against specific tissue antigens: surface receptors, intracellular enzymes and secreted proteins (hormones).

Animal Models/Main Features

The transgenic mouse models display a variable combination of autoimmune endocrine tissue destruction, mucocutaneous candidiasis and ectodermal dystrophies [1031].

Clinical Description

Main features
Polyendocrinopathy Disorder Subdivisions

- PGA I polyglandular autoimmune syndrome, type I
- PGA II polyglandular autoimmune syndrome, type II
- PGA III polyglandular autoimmune syndrome, type III

Autoimmune disease affecting one gland is frequently followed by the impairment of other glands. In this syndrome, two major types of failure have been described.

Type I affects children and adults under age 35 and is characterized by hypoparathyroidism (79%) and hypoadrenocorticism (72%). About 60% of affected women and about 15% of men fail to mature sexually (hypogonadism). Mucocutaneous candidiasis is common and chronic. Type II more frequently strikes adults, with a peak incidence at about 30 years. Almost invariably it involves the adrenal cortex, with thyroid involvement somewhat less frequently. It may also involve the pancreatic islets, producing insulin-dependent diabetes mellitus.

Cranio-oro-dental features

- Enamel hypoplasia in the form of pits, missing enamel and grooves which defects occurred in a chronological pattern and were linked to the hypoparathyroidism

- Candidiasis
- Susceptibility to oral or oesophageal squamous cell carcinoma (SCC).

Management/Oral Health

- Scaling and root planing
- Oral hygiene and health instruction
- Candidiasis necessitates antimycotic treatment. Hypoparathyroidism benefits from calcium and vitamin D therapy. Adrenal failure requires hydrocortisone replacement. Vitamin and mineral replacement can be useful to complement hormone replacement.

References

[1038–1051]

6.2.4 Albright Syndrome (Pseudohypoparathyroidism Type IA, PHP1A)

Definition

Albright hereditary osteodystrophy (AHO) is characterized by obesity, round faces, subcutaneous ossifications and skeletal anomalies resulting in short stature and brachydactyly. Some patients have learning disability. AHO is often associated with pseudohypoparathyoidism type Ia (PHP Ia), hypocalcaemia and elevated parathyroid hormone (PTH) levels. Pseudohypoparathyroidism is a genetic disorder that resembles hypoparathyroidism (lowered levels of PTH) but is caused by a lack of response to PTH rather than a deficiency in it.

OMIM Number

#103580

Prevalence

0,72 in 100,000 [259].

Inheritance

Autosomal dominant with a parental origin effect.

Aetiology – Molecular Basis – Gene

Pseudohypoparathyroidism type Ia (PHP Ia) is caused by a mutation resulting in loss of function of the Gs-alpha isoform of the *GNAS* gene (20q13.2) on the maternal allele. This results in expression of the Gs-alpha protein only from the paternal allele.

Pseudopseudohypoparathyroidism (PPHP) is caused by a mutation resulting in loss of function of the Gs-alpha isoform of the *GNAS* gene on the paternal allele. This results in expression of the Gs-alpha protein only from the maternal allele.

Activating or gain-of-function *GNAS1* mutations in patients result in the McCune–Albright syndrome #174800.

Animal Models/Main Features

Gene targeting of specific *Gnas* transcripts demonstrates that heterozygous mutation of Gs-alpha on the maternal (but not the paternal) allele leads to early lethality, perinatal subcutaneous edema, severe obesity and multihormone resistance, while the paternal mutation leads to only mild obesity and insulin resistance [1052].

Clinical Description

Main features

- Short stature
- Obesity
- Brachydactyly
- Subcutaneous ossifications
- Hypocalcemia (seizures, muscular hyperexitability, spasms)
- PTH/TSH resistance
- Pubertal delay
- A syndrome very similar to hypoparathyroidism develops with low blood calcium and high phosphate levels.
- Maternally derived mutations in *GNAS* are usually associated with resistance to parathyroid hormone – termed *pseudohypoparathyroidism type Ia* (Albright syndrome AHO).
- Paternally derived mutations are associated with AHO but usually normal hormone responsiveness, known as *pseudo-pseudohypoparathyroidism.*

Type Ib is characterized by resistance to PTH confined to the kidney. As a result, the calcium and phosphate problems are seen but not the rest of the clinical manifestations of the syndrome. However, rarely some features of AHO are seen. It is very similar to type I in its clinical features, but the underlying mechanism in the kidney is different.

Cranio-oro-dental features

- Round face
- Enamel hypoplasia
- Delayed tooth eruption

Other dental manifestations reported in the literature:

- Hypodontia
- Microdontia
- Malformed roots
- Enlarged pulp chambers
- Pulp calcifications
- Ankylosis and an enlarged frontal sinus have also been reported.

Management/Oral Health

- Multidisciplinary management:
- Scaling and root planing
- Oral hygiene and health instruction

References

[259,1052–1071]

6.2.5 Oculodentodigital Dysplasia (ODDD)

Definition

This syndrome, also known as oculodentodigital (ODD) or oculodentoosseous dysplasia (ODOD), is characterized by a thin nasal ridge with underdeveloped alae and narrow nostrils, microcornea with iris anomalies, syndactyly and camptodactyly of the fourth and fifth fingers and enamel hypoplasia similar to amelogenesis imperfecta.

OMIM Number

#164200

Frequency

ODDD is rare; only about 150 cases have been reported.

Inheritance

- ODD has autosomal dominant inheritance with variable expressivity.
- An autosomal recessive form has also been described.
- Around 50% of cases represent new mutations.

Aetiology – Molecular Basis – Gene

The gene responsible for ODDD is considered to be *connexin-43* (*CX43* also symbolized as GJA1 gap junction protein, alpha-1), and it maps to the long arm of chromosome 6 (6q21–23.2).

Animal Models/Main Features

In addition to the classic features of ODDD, these mutant mice also showed decreased bone mass and mechanical strength, as well as altered haematopoietic stem cell and progenitor populations [1072,1073].

Clinical Description

Main features

- Spastic paraplegia
- Somatic growth and mental development within normal limits
- Syndactyly of the fourth and fifth fingers

- Camptodactyly and ulnar clinodactyly
- Middle phalanx of these fingers shown in radiographs to be cube shaped
- Aplasia or hypoplasia of the middle phalanx of one or more toes
- Occasional finding of syndactyly of the third and fourth toes
- Mild broadening of the metaphyseal area of the femur
- Generally osteoporotic hand and feet bones
- Dry, thin and sparse hair that fails to grow to normal length noted in several patients

Cranio-oro-dental features

- Several patients have proved to be microcephalic.
- Characteristic face with small, sunken eyes.
- Hypertelorism.
- Epicanthal folds are frequent.

Eye:
  - In most cases, vision is essentially normal.
  - Microcornea is the most constant eye anomaly, the corneal diameter generally being less than 10 mm.
  - Microphthalmos
  - Secondary glaucoma
  - Congenital cataract
  - Persistence of pupillary membrane
  - Disk coloboma
  - Synechial strabismus
  - Optic atrophy
  - Eyelid aperture reduced in width to about 24–25 mm
- The nasal alae are underdeveloped, and the nostrils are slit-like.
- The ears usually are malformed.
- Conductive deafness has been noted in some patients.
- Cleft lip and palate has been reported.
- Highly arched palate.
- Small chin.

Dental findings:

- Both dentitions may be affected by a generalized enamel hypoplasia, similar to that seen in amelogenesis imperfecta.
- Teeth are usually yellow on eruption.
- Tooth agenesis is present.
- Microdontia occurs.
- Premature loss of teeth has also been reported.
- Thickening of the mandibular alveolar ridges has also been reported.

Management/Oral Health

Multidisciplinary management:

- Oral hygiene and health protocol.
- Aesthetic dental procedures to improve appearance of, especially anterior teeth, by means of ceramic or porcelain crowns.
- Surgical correction of the cleft lip, if present.
- Orthodontic treatment, if indicated.

References

[1073–1083]

Pictures Illustrating the Oro-Dental Features

Figure 6.13

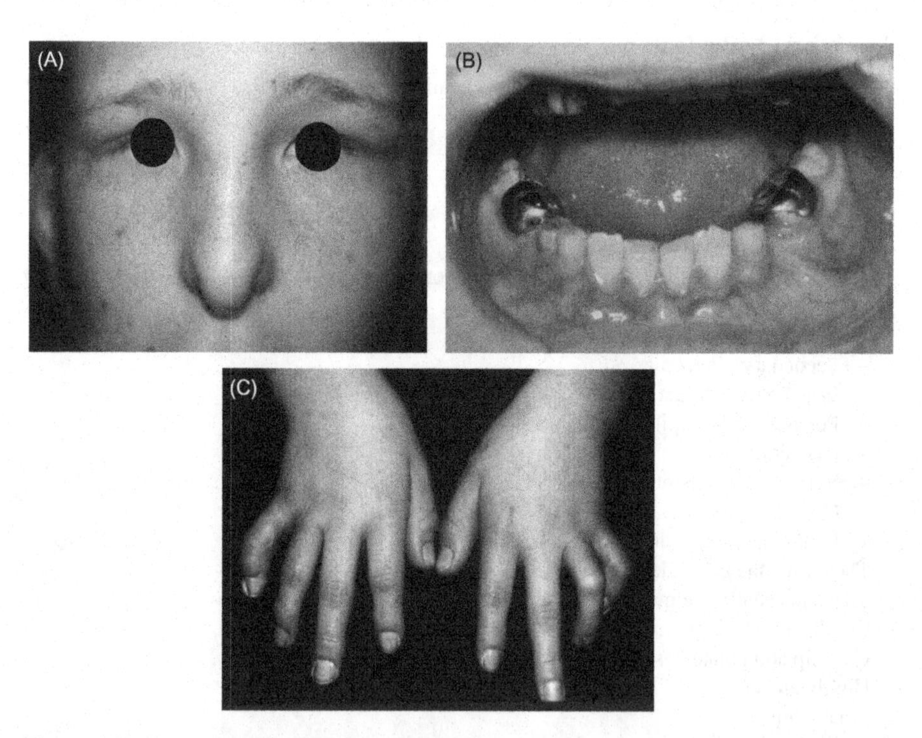

Figure 6.13 Oculodentodigital dysplasia (ODDD). (A) Typical facial appearance of a patient with oculo-dento-digital dysplasia. Note the thin aquiline nose with underdeveloped alae nasi; also note epicanthic folds. (B) Enamel hypoplasia, as seen here, is a constant finding in patients with ODD. (C) Note the ulnar deviation of the fourth and fifth fingers of both hands. Soft tissue syndactyly was present in those fingers that were separated by surgery.

6.2.6 Pycnodysostosis (PKND; PYCD; Stanesco's Dysostosis Syndrome)

Definition

The syndrome consists of short stature, increased bone density (osteosclerosis/osteopetrosis), underdevelopment of the tips of fingers with absent or small nails, fragile bones, abnormal or absent clavicle and delayed skull suture closure.

OMIM Number

#265800

Prevalence

Not yet determined.

Inheritance

Autosomal recessive.

Aetiology – Molecular Basis – Gene

Mutations in the *CTSK* (cathepsin K) gene situated at 1q21 that codes for cathepsin K – a lysosomal cysteine protease. Mutations in this gene disturb bone resorption and remodelling.

Animal Models/Main Features

Mouse models of Cathepsin K deficiency recapitulate the osteopetrosis of human pyknodysostosis [1084,1085].

Clinical Description

Main features

- Open cranial sutures and fontanelles
- Wormian bones
- Dolichocephaly
- Sclerotic vertebrae
- Fractured long bones
- Partial agenesis/aplasia of terminal phalanges causing short, stubby hands
- Developmental delay and mild dwarfism (acromelic type) in a minority of patients

Cranio-oro-dental features

- The nasal ridge is convex
- The jaw is small with an obtuse mandibular angle, micrognathia
- Narrow palate
- Hypodontia
- Enamel hypoplasia
- Delayed eruption of primary and permanent teeth
- Retention of primary teeth

Management/Oral Health

Multidisciplinary management:

- Scaling and root planing
- Oral hygiene and health instructions

References

[1086–1099]

Pictures Illustrating the Oro-Dental Features

Figure 6.14

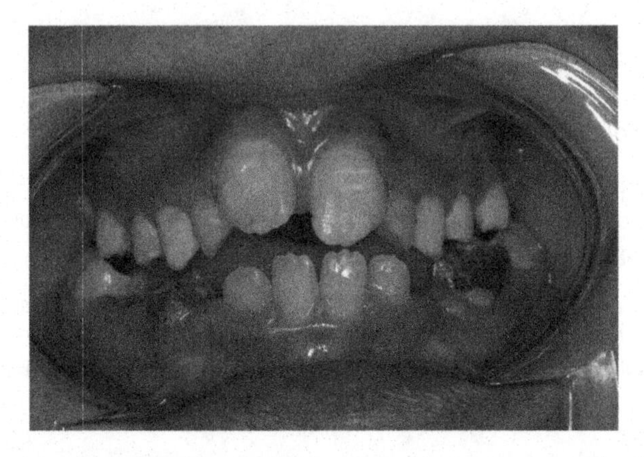

Figure 6.14 Pyknodysostosis syndrome. The jaw is small with an obtuse mandibular angle, and the palate is narrow. Enamel hypoplasia is visible.

6.2.7 Vitamin D-Dependent Rickets or Pseudo-Deficiency Rickets

Definition

Vitamin D-dependent rickets is characterized by severe hypocalcaemia with secondary hyperparathyroidism and is subdivided into type I (VDDR I or 1-alpha-hydroxylase deficiency) and type II (VDDR II). The two types are similar, but in type II there is usually also total absence of scalp hair (*alopecia totalis*). The VDDR II patient may, however, present without alopecia and with normal vitamin D receptor.

OMIM Number

#264700 type I
#277440 type II

Prevalence

Highest prevalence is 1 in 2358 births in the Saguenay-Lac St. Jean region of Quebec, where the carrier rate is 1 in 26 people.

Inheritance

Autosomal recessive.

Aetiology – Molecular Basis – Gene

VDDR I is caused by mutation in the gene encoding 25-hydroxyvitamin D3-1-alpha-hydroxylase (*CYP27B1*; 609506) (12q13.1–q13.3), a renal proximal tubular enzyme that catalyses the hydroxylation and activation of 25-hydroxyvitamin D3 into the active hormone 1,25-dihydroxyvitamin D3.

VDDR II is caused by mutation in the gene encoding the vitamin D receptor (*VDR*; 601769) located in 12q12–q14.

Animal Models/Main Features

Serum analysis of homozygous mutant animals for the 25-hydroxyvitamin D(3)-1-alpha-hydroxylase gene confirmed that they were hypocalcaemic, hypophosphataemic and hyperparathyroidic, and that they had undetectable 1-alpha, 25-dihydroxyvitamin D(3). Histological analysis of the bones from 3-week-old mutant animals confirmed rickets. Treatment with 1,25-dihydroxyvitamin D3 can rescue the phenotype [1100–1102].

Clinical Description

Main features

- Growth failure, hypotonia, weakness, rachitic rosary, convulsions, tetany and pathologic fractures
- Bone deformities affecting all long bones
- Possibility of secondary hyperparathyroidism, hypocalcaemia and variable serum phosphate levels due to raised levels of serum alkaline phosphatase

Cranio-oro-dental features

- Open fontanelles
- Frontal bossing
- Enamel hypoplasia
- Enamel alteration as a consequence of the abnormal metabolism of calcium and phosphorus which are a direct result of the inherited defect
- Delayed tooth eruption
- Large pulp chamber, thin dentine
- Short roots
- Marked anterior open bite

Management/Oral Health

Multidisciplinary management:

- Patients with vitamin D-dependent rickets respond to treatment with vitamin D3 at doses of 5000 to 10,000 units per day. Lifelong treatment is required to totally prevent or cure the bone, radiological and biological features of rickets.
- Type II rickets is only partially responsive and may be non-responsive even to high-dose treatment with vitamin D derivatives and needs parenteral calcium supplementation. The alopecia is not responsive to either treatment.

References

[1100–1109]

Pictures Illustrating the Oro-Dental Features

Figure 6.15

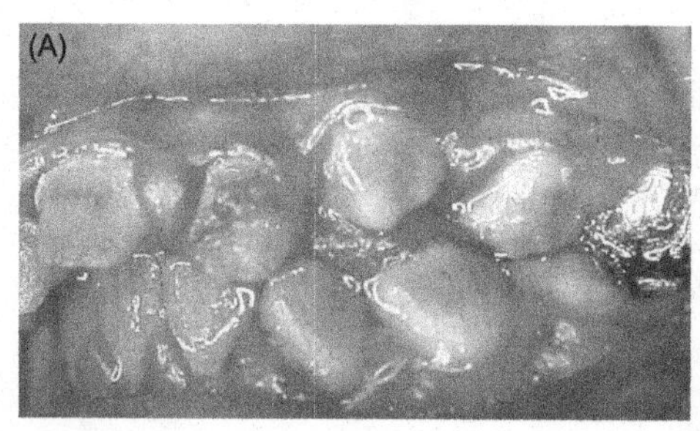

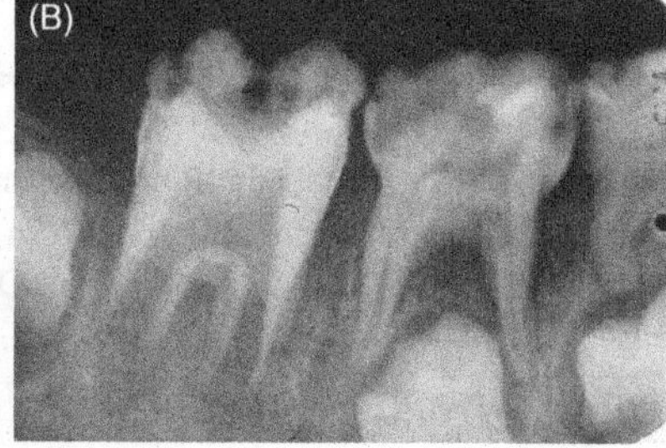

Figure 6.15 Vitamin D-dependent rickets or pseudo-deficiency rickets. (A) These are the teeth of a 10-year-old girl with vitamin D-dependent rickets type 1. Note that the appearance of the enamel is similar to that seen in the hypocalcified type of amelogenesis imperfecta. (B) An intraoral X-ray from another patient with the same syndrome, showing the enamel defect and large pulp chambers.

7 Anomalies of Teeth Eruption and/or Resorption

Dento/Oro/Craniofacial Anomalies and Genetics. DOI: 10.1016/B978-0-12-416038-5.00007-X

Syndrome	Associated Features	Gene	Molecules	Inheritance	Locus
Sotos syndrome #117550	Premature tooth eruption Rapid growth, acromegalic features typical face Learning disability	*NSD1*	Nuclear receptor-binding SET domain protein-1	Isolated cases	5q35
Hypophosphataemic vitamin D-resistant rickets #307800	Very short stature Rachitic rosary Bowing of the extremities Delayed eruption of teeth Dentine defects Abnormal cement	*PHEX*	Membrane-bound endoprotease (Zn-metaloendopeptidase proteolytic enzyme)	X	Xp22.2–1
Pycnodysostosis #265800	Short stature Deformity of the skull (including wide sutures), maxilla and phalanges (acroosteolysis), osteosclerosis and fragility of bone Micrognathia Narrow palate Delayed eruption of deciduous teeth Persistence of deciduous teeth Delayed eruption of permanent teeth Hypodontia Caries, enamel hypoplasia	*CTSK*	Cathepsin K	AR	1q21
Primary failure of tooth eruption #125350 [1110] Figure 7.1	Hypodontia Posterior open bite Ankylosis of deciduous teeth Not responsive to orthodontic treatment	*PTHR1*	Parathyroid hormone receptor-1 gene	AD	3p22–p21.1
Osteopetrosis #259710 #607649	Osteosclerosis Delay or failure of tooth eruption	*TNFSF11* *TCIRG1 subunit*	– Tumour necrosis factor ligand superfamily, member 11 RANKL – T-cell immune regulator 1	AR AR	13q14 11q13.4–q13.5

Hypomyelinating leukoencephalopathy 4H syndrome/ADDH #607694 [1111]	Neurodegenerative disorder Delay in primary tooth eruption, complete retention of the primary maxillary central incisors and shape abnormalities of the permanent maxillary central incisors, oligodontia	*POLR3A* [1061]	RNA polymarase III subunit C	AR	10q22.3
Hypophosphatasia, infantile childhood, adult #241500, #241510, #146300	Skeletal abnormalities, seizures Premature tooth loss	*ALPL* *Alkaline phosphatase*	Enzyme	AR, AD	1p36.1–p34
Papillon–Lefèvre syndrome #245000	Hyperkeratosis of palms and soles Premature tooth loss Severe periodontitis	*CTSC* *Cathepsin C gene*	Enzyme Lysosomal protease	AR	11q14.1–q14.3
Haim–Munk syndrome #245010	Keratosis palmoplantaris Premature tooth loss Severe periodontitis Onychogryposis Arachnodactyly, acroosteolysis	*CTSC* *Cathepsin C gene*	Enzyme	AR	11q14.1–q14.3
Aggressive periodontitis #170650		CTSC Cathepsin C	Enzyme Lysosomal protease	AR	11q14.1–q14.3
Ehlers-Danlos type VIII %130080 [1113]	Joint hypermobility, normal scar formation but eventual scar atrophy and severe periodontal disease	?	?	AD	12p13 Other locus
Neutropenia chronic familial %162700	Neutropenia, hyperglobulinemia, oral ulcers, gingivitis, periodontitis, early tooth loss	?	?	AD, AR	?
Familial expansile osteolysis #174810 Expansile skeletal hyperphosphatasia	Root resorption	*TNFRSF11A*	RANK Receptor Activator of NF-kappaB	AD	18q21.1–22

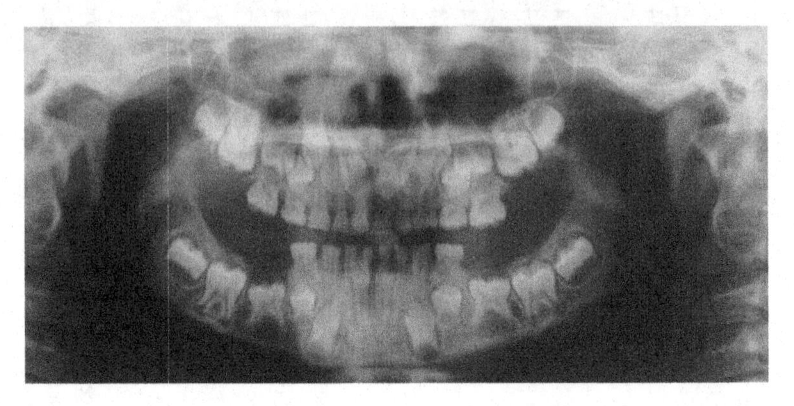

Figure 7.1 Primary failure of teeth eruption with ankylosis of deciduous teeth (9-year-old child) Courtesy of GJ Roberts, EDI, UCL.

Tooth eruption depends on the presence of the dental follicules, of osteoclasts to create an eruption pathway through the alveolar bone and of osteoblasts to form new alveolar bone [1114]. In diseases where osteoclast formation or function is reduced, such as the various types of osteopetrosis, tooth eruption is affected and can be abnormal or absent. Diseases in which osteoclast formation or activity is increased, such as in familiar expansile osteolysis and Paget disease of bone, are associated with dental abnormalities such as hypercementosis, root resorption and premature tooth loss [96]. Premature tooth loss can also originate from abnormal periodontium and cementum formation, as possibly seen in hypophosphatasia and Papillon–Lefèvre syndrome [1115]. In Papillon–Lefèvre syndrome, cathepsin C is involved in immune and inflammatory processes and associated destructive periodontal disease results. In hypophosphatasia, however, early loss of teeth occurs without any periodontal inflammation [1116]. Recently *PTHR1* loss-of-function mutations were discovered in familial, non-syndromic primary failure of tooth eruption [1110]. *PTHR1* encodes the G protein-coupled receptor for parathyroid hormone and parathyroid hormone-like hormone (PTHR1).

7.1 Sotos Syndrome (Cerebral Gigantism)

Definition

The syndrome consists of excessive growth, distinctive craniofacial features, dolichocephaly, advanced bone age and learning disability. The syndrome may also be associated with an increased risk of tumours (Wilms tumour, hepatocarcinoma).

OMIM Number

#117550

Prevalence

At birth, 7 in 100,000.

Inheritance

Most cases are sporadic mutations but autosomal dominant inheritance occurs.

Aetiology – Molecular Basis – Gene

Intragenic mutations and microdeletions of the *NSD1* gene (nuclear receptor-binding Su-var, enhancer of zeste and trithorax domain protein-1 gene located at chromosome 5q35 and coding for a histone methyltransferase implicated in transcriptional regulation) are responsible for more than 75% of cases.

Haploinsufficiency is the major cause: the single functional copy of the gene does not produce enough of a gene product to bring about a wild-type condition. A heterozygote displays a phenotypic effect.

Sotos syndrome may be allelic to some patients with Weaver syndrome (**#277590**) as some patients with Weaver syndrome also have a mutation in the *NSD1* gene. The phenotypes can be overlapping.

Animal Models/Main Features

Homozygous mutant *Nsd1* embryos, which initiate mesoderm formation, display a high incidence of apoptosis and fail to complete gastrulation [1117].

Clinical Description

Main features

- Growth perhaps accelerated, but other aspects of development often delayed
- Excessive growth, especially in early childhood, leading to acromegalic features
- Learning disability
- Language impairments (particularly expressive)
- Speech sound production impairments
- Voice impairments
- Dysfluencies (stuttering)
- Blood coagulation factor XII deficiency

Cranio-oro-dental features

- Head large and dolicocephalic, with a rounded and prominent forehead, accentuated by frontoparietal balding
- Narrowing at the temples, fullness of the cheeks and tapering to a pointed chin
- Mandibule long and narrow inferiorly, square or pointed, but prognathism is rare
- Hypertelorism, antimongoloid slant of eyes
- Highly arched narrow palate
- Premature tooth eruption
- Hypodontia
- Enamel hypoplasia

Management/Oral Health

Multidisciplinary management.

References

[1118–1135]

Pictures Illustrating the Oro-Dental Features

Figure 7.2

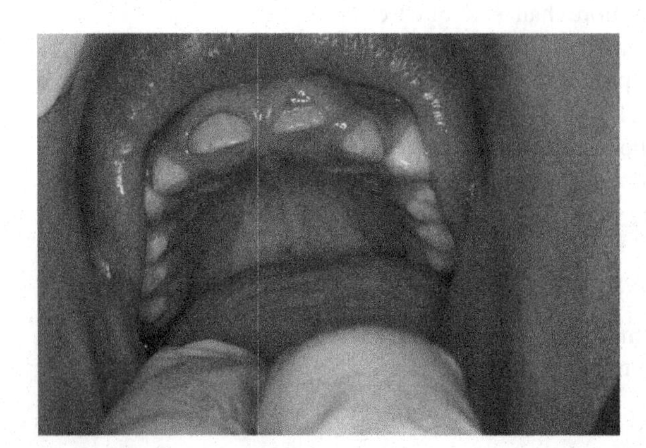

Figure 7.2 Signs of premature eruption (upper permanent incisors) and enamel hypoplasia in this 6-year-old boy with Sotos syndrome.

7.2 Hypophosphataemic Vitamin D-Resistant Rickets, Hypophosphataemia X-Linked

Definition

It is characterized by hypophosphataemia (diminished amount of serum phosphate) associated with diminished renal tubular resorption of inorganic phosphate, normocalcaemia and increased excretion of phosphate by the renal tubules.

OMIM Number

#307800

Prevalence

Not determined yet.

Inheritance

X-linked dominant.

Aetiology – Molecular Basis – Gene

Mutations in the *PHEX* gene (phosphate regulating endopeptidase homologue, X-linked, gene) located in the X chromosome (Xp22.2–1) are responsible for this disorder. The *PHEX* gene codes for a membrane-bound endoprotease (Zn-metaloendopeptidase proteolytic enzyme) which is predominantly expressed in osteoblasts that regulates phosphates.

PHEX regulates the function of fibroblast growth factor 23 (FGF-23) – the most important factor in the pathophysiology of hyperphosphaturic disturbances. Defective PHEX function leaves phosphaturic peptides such as FGF-23 uncleaved, enabling these phosphatonins to exert their phosphaturic activities in the renal proximal tubule.

FGF23 (12p13.3) is responsible for an autosomic dominant form of the disease (ADHR #193100), *DMP1* for an autosomic recessive form (ARHR #241520) and *ClCN5* for an X-linked recessive form (#300554).

Animal Models/Main Features

The hyp-mouse, a model of XLH characterized by a deletion in the *Phex* gene, manifests hypophosphataemia, renal phosphate wasting and rickets/osteomalacia [1136–1140].

Clinical Description

Main features

- Short stature
- Rachitic rosary
- No response to vitamin D therapy
- Bowing of the extremities
- No associated tetany or seizures
- Raised serum alkaline phosphatase levels

Cranio-oro-dental features

- Frontal bossing
- Delayed tooth eruption
- Enamel hypomineralization
- Abnormally large pulp chambers and pulp horns with microtracts extending into the dentine that cause communication of the pulp with the oral cavity, which eventually produces pulpitis and consequent pulp necrosis with periapical pathology
- Globular dentine which is hypocalcified, presenting clefts and tubular defects
- Abnormal cementum
- No susceptibility to gingivitis and/or periodontitis

Management/Oral Health

- Oral hygiene protocol

- the application of fluid resin composites with a self-etching primer bonding system to all primary/permanent teeth to prevent abscess formation
- Drainage of abscesses when present, as well as antibiotic therapy
- Replacement of lost teeth by prostheses or implants

Treatment of hypophosphataemic children, starting in early childhood, with 1-(OH) vitamin D and oral phosphate ensures good dentine development and mineralization and prevents clinical anomalies such as the dental necrosis classically associated with the disease.

References

[1140–1150]

Pictures Illustrating the Oro-Dental Features

Figure 7.3

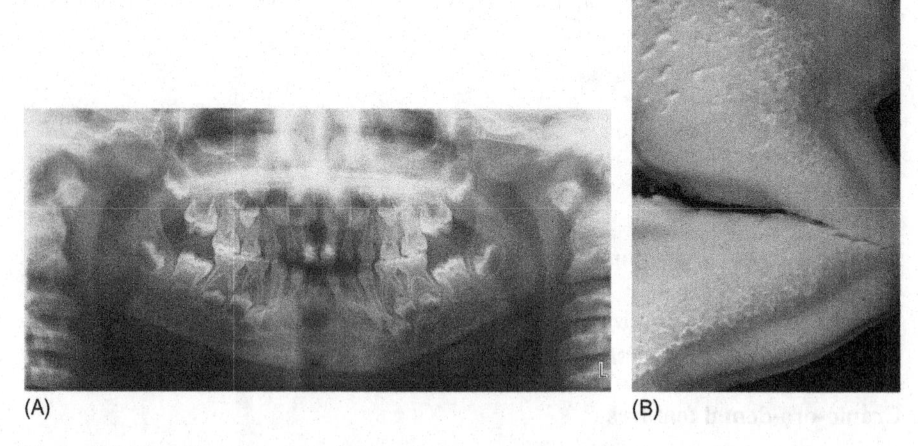

Figure 7.3 Hypophosphataemic vitamin D-resistant rickets. (A) Abnormally large pulp chambers and pulp horns extending into the dentine (panoramic radiograph of a 6-year-old child). Tooth eruption is also delayed the lower incisors and first permanent molars have not erupted yet. (B) This gross section of a non-decalcified molar shows a markedly globular, abnormal dentine with large dentinal tubules and a crack that extends to the enamel. Such cracks are the initial point of pulpitis and eventual pulpal necrosis.

7.3 Osteopetrosis

Definition

Osteopetrosis ('marble bone disease') is a descriptive term that refers to a group of heterogeneous rare genetic diseases of the skeleton characterized by increased bone density on radiograph. The name is derived from the Greek *osteo*, meaning 'bone', and *petros*, meaning 'stone'. The autosomal dominant form of the disease is also known as Albers-Schönberg disease.

Osteopetrosis ranges in severity from asymptomatic to fatal in infancy.

OMIM Number

#307800

Prevalence

From 0.75 in 100,000 (prevalence at birth) for the malignant, autosomal recessive form, to around 50 cases described in the literature.

Inheritance

Autosomal dominant (ADO), autosomal recessive (ARO), X-linked osteopetrosis. Dominant forms are usually more benign.

Aetiology – Molecular Basis – Gene

Osteopetrosis is caused by failure of osteoclast development or function, and causative mutations in at least 10 genes have been identified, accounting for 70% of all cases. The known genes involved are listed here:

- *TNFSF11* gene or *RANKL*, 13q14, ARO, Dental anomalies #259710
- *TNFRSF11A* or *RANK* gene severe associated with hypogammaglobulinemia, 18q22.1, ARO #612301
- *CLCN7* (chloride channel 7) gene, 16p13, ADO, ARO #611490, #166600
- *TCIRG1* proton pump vacuolar ATPase (V-ATPase) gene, 11q13.4–q13.5, ARO, dental anomalies #259700
- *OSTM1* osteopetrosis-associated transmembrane protein-1 gene, 6q21, ARO #259720
- *PLEKHM1* pleckstrin homology domain-containing protein, family m, member 1 gene, 17q21.3, ARO #611497
- *Carbonic anhydrase II* gene, 8q22, ARO with renal tubular acidosis, dental anomalies #259730
- *LRP5* low-density lipoprotein receptor-related protein 5 gene, 11q13, ADO #607634
- *IKBKG (NEMO)*, Xq28, ARO with with immune system dysfunction in ectodermal dysplasia, anhidrotic, with immunodeficiency, osteopetrosis and lymphedema; oledaid #300301
- *CalDAG-GEF1*, ARO with immune system dysfunction
- *FERMT3* or *KINDLIN-3*, 11q12, ARO with immune system dysfunction (leucocyte adhesion deficiency-3 [LAD3])
- *CTSK cathepsin K* gene, 1q21, mutations in the human cathepsin K gene leading to pycnodysostosis
- *LEMD3* gene, 12q14, which codes for an integral protein of the inner nuclear membrane result in osteopoikilosis, Buschke–Ollendorff syndrome and melorheostosis
- *PORCN* (Xp11.23) and *WTX* (Xq11.1) genes, associated with hyperostotic phenotypes (osteopathia striata in Goltz-Gorlin syndrome and osteopathia striata with cranial stenosis [OSCS], respectively)

Animal Models/Main Features

- *M-CSF*, 1p21–p13, this mouse model of osteopetrosis (op/op) demonstrates osteopetrosis, with a highly sclerotic skeleton, lack of marrow spaces, failure of tooth eruption and other pathologies.
- *SLC4A2*, solute carrier family 4, anion exchanger, member 2, gene Ae2: this osteopetrotic mouse model also shows a similar phenotype.
- Cathepsin K(–/–) mice exhibit many characteristics of the human pycnodysostosis-like phenotype.

Clinical Description

Main features

- Short stature
- Fractures
- Osteosclerosis relatively diffuse affecting the skull, spine, pelvis and appendicular bones
- Focal sclerosis of the skull base, pelvis and vertebral end plates
- Bone modelling defects
- Osteomyelitis, especially of the mandible (*TNFS11*)
- Compressive neuropathies
- Optic nerve atrophy, compression, visual impairment
- Conductive hearing loss
- Hypocalcaemia with attendant tetanic seizures
- Bone marrow failure
- Hepatosplenomegaly
- Anaemia
- Thrombocytopenia
- Life-threatening pancytopenia

Cranio-oro-dental features

- Hydrocephaly
- Facial nerve palsy
- Cranial hyperostosis
- Frontal bossing
- Altered craniofacial morphology
- Macrocephaly
- Prognathism
- Micrognathia in pycnodysostosis
- Dental malocclusion
- Dental anomalies
- Delay or failure of tooth eruption
- Deciduous teeth retention
- Tooth crown malformation
- Distorted primary molars
- Root anomalies with even sometime absence of root development
- Dental caries

Management/Oral Health

Bone marrow transplantation induces normalization of osteoclast function, which is a prerequisite for normal dental development and eruption of teeth.

References

[96,428,1051–1056,1084,1086,1088,1097]

Pictures Illustrating the Oro-Dental Features

Figure 7.4

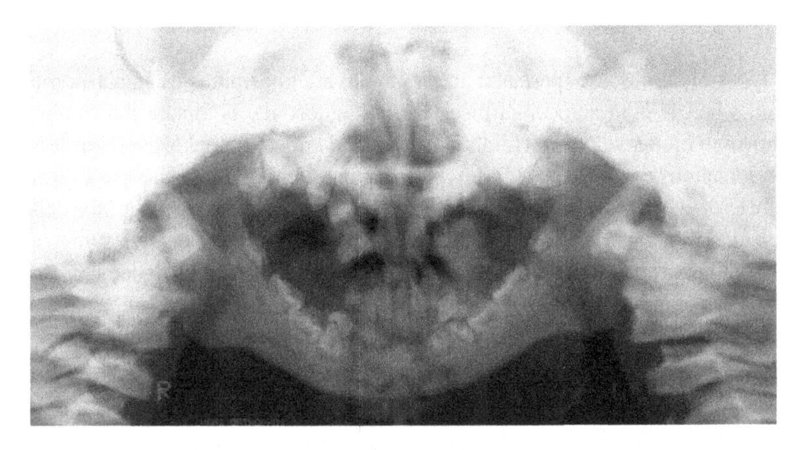

Figure 7.4 Osteopetrosis. Failure of tooth eruption and impacted teeth visible on this panoramic radiograph of a patient affected by osteopetrosis. Root development is impaired.

7.4 Hypophosphatasia

Definition

Hypophosphatasia (HP) is a rare inherited disorder characterized by a wide spectrum of defects in mineralized tissues (e.g. bone and teeth) and caused by deficiency in the tissue-non-specific alkaline phosphatase gene (*ALPL*). It manifests with reduced serum alkaline phosphatase (AP) levels and the presence in the urine of phosphoethanolamine. Six clinical types of HP are currently recognized, depending on the age of onset, are highly variable in their clinical expression and relate to different mutations in the *ALPL* gene. The first clinical sign of the disease is often a premature loss of primary teeth, mostly in the moderate types.

In addition to clinical and radiographical examinations, laboratory assays are indicated: total serum alkaline phosphatatase activity is markedly reduced. Molecular biology alone can confirm the diagnosis. The detection rate is about 95% in the severe forms.

OMIM Number

Lethal perinatal (**#241500**)
Prenatal (benign), infantile (**#241500**)
Childhood (**#241510**)
Adult and odontohypophosphatasia (**#146300**)

Prevalence

The birth prevalence of the severe form of hypophosphatasia has been estimated at 1 in 100,000 live births. It has been estimated that 1 in 25 Manitoba Mennonites are carriers of infantile hypophosphatasia. The incidence of moderate forms has never been estimated but is probably much higher and is estimated in the European population to be 1 in 6370.

Inheritance

Severe types of the disease (perinatal and infantile) are transmitted as an autosomal recessive trait, whereas both autosomal recessive and autosomal dominant transmission have been shown in moderate types (childhood, adult and odontohypophosphatasia). Moreover, *de novo* mutations can be seen. Mild hypophosphatasia not only results partly from compound heterozygosity for severe and moderate mutations but also in a large part from heterozygous mutations with a dominant negative effect. In patients carrying only one mutation, although this mutation is known to be severe, there is a moderately decreased serum AP value and a moderate form of the disease. This is due to the normal allele that provides enough AP activity to limit the dominant negative effect of the mutation.

Aetiology – Molecular Basis – Gene

Hypophosphatasia is the result of defects in the *ALPL* gene (1p36.1–p34) – encoding for tissue-non-specific alkaline phosphatase (TNSALP). More than 190 mutations are currently described worldwide in patients. This diversity of mutations explains the variable clinical expressions, the heterogeneity of the phenotype and the overlap of the clinical subtypes.

TNSALP is also known as liver/bone/kidney alkaline phosphatase, and in humans it hydrolyses phosphate from pyridoxal-5′-phosphate, pyrophosphate and phosphoethanolamine. The resulting inorganic phosphate is utilized in the production of hydroxyapatite crystals which are necessary for bone, enamel, dentine and cementum formation.

Animal Models/Main Features

Alkaline phosphatase knockout mice recapitulate the metabolic and skeletal defects of infantile hypophosphatasia [1157–1163].

Clinical Description

Distinction of infantile from childhood hypophosphatasia is somewhat arbitrary and the phenotype is more a continuum than a precise subtype, even though, in the literature, these subtypes are often cited.

Main features
Infantile form:

- Craniosynostosis
- Severe rickets
- Hypercalcaemia

- Failure to thrive
- Short stature
- Seizures
- Gastrointestinal and renal problems

Childhood form:

- Craniosynostosis
- Bone deformities
- Short stature
- Delay in walking: waddling gait
- Costochondral junction enlargement (rachitic rosary)

Adult form:

- Includes odontohypophosphatasia
- Spontaneous fractures and bone changes similar to those seen in rickets but to a minor degree

Cranio-oro-dental features

- Spontaneous and premature loss of fully rooted primary teeth before 3 years of age. This sign is a trigger sign especially in the milder forms to the diagnosis.
- Almost always, only the primary uniradicular teeth are shed, signifying moderate forms of the disease.
- Multiradicular teeth as well as permanent teeth may also be lost.
- Severe types manifest with premature loss of anterior teeth before 1 year of age and exfoliation of posterior teeth.
- A prognosis for the permanent dentition in the light of the changes in the primary one cannot be established.
- There is no periodontal or gingival involvement.
- The pulp chamber of all teeth is larger than normal.
- Teeth roots are not resorbed.
- Absence of cementum formation.
- There is horizontal atrophy of alveolar bone because there is no cementum to provide periodontal ligament attachment and mechanical stimulation of the alveolus.
- Dental defects such as abnormalities of tooth shape, structure or eruption occur mainly in the infantile type.
- Some patients with the adult type or odontohypophosphatasia present only dental manifestations similar to those seen in the childhood variety: those cases have been reported under the name of odontohypophosphatasia.

Management/Oral Health

- It is important to recognize that premature loss of primary teeth can be a diagnostic sign for this disease and to refer the patient and family to the paediatrician and genetician.
- Oral hygiene and health protocol are needed to avoid any periodontal disease.
- Removable prostheses for children in the primary dentition and for adults are necessary.
- Implants might be considered.

A recent treatment by enzyme replacement therapy (Enobia *ENB-0040*) prevents dental defects in a mouse model of hypophosphatasia and improves survival and bone defects in treated patients. Promising observations describe an improvement in dental defects.

References

[1116,1164–1179]

Pictures Illustrating the Oro-Dental Features

Figure 7.5

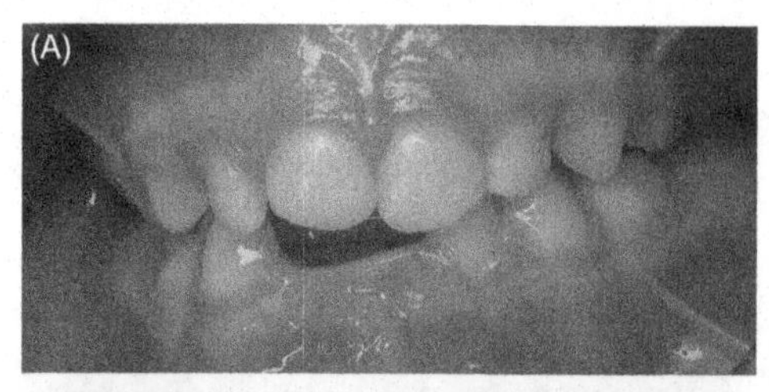

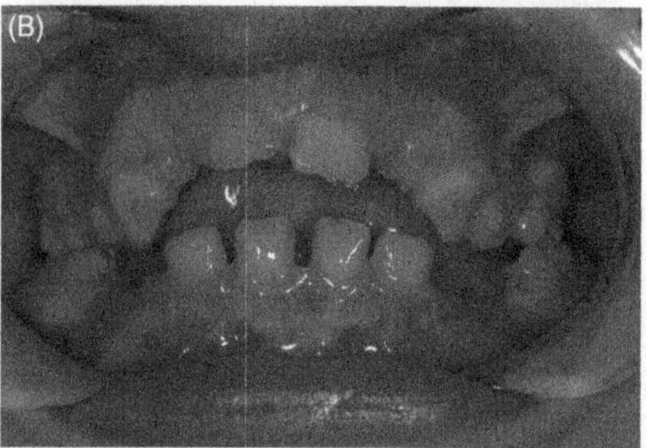

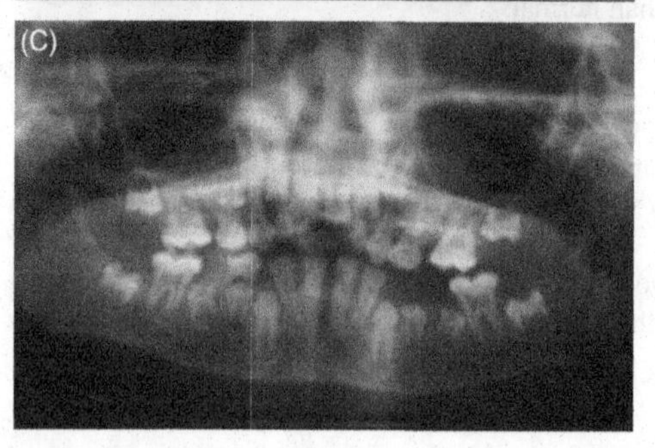

Figure 7.5 Hypophosphatasia. (A) Intraoral view at 2½ years of age (childhood form of hypophosphatasia). The lower incisors (71, 72 and 81) were lost spontaneously without gingivitis. (B) Intraoral view at 8 years of age (infantile form of hypophosphatasia). The clinical examination reveals midcrown linear enamel hypoplasia of the maxillary canines (53 and 63) and delay in the eruption of the maxillary right central incisor (11), whereas the left maxillary permanent central incisor (21) had erupted at 6½ years of age. The colour of the permanent teeth is dark yellow. (C) Panoramic radiograph of patient (B). Wide pulp chambers and enamel and dentine defects are visible.

7.5 Papillon–Lefèvre Syndrome (PLS)

Definition

The syndrome consists of palmoplantar hyperkeratosis and marked periodontoclasia (periodontal destruction) in both dentitions, with premature loss of teeth.

OMIM Number

#24500

Prevalence

1 in 4,000,000 at birth.

Inheritance

Autosomal recessive, 40% consanguinity.

Aetiology – Molecular Basis – Gene

Mutations of the lysosomal protease *cathepsin C* gene (*CTSC*, 11q14.1–14.30) have been associated with PLS.

Cathepsin C is a lysosomal cysteine protease involved in the activation of other proteases within bone marrow derived cells which participate in cell-mediated phagocytosis, cytotoxicity and activation of various inflammatory mediators.

It has been suggested that the genetic component of PLS is a predisposition rather than the main determinant of periodontal disease.

Animal Models/Main Features

De Haar et al., 2006 [1180], did not find major changes in the periodontium of *cathepsin C* (–/–) deficient mice.

Clinical Description

Main features

- Reddening of palms and soles is present at birth.
- Red, scaly keratosis of palms and soles is concurrent with the oral lesions.
- Occasionally, keratosis of the dorsal surfaces of hand and feet is present.
- Natural killer (NK) cell cytotoxicity has been reported consistently and severely depressed in patients with PLS.
- The impaired NK cell cytotoxicity might contribute to the pathogenesis of PLS-associated periodontitis.

Cranio-oro-dental features

- Teeth erupt in normal sequence.
- Marked gingivo-periodontal inflammation starts at around 1½ to 2 years of age with gingival edema and bleeding, marked alveolar bone resorption and vertical pockets on radiographs, and eventual tooth mobility with consequent exfoliation.
- Teeth are lost in eventually the same sequence in which they erupted.
- After the primary teeth are lost, the gingiva acquires a normal appearance.

- Permanent teeth erupt at the correct time, but at around 8–9 years of age, the gingivo-periodontal destruction is repeated in the same manner as in the primary dentition; all permanent teeth are lost before 14 years of age and the gingivae then resume their normal appearance.
- Marked halitosis is present.

Management/Oral Health

- Scaling and root planning are required.
- Oral hygiene instruction is needed.
- Some authors have used systemic amoxicillin–metronidazole therapy (250 mg of each/3 times daily/10 days) which, based on follow-up microbiological testing, can be repeated after 4 months.
- Recently, treatment with acitretin, 10 mg oral daily, and trimethoprim-sulfamethoxazole has proven to be effective.
- Prevention of tooth loss in some patients with the syndrome has been partially obtained with initial periodontal treatment followed by a carefully monitored oral hygiene programme and the use of ultrasonic scaling.
- Edentulous patients tolerate full dentures well.
- Osteo-integrated implants have been utilized successfully in a few cases.
- The skin manifestations remain for life, but treatment with retinoids (oral etretinate) can control the hyperkeratinization.

References

[763,1181–1195]

Pictures Illustrating the Oro-Dental Features

Figure 7.6

7.6 Haim–Munk Syndrome (Cochin Jewish Disorder, Congenital Keratosis Palmoplantaris)

Definition

The syndrome consists of long pointy fingers, periodontitis and callous patches on the palms and soles.

OMIM Number

#245010

Prevalence

The estimated occurrence of Papillon–Lefèvre syndrome, of which Haim–Munk is an extremely rare variant, is considered to be 1 or 2 persons in 1,000,000; however,

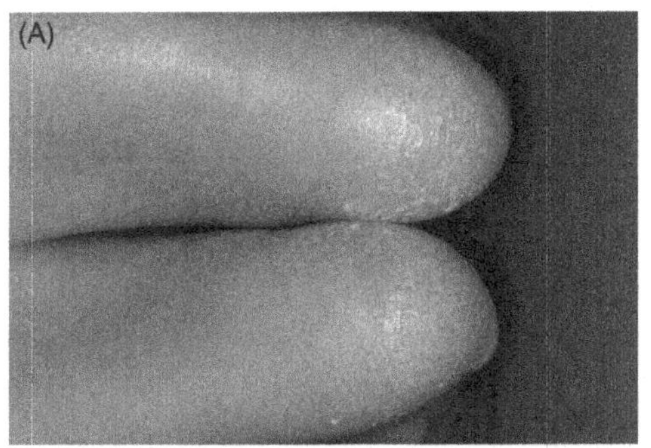

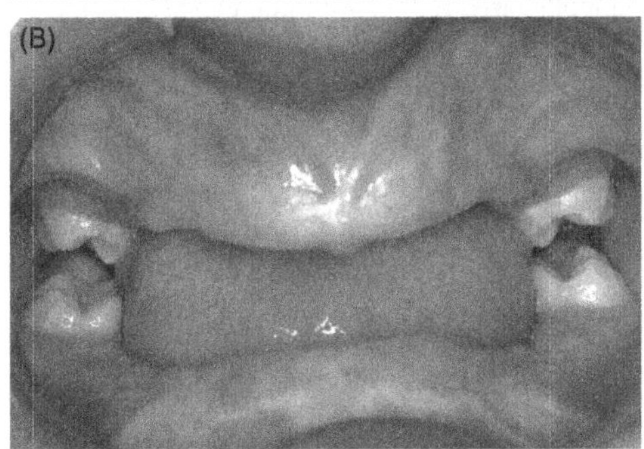

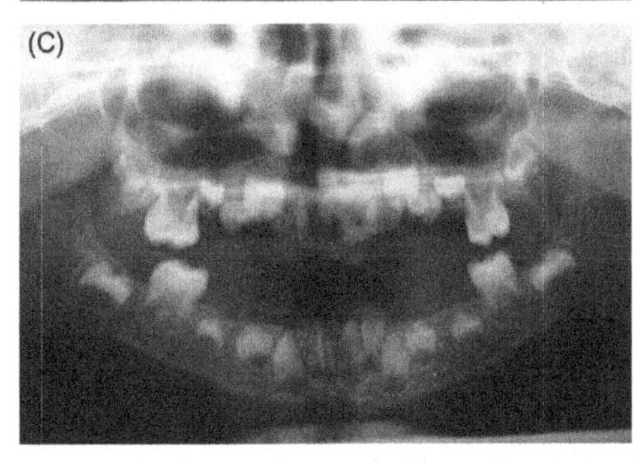

Figure 7.6 Papillon–Lefèvre syndrome. (A) Five-year-old boy presenting discrete plantar keratosis in the context of Papillon–Lefèvre syndrome, with an R272P homozygous mutation in the *cathepsin C* gene. (B) Intraoral picture demonstrating the loss of all primary teeth in patient (A). (C) Panoramic radiograph of patient (A).

although Papillon–Lefèvre syndrome cases have been identified worldwide, Haim–Munk syndrome has only been described among descendants of an inbred Jewish family originally from Cochin, India, who migrated to Israel.

Inheritance

Autosomal recessive.

Aetiology – Molecular Basis – Gene

A mutation of *cathepsin C* lysosomal protease (*CTSC*: exon 6, 2127A*G) (11q14.1–q14.3) has been identified. Haim–Munk syndrome and Papillon–Lefèvre syndrome are allelic mutations in *cathepsin C*, with variation in phenotype.

Animal Models/Main Features

None reported in the literature.

Clinical Description

Main features

- Onychogryphosis (overgrowth of the fingernails and toenails)
- Arachnodactyly (long, thin fingers)
- Acro-osteolysis (bone loss in the fingers or toes)
- Palmoplantar hyperkeratosis (keratosis palmoplantaris; red, scaly, thick patches of skin on the palms of the hands and soles of the feet, apparent at birth)
- Pyogenic skin infections
- Periodontitis beginning in childhood

Cranio-oro-dental features

- Severe periodontitis
- Premature tooth loss, in both dentitions

Management/Oral Health

- Oral hygiene protocol
- Periodontal scaling and surgery, when indicated, to preserve teeth as long as possible
- Extractions of mobile teeth
- Implants or prostheses

References

[1196–1202]

7.7 Familial Expansile Osteolysis Syndrome

Definition

The syndrome consists of familial expansile osteolysis.

Synonyms

Polyostotic osteolytic dysplasia, hereditary expansile, McCabe disease, expansile osteolysis, familial.

OMIM Number

#174810

Prevalence

Familial expansile osteolysis is a rare syndrome which worldwide has been observed in only four kinships and in two unrelated American individuals and a family in the Czech Republic.

Inheritance

Autosomal dominant.

Aetiology – Molecular Basis – Gene

Mutations in *TNFRSF11A* gene (tumour necrosis factor receptor superfamily, member 11a) (18q22.1) that encodes RANK [receptor activator of NF-kappa B (nuclear factor-kB)]. This receptor can interact with various TRAF family proteins, through which this receptor induces the activation of NF-kappa B pathway and MAPK8/JNK. This receptor and its ligand are important regulators of the interaction between T cells and dendritic cells. This receptor is also an essential mediator for osteoclast and lymph node development.

Familial expansile osteolysis, early-onset familial Paget disease of Bone PDB (#602080), osteopetrosis, autosomal recessive 7 (#612301) and expansile skeletal hyperphosphatasia are related disorders all caused by mutations affecting the *TNFRSF11A* gene.

Animal Models

Animals genetically deficient in *RANK* or the cognate *RANK ligand* are profoundly osteopetrotic because of the lack of bone resorption and remodelling [1152].

Clinical Description

Main features

- Familial expansile osteolysis
  - Severe bone pain
  - Abnormal bone modelling
  - Fractures
  - Hearing loss.

Cranio-oro-dental features

- Root resorption
- Tooth mobility
- Early loss of teeth

Management/Oral Health

Therapy using a bisphosphonate such as alendronate, a potent inhibitor of osteoclastic activity, has reduced alkaline phosphatase levels and bone pain, and may offer an effective strategy to prevent tooth root resorption.

References

[96,1203–1219]

Pictures Illustrating the Oro-Dental Features

Figure 7.7

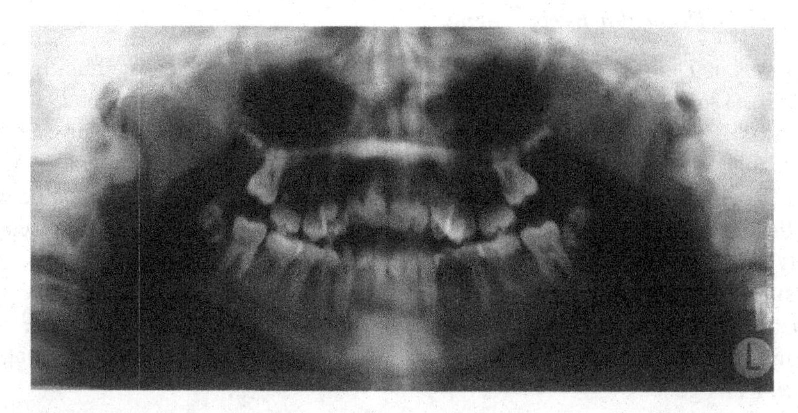

Figure 7.7 Familial expansile osteolysis syndrome. Extensive tooth resorption seen in familial expansile osteolysis syndrome.

8 Pathology and Dental Anomalies

Syndrome	Associated Features	Gene	Molecules	Inheritance	Locus
Pathology					
Gorlin syndrome #109400	Odontogenic keratocysts (keratocystic odontogenic tumours) Basal cell naevi Bifid ribs, calcified falx, palmo-plantar pits	*PTCH1* *PTCH2* *SUFU*	Transmembrane protein receptor of SHH	AD	9q22.3 1p32 10q24–q25
Cherubism #118400 [1220]	Osseous dysplasia Teeth agenesis (second and third molars) Malposition Gum hypertrophy Large multilocular, radiolucent lesions of jaws Deformation of lower part of the face Regression of the lesions during the third decade of life	*SH3BP2*	SH3-binding protein Participate to the regulation of osteoclastogenesis	AD	4p16.3

8.1 Naevoid Basal Cell Carcinoma Syndrome (NBCCS; Gorlin Syndrome)

Definition

This syndrome, also known as Gorlin syndrome or Gorlin–Goltz syndrome, is characterized by multiple basal carcinomas, odontogenic keratocysts (keratocystic odontogenic tumours), neoplasms that arise in different tissues and various skeletal and developmental anomalies.

OMIM Number

#109400

Dento/Oro/Craniofacial Anomalies and Genetics. DOI: 10.1016/B978-0-12-416038-5.00008-1

Prevalence

The estimated prevalence varies from 1 in 57,000 to 1 in 256,000, with a male-to-female ratio of 1:1. The syndrome represents 0.4% of all cases of basal cell carcinomas.

Inheritance

Autosomal dominant with complete penetrance and variable expressivity – 50% are new mutations.

Aetiology – Molecular Basis – Gene

It is caused by mutations in the *PTCH1* gene (9q22) or the *PTCH2* gene (1p32) or even the *SUFU* gene (10q24–q25) which may be deletions, insertions, splice site alterations, nonsense or missense mutations.

PTCH1 (Patched) is a transmembrane protein, the principal receptor for SHH, that represses, cell selectively, transcription of genes from the TGFβ and WNT families. *PTCH1* is a tumour suppressor gene.

The HH signal is received and transduced via a specific receptor complex composed of PTCH and smoothened (SMOH) transmembrane proteins. Abnormalities in this signalling cascade have been found in various developmental pathologies and neoplasms such as basal cell carcinoma. Excessive HH signalling is associated with an inherited cancer predisposition syndrome (Gorlin syndrome).

PTCH2 encodes a 1203 amino acid putative transmembrane protein that is highly homologous to the PTCH1 product.

PTCH2 is able to modulate tumourigenesis associated with PTCH1 haploinsufficiency.

Recently a *SUFU* germline mutation was identified in a family with Gorlin syndrome.

SUFU (suppressor of fused) encodes another component of the SHH/PTCH signalling pathway.

Animal Models/Main Features

These animals show close phenotypic resemblance to the human disease [1221–1226].

Clinical Description

Main features

Dermatologic findings:

- Basal cell carcinomas appear early in life over the nose, eyelids, cheeks, neck, arms and trunk, 40% in blacks, 90% in whites. The carcinomas can be flesh coloured or pale brown, single or in clusters.
- Small pits on palms and soles that become filled with dirt and appear as dark spots are present in 65–80% of patients.
- Hair tufts occur on various dermatologic locations.

Neoplasms and cysts:

- Various different neoplasms exist, such as medulloblastoma, meningioma, cardiac fibroma, foetal rhabdomyoma and hamartomas.
- Less frequent findings include calcified ovarian fibromas and mesenteric cysts.

Skeletal findings:

- Bifurcation or splaying of one or more ribs
- Brachydactyly (shortening of the metacarpals)
- Spina bifida occulta
- Scoliosis or kyphoscoliosis
- Ectopic calcification of falx cerebri and other skull–brain locations

Cranio-oro-dental features

Craniofacial findings:

- Mild hypertelorism
- Frontal and parietal enlargement
- Broad nasal bridge
- Mild prognathism

Ocular findings (10–25%):

- Strabismus
- Microphthalmia
- Congenital cataracts
- Iris coloboma
- Milia on the palpebral conjunctivae (40%)

Oro-dental findings:

- Highly arched palate (50%)
- Cleft lip and palate (2–8%)
- Open bite
- Multiple keratocysts odontogenic tumours of the jaws which may develop as early as 5–6 years of age and interfere with normal development of the jaw bones and teeth; rarely arising in these cysts, an ameloblastoma
- Teeth agenesis
- Impacted teeth

Management/Oral Health

- Oral hygiene protocol
- Surgical treatment of the keratocystic odontogenic tumours
- Surgical correction of the cleft, if present
- Orthodontic treatment, if needed
- Prosthodontic rehabilitation, if needed

References

[394,1227–1240]

Pictures Illustrating the Oro-Dental Features

Figure 8.1

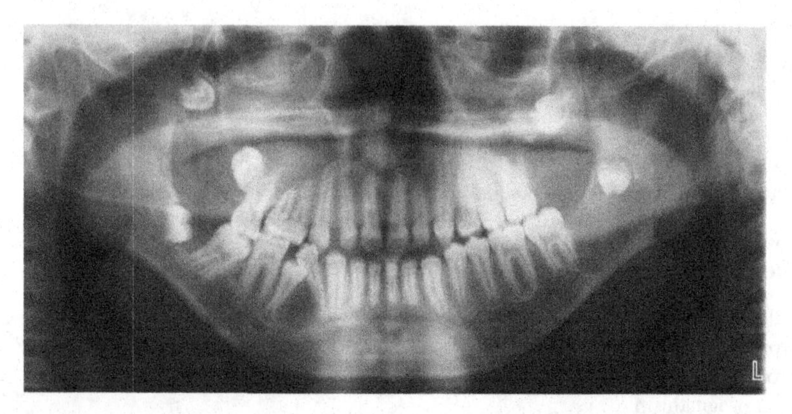

Figure 8.1 Gorlin syndrome. Multiple keratocysts visible on this panoramic radiograph. Gorlin syndrome, Courtesy of AL Garret, Reference Centre for Orodental Manifestations of Rare Diseases, Pôle de Médecine et Chirurgie Bucco-dentaires, Hôpitaux Universitaires de Strasbourg, Université de Strasbourg.

8.2 Cherubism

Definition

Cherubism is a rare non-neoplastic, hereditary disorder characterized by bilateral, multilocular, radiolucent lesions producing expansion of the mandible and occasionally of the maxilla, with a resulting, characteristic facial appearance.

OMIM Number

#118400

Prevalence

More than 250 cases described.

Inheritance

Autosomal dominant, with almost complete penetrance in males and incomplete penetrance in females, and variable expressivity. Familial occurrence is reported in most cases.

Aetiology – Molecular Basis – Gene

Cherubism is caused by gain of function mutations in the *SH3BP2* gene (4p16.3) coding for SH3-binding protein 2 potentially involved in signal transduction and participating to the regulation of osteoclastogenesis. Most missense mutations identified to date are in exon 9. Cherubism has been reported in rare cases of Noonan syndrome (a developmental disorder characterized by unusual face, short stature, and heart defects).

Animal Models/Main Features

Sh3bp2 −/− mice developed osteoporosis as a result of reduced bone formation despite the fact that bone resorption is impaired.

Clinical Description

Main features
Cranio-oro-dental features

- Symmetrical enlargement of the mandible due to
  - Bilateral multilocular areas of diminished density in the mandible on X-rays
  - Those areas are occupied by connective tissue with multinucleated cells.
- Bilateral swelling at the angle of the mandible in children
- Possible problems associated with respiration, speech, swallowing and vision
- Onset between ages 2 and 5
- Progression of the lesions until puberty
- Tendency for condition to regress in adults
- Agenesis of teeth (molars)
- Premature exfoliation of deciduous teeth
- Delayed tooth eruption
- Malocclusion
- Displacement of primary and permanent teeth

Management/Oral Health

- Multidisciplinary management is required.
- Surgery (curettage with eventually bone grafting) might be needed for aggressive lesions.
- Orthodontic treatment may reduce the risk of airway obstruction and tooth displacement.
- Ophthalmologic treatment is needed for displacement of the globe or vision loss.
- Long-term follow-up is needed.
- Lesions might regress spontaneously.

References

[1220,1241,1242]

Pictures Illustrating the Oro-Dental Features

Figure 8.2

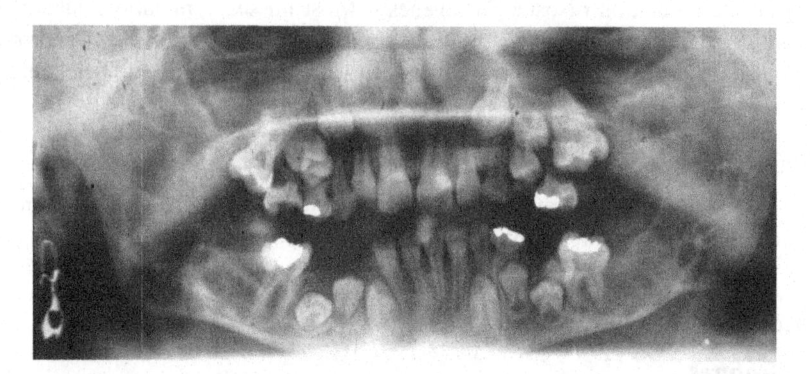

Figure 8.2 Cherubism. Bilateral multilocular areas of diminished density in the mandible on this panoramic radiograph.

Bibliography

[1] J. London, H. Birkedal-Hansen, Opportunities in dental, oral, and craniofacial research, Compend. Contin. Educ. Dent. 21 (9) (2000) 760–762, 764, 766.

[2] H.C. Slavkin, Molecular biology experimental strategies for craniofacial-oral-dental dysmorphology, Connect. Tissue Res. 32 (1–4) (1995) 233–239.

[3] H. Slavkin, Craniofacial-oral-dental research in the 21st century, J. Dent. Res. 76 (2) (1997) 628–630.

[4] R.D. Knight, T.F. Schilling, Cranial neural crest and development of the head skeleton, Adv. Exp. Med. Biol. 589 (2006) 120–133.

[5] M.T. Cobourne, T. Mitsiadis, Neural crest cells and patterning of the mammalian dentition, J. Exp. Zoolog. B Mol. Dev. Evol. 306 (3) (2006) 251–260.

[6] D.M. Noden, R.A. Schneider, Neural crest cells and the community of plan for craniofacial development: historical debates and current perspectives, Adv. Exp. Med. Biol. 589 (2006) 1–23.

[7] J. Jernvall, S.V. Keranen, I. Thesleff, From the cover: evolutionary modification of development in mammalian teeth: quantifying gene expression patterns and topography, Proc. Natl. Acad. Sci. U.S.A. 97 (26) (2000) 14444–14448.

[8] I. Salazar-Ciudad, J. Jernvall, A gene network model accounting for development and evolution of mammalian teeth, Proc. Natl. Acad. Sci. U.S.A. 99 (12) (2002) 8116–8120.

[9] M.V. Plikus, M. Zeichner-David, J.A. Mayer, J. Reyna, P. Bringas, J.G. Thewissen, et al., Morphoregulation of teeth: modulating the number, size, shape and differentiation by tuning Bmp activity, Evol. Dev. 7 (5) (2005) 440–457.

[10] I. Thesleff, T. Aberg, Molecular regulation of tooth development, Bone 25 (1) (1999) 123–125.

[11] A.S. Tucker, P.T. Sharpe, Molecular genetics of tooth morphogenesis and patterning: the right shape in the right place, J. Dent. Res. 78 (4) (1999) 826–834.

[12] H. Peters, R. Balling, Teeth. Where and how to make them, Trends Genet. 15 (2) (1999) 59–65.

[13] A. Tucker, P. Sharpe, The cutting-edge of mammalian development; how the embryo makes teeth, Nat. Rev. Genet. 5 (7) (2004) 499–508.

[14] I. Thesleff, Epithelial–mesenchymal signalling regulating tooth morphogenesis, J. Cell Sci. 116 (9) (2003) 1647–1648.

[15] Y. Chai, X. Jiang, Y. Ito, P. Bringas, Jr., J. Han, D.H. Rowitch, et al., Fate of the mammalian cranial neural crest during tooth and mandibular morphogenesis, Development 127 (8) (2000) 1671–1679.

[16] I. Miletich, P.T. Sharpe, Neural crest contribution to mammalian tooth formation, Birth Defects Res. C Embryo. Today 72 (2) (2004) 200–212.

[17] P. Nieminen, M. Pekkanen, T. Aberg, I. Thesleff, A graphical WWW-database on gene expression in tooth, Eur. J. Oral Sci. 106 (Suppl 1) (1998) 7–11.

[18] X. Nie, K. Luukko, P. Kettunen, BMP signalling in craniofacial development, Int. J. Dev. Biol. 50 (6) (2006) 511–521.

[19] X. Nie, K. Luukko, P. Kettunen, FGF signalling in craniofacial development and developmental disorders, Oral Dis. 12 (2) (2006) 102–111.

[20] Z. Hardcastle, C.C. Hui, P.T. Sharpe, The Shh signalling pathway in early tooth development, Cell. Mol. Biol. (Noisy-le-grand) 45 (5) (1999) 567–578.

[21] H.R. Dassule, P. Lewis, M. Bei, R. Maas, A.P. McMahon, Sonic hedgehog regulates growth and morphogenesis of the tooth, Development 127 (22) (2000) 4775–4785.

[22] J. Pispa, I. Thesleff, Mechanisms of ectodermal organogenesis, Dev. Biol. 262 (2) (2003) 195–205.

[23] M.T. Cobourne, P.T. Sharpe, Sonic hedgehog signaling and the developing tooth, Curr. Top. Dev. Biol. 65 (2005) 255–287.

[24] M. Tummers, I. Thesleff, The importance of signal pathway modulation in all aspects of tooth development, J. Exp. Zool. B Mol. Dev. Evol. 312B (4) (2009) 309–319.

[25] I. Thesleff, Developmental biology and building a tooth, Quintessence Int. 34 (8) (2003) 613–620.

[26] I. Thesleff, M. Mikkola, The role of growth factors in tooth development, Int. Rev. Cytol. 217 (2002) 93–135.

[27] I. Thesleff, A. Vaahtokari, A.M. Partanen, Regulation of organogenesis. Common molecular mechanisms regulating the development of teeth and other organs, Int. J. Dev. Biol. 39 (1) (1995) 35–50.

[28] E. Sonnenberg, D. Meyer, K.M. Weidner, C. Birchmeier, Scatter factor/hepatocyte growth factor and its receptor, the c-met tyrosine kinase, can mediate a signal exchange between mesenchyme and epithelia during mouse development, J. Cell Biol. 123 (1) (1993) 223–235.

[29] M.J. Tabata, K. Kim, J.G. Liu, K. Yamashita, T. Matsumura, J. Kato, et al., Hepatocyte growth factor is involved in the morphogenesis of tooth germ in murine molars, Development 122 (4) (1996) 1243–1251.

[30] X. Xu, P. Bringas, Jr., P. Soriano, Y. Chai, PDGFR-alpha signaling is critical for tooth cusp and palate morphogenesis, Dev. Dyn. 232 (1) (2005) 75–84.

[31] T.A. Mitsiadis, M. Salmivirta, T. Muramatsu, H. Muramatsu, H. Rauvala, E. Lehtonen, et al., Expression of the heparin-binding cytokines, midkine (MK) and HB-GAM (pleiotrophin) is associated with epithelial–mesenchymal interactions during fetal development and organogenesis, Development 121 (1) (1995) 37–51.

[32] T.A. Mitsiadis, T. Muramatsu, H. Muramatsu, I. Thesleff, Midkine (MK), a heparin-binding growth/differentiation factor, is regulated by retinoic acid and epithelial–mesenchymal interactions in the developing mouse tooth, and affects cell proliferation and morphogenesis, J. Cell Biol. 129 (1) (1995) 267–281.

[33] T.A. Mitsiadis, K. Luukko, Neurotrophins in odontogenesis, Int. J. Dev. Biol. 39 (1) (1995) 195–202.

[34] J. Jernvall, I. Thesleff, Reiterative signaling and patterning during mammalian tooth morphogenesis, Mech. Dev. 92 (1) (2000) 19–29.

[35] R. Maas, M. Bei, The genetic control of early tooth development, Crit. Rev. Oral Biol. Med. 8 (1) (1997) 4–39.

[36] A.S. Tucker, G. Yamada, M. Grigoriou, V. Pachnis, P.T. Sharpe, Fgf-8 determines rostral-caudal polarity in the first branchial arch, Development 126 (1) (1999) 51–61.

[37] M.T. Cobourne, P.T. Sharpe, Tooth and jaw: molecular mechanisms of patterning in the first branchial arch, Arch. Oral Biol. 48 (1) (2003) 1–14.

[38] I. Satokata, L. Ma, H. Ohshima, M. Bei, I. Woo, K. Nishizawa, et al., Msx2 deficiency in mice causes pleiotropic defects in bone growth and ectodermal organ formation, Nat. Genet. 24 (4) (2000) 391–395.

[39] B.L. Thomas, A.S. Tucker, M. Qui, C.A. Ferguson, Z. Hardcastle, J.L. Rubenstein, et al., Role of *Dlx-1* and *Dlx-2* genes in patterning of the murine dentition, Development 124 (23) (1997) 4811–4818.

[40] M.L. Mikkola, S.E. Millar, The mammary bud as a skin appendage: unique and shared aspects of development, J. Mammary Gland Biol. Neoplasia 11 (3–4) (2006) 187–203.

[41] H.S. Jung, P.H. Francis-West, R.B. Widelitz, T.X. Jiang, S. Ting-Berreth, C. Tickle, et al., Local inhibitory action of BMPs and their relationships with activators in feather formation: implications for periodic patterning, Dev. Biol. 196 (1) (1998) 11–23.

[42] S.E. Millar, Molecular mechanisms regulating hair follicle development, J. Invest. Dermatol. 118 (2) (2002) 216–225.

[43] J. Laurikkala, M. Mikkola, T. Mustonen, T. Aberg, P. Koppinen, J. Pispa, et al., TNF signaling via the ligand–receptor pair ectodysplasin and edar controls the function of epithelial signaling centers and is regulated by Wnt and activin during tooth organogenesis, Dev. Biol. 229 (2) (2001) 443–455.

[44] M.L. Mikkola, I. Thesleff, Ectodysplasin signaling in development, Cytokine Growth Factor Rev. 14 (3–4) (2003) 211–224.

[45] A. Smahi, G. Courtois, S.H. Rabia, R. Doffinger, C. Bodemer, A. Munnich, et al., The NF-kappaB signalling pathway in human diseases: from incontinentia pigmenti to ectodermal dysplasias and immune-deficiency syndromes, Hum. Mol. Genet. 11 (20) (2002) 2371–2375.

[46] T. Rinne, H.G. Brunner, H. van Bokhoven, p63-associated disorders, Cell Cycle 6 (3) (2007) 262–268.

[47] I. Thesleff, J. Jernvall, The enamel knot: a putative signaling center regulating tooth development, Cold Spring Harb. Symp. Quant. Biol. 62 (1997) 257–267.

[48] A. Gritli-Linde, M. Bei, R. Maas, X.M. Zhang, A. Linde, A.P. McMahon, Shh signaling within the dental epithelium is necessary for cell proliferation, growth and polarization, Development 129 (23) (2002) 5323–5337.

[49] J. Jernvall, T. Aberg, P. Kettunen, S. Keranen, I. Thesleff, The life history of an embryonic signaling center: BMP-4 induces p21 and is associated with apoptosis in the mouse tooth enamel knot, Development 125 (2) (1998) 161–169.

[50] M. Bei, K. Kratochwil, R.L. Maas, BMP4 rescues a non-cell-autonomous function of Msx1 in tooth development, Development 127 (21) (2000) 4711–4718.

[51] A. Bloch-Zupan, T. Leveillard, P. Gorry, J.L. Fausser, J.V. Ruch, Expression of p21(WAF1/CIP1) during mouse odontogenesis, Eur. J. Oral Sci. 106 (Suppl. 1) (1998) 104–111.

[52] K. Kratochwil, M. Dull, I. Farinas, J. Galceran, R. Grosschedl, Lef1 expression is activated by BMP-4 and regulates inductive tissue interactions in tooth and hair development, Genes Dev. 10 (11) (1996) 1382–1394.

[53] P. Kettunen, I. Thesleff, Expression and function of FGFs-4, −8, and −9 suggest functional redundancy and repetitive use as epithelial signals during tooth morphogenesis, Dev. Dyn. 211 (3) (1998) 256–268.

[54] X. Wang, I. Thesleff, Tooth development, in: K. Unsicker, K. Krieglstein, (Eds.), Cell Signalling and Growth Factors in Development, Wiley-VCH, Weinheim, 2006, pp. 719–754.

[55] S. Vainio, I. Karavanova, A. Jowett, I. Thesleff, Identification of BMP-4 as a signal mediating secondary induction between epithelial and mesenchymal tissues during early tooth development, Cell 75 (1) (1993) 45–58.

[56] M. Bei, R. Maas, FGFs and BMP4 induce both Msx1-independent and Msx1-dependent signaling pathways in early tooth development, Development 125 (21) (1998) 4325–4333.

[57] H. Peters, A. Neubuser, R. Balling, Pax genes and organogenesis: Pax9 meets tooth development, Eur. J. Oral Sci. 106 (Suppl. 1) (1998) 38–43.

[58] T. Aberg, X.P. Wang, J.H. Kim, T. Yamashiro, M. Bei, R. Rice, et al., Runx2 mediates FGF signaling from epithelium to mesenchyme during tooth morphogenesis, Dev. Biol. 270 (1) (2004) 76–93.

[59] R.N. D'Souza, T. Aberg, J. Gaikwad, A. Cavender, M. Owen, G. Karsenty, et al., Cbfa1 is required for epithelial–mesenchymal interactions regulating tooth development in mice, Development 126 (13) (1999) 2911–2920.

[60] T. Aberg, A. Cavender, J.S. Gaikwad, A.L. Bronckers, X. Wang, J. Waltimo-Siren, et al., Phenotypic changes in dentition of Runx2 homozygote-null mutant mice, J. Histochem. Cytochem. 52 (1) (2004) 131–140.

[61] A.S. Tucker, D.J. Headon, P. Schneider, B.M. Ferguson, P. Overbeek, J. Tschopp, et al., Edar/Eda interactions regulate enamel knot formation in tooth morphogenesis, Development 127 (21) (2000) 4691–4700.

[62] J.M. Courtney, J. Blackburn, P.T. Sharpe, The Ectodysplasin and NFkappaB signalling pathways in odontogenesis, Arch. Oral. Biol. 50 (2) (2005) 159–163.

[63] E. Jarvinen, I. Salazar-Ciudad, W. Birchmeier, M.M. Taketo, J. Jernvall, I. Thesleff, Continuous tooth generation in mouse is induced by activated epithelial Wnt/beta-catenin signaling, Proc. Natl. Acad. Sci. U.S.A. 103 (49) (2006) 18627–18632.

[64] S.W. Cho, J.Y. Kim, J. Cai, J.M. Lee, E.J. Kim, H.A. Lee, et al., Temporospatial tissue interactions regulating the regeneration of the enamel knot in the developing mouse tooth, Differentiation 75 (2) (2007) 158–165.

[65] S.W. Cho, H.A. Lee, J. Cai, M.J. Lee, J.Y. Kim, H. Ohshima, et al., The primary enamel knot determines the position of the first buccal cusp in developing mice molars, Differentiation 75 (5) (2007) 441–451.

[66] E. Matalova, G.S. Antonarakis, P.T. Sharpe, A.S. Tucker, Cell lineage of primary and secondary enamel knots, Dev. Dyn. 233 (3) (2005) 754–759.

[67] J. Jernvall, P. Kettunen, I. Karavanova, L.B. Martin, I. Thesleff, Evidence for the role of the enamel knot as a control center in mammalian tooth cusp formation: non-dividing cells express growth stimulating Fgf-4 gene, Int. J. Dev. Biol. 38 (3) (1994) 463–469.

[68] A. Vaahtokari, T. Aberg, J. Jernvall, S. Keranen, I. Thesleff, The enamel knot as a signaling center in the developing mouse tooth, Mech. Dev. 54 (1) (1996) 39–43.

[69] E. Matalova, A.S. Tucker, P.T. Sharpe, Death in the life of a tooth, J. Dent. Res. 83 (1) (2004) 11–16.

[70] H. Magloire, M.L. Couble, A. Romeas, F. Bleicher, Odontoblast primary cilia: facts and hypotheses, Cell Biol. Int. 28 (2) (2004) 93–99.

[71] W.T. Butler, J.C. Brunn, C. Qin, Dentin extracellular matrix (ECM) proteins: comparison to bone ECM and contribution to dynamics of dentinogenesis, Connect. Tissue Res. 44 (Suppl. 1) (2003) 171–178.

[72] M. MacDougall, J. Dong, A.C. Acevedo, Molecular basis of human dentin diseases, Am. J. Med. Genet. A 140 (23) (2006) 2536–2546.

[73] L.W. Fisher, N.S. Fedarko, Six genes expressed in bones and teeth encode the current members of the SIBLING family of proteins, Connect. Tissue Res. 44 (Suppl. 1) (2003) 33–40.

[74] M. Goldberg, O. Rapoport, D. Septier, K. Palmier, R. Hall, G. Embery, et al., Proteoglycans in predentin: the last 15 micrometers before mineralization, Connect. Tissue Res. 44 (Suppl. 1) (2003) 184–188.

[75] G. Embery, R. Hall, R. Waddington, D. Septier, M. Goldberg, Proteoglycans in dentinogenesis, Crit. Rev. Oral Biol. Med. 12 (4) (2001) 331–349.
[76] M. Goldberg, D. Septier, S. Lecolle, L. Vermelin, P. Bissila-Mapahou, J.P. Carreau, et al., Lipids in predentine and dentine, Connect. Tissue Res. 33 (1–3) (1995) 105–114.
[77] M. Goldberg, D. Septier, Phospholipids in amelogenesis and dentinogenesis, Crit. Rev. Oral Biol. Med. 13 (3) (2002) 276–290.
[78] N.S. Fedarko, A. Jain, A. Karadag, L.W. Fisher, Three small integrin binding ligand N-linked glycoproteins (SIBLINGs) bind and activate specific matrix metalloproteinases, FASEB J. 18 (6) (2004) 734–736.
[79] J.P. Simmer, J.C. Hu, Expression, structure, and function of enamel proteinases, Connect. Tissue Res. 43 (2–3) (2002) 441–449.
[80] J.D. Bartlett, B. Ganss, M. Goldberg, J. Moradian-Oldak, M.L. Paine, M.L. Snead, et al., 3. Protein–protein interactions of the developing enamel matrix, Curr. Top. Dev. Biol. 74 (2006) 57–115.
[81] K. Iwasaki, E. Bajenova, E. Somogyi-Ganss, M. Miller, V. Nguyen, H. Nourkeyhani, et al., Amelotin – a novel secreted, ameloblast-specific protein, J. Dent. Res. 84 (12) (2005) 1127–1132.
[82] J.V. Ruch, H. Lesot, Y. Cam, J.M. Meyer, A. Bloch-Zupan, C. Begue-Kirn, Control of odontoblast differentiation: current hypothesis. Proceedings of the International Conference on Dentin/Pulp Complex 1995, Japan, 1996, pp. 105–111.
[83] J.V. Ruch, Odontoblast commitment and differentiation, Biochem. Cell Biol. 76 (6) (1998) 923–938.
[84] A.G. Fincham, J. Moradian-Oldak, J.P. Simmer, The structural biology of the developing dental enamel matrix, J. Struct. Biol. 126 (3) (1999) 270–299.
[85] H. Lesot, S. Lisi, R. Peterkova, M. Peterka, V. Mitolo, J.V. Ruch, Epigenetic signals during odontoblast differentiation, Adv. Dent. Res. 15 (2001) 8–13.
[86] C. Begue-Kirn, A.J. Smith, J.V. Ruch, J.M. Wozney, A. Purchio, D. Hartmann, et al., Effects of dentin proteins, transforming growth factor beta 1 (TGF beta 1) and bone morphogenetic protein 2 (BMP2) on the differentiation of odontoblast *in vitro*, Int. J. Dev. Biol. 36 (4) (1992) 491–503.
[87] K. Tompkins, Molecular mechanisms of cytodifferentiation in mammalian tooth development, Connect. Tissue Res. 47 (3) (2006) 111–118.
[88] X.P. Wang, M. Suomalainen, C.J. Jorgez, M.M. Matzuk, S. Werner, I. Thesleff, Follistatin regulates enamel patterning in mouse incisors by asymmetrically inhibiting BMP signaling and ameloblast differentiation, Dev. Cell 7 (5) (2004) 719–730.
[89] E. Koyama, C. Wu, T. Shimo, M. Iwamoto, T. Ohmori, K. Kurisu, et al., Development of stratum intermedium and its role as a Sonic hedgehog-signaling structure during odontogenesis, Dev. Dyn. 222 (2) (2001) 178–191.
[90] A. Berdal, P. Papagerakis, D. Hotton, I. Bailleul-Forestier, J.L. Davideau, Ameloblasts and odontoblasts, target-cells for 1,25-dihydroxyvitamin D3: a review, Int. J. Dev. Biol. 39 (1) (1995) 257–262.
[91] L. Ye, T.Q. Le, L. Zhu, K. Butcher, R.A. Schneider, W. Li, et al., Amelogenins in human developing and mature dental pulp, J. Dent. Res. 85 (9) (2006) 814–818.
[92] S.N. White, M.L. Paine, A.Y. Ngan, V.G. Miklus, W. Luo, H. Wang, et al., Ectopic expression of dentin sialoprotein during amelogenesis hardens bulk enamel, J. Biol. Chem. 282 (8) (2007) 5340–5345.
[93] T. Yokohama-Tamaki, H. Ohshima, N. Fujiwara, Y. Takada, Y. Ichimori, S. Wakisaka, et al., Cessation of Fgf10 signaling, resulting in a defective dental epithelial stem cell compartment, leads to the transition from crown to root formation, Development 133 (7) (2006) 1359–1366.

[94] X. Luan, Y. Ito, T.G. Diekwisch, Evolution and development of Hertwig's epithelial root sheath, Dev. Dyn. 235 (5) (2006) 1167–1180.

[95] B.L. Foster, T.E. Popowics, H.K. Fong, M.J. Somerman, Advances in defining regulators of cementum development and periodontal regeneration, Curr. Top. Dev. Biol. 78 (2007) 47–126.

[96] M.H. Helfrich, Osteoclast diseases and dental abnormalities, Arch. Oral Biol. 50 (2) (2005) 115–122.

[97] G.E. Wise, S.J. Lumpkin, H. Huang, Q. Zhang, Osteoprotegerin and osteoclast differentiation factor in tooth eruption, J. Dent. Res. 79 (12) (2000) 1937–1942.

[98] W.M. Philbrick, B.E. Dreyer, I.A. Nakchbandi, A.C. Karaplis, Parathyroid hormone-related protein is required for tooth eruption, Proc. Natl. Acad. Sci. U.S.A. 95 (20) (1998) 11846–11851.

[99] S. Camilleri, F. McDonald, Runx2 and dental development, Eur. J. Oral Sci. 114 (5) (2006) 361–373.

[100] S. Alaluusua, P.L. Lukinmaa, Developmental dental toxicity of dioxin and related compounds – a review, Int. Dent. J. 56 (6) (2006) 323–331.

[101] S. Alaluusua, [Amoxicillin may be a cause of enamel hypomineralization], Duodecim 122 (5) (2006) 491–492.

[102] S. Alaluusua, P.L. Lukinmaa, J. Torppa, J. Tuomisto, T. Vartiainen, Developing teeth as biomarker of dioxin exposure, Lancet 353 (9148) (1999) 206.

[103] G. Koch, Prevalence of enamel mineralisation disturbances in an area with 1-1.2 ppm F in drinking water. Review and summary of a report published in Sweden in 1981, Eur. J. Paediatr. Dent. 4 (3) (2003) 127–128.

[104] K.L. Weerheijm, Molar incisor hypomineralisation (MIH), Eur. J. Paediatr. Dent. 4 (3) (2003) 114–120.

[105] A. Berdal, [Gene/environment relations in the development of tooth anomalies], Arch. Pediatr. 10 (Suppl. 1) (2003) 16s–18s.

[106] A. Bloch-Zupan, Odonto-génétique: une nouvelle facette de notre profession!, Le Chirurgien Dentiste de France 1182 (2004) 77–86.

[107] I. Thesleff, The genetic basis of tooth development and dental defects, Am. J. Med. Genet. A 140 (23) (2006) 2530–2535.

[108] M.J. Aldred, R. Savarirayan, P.J. Crawford, Amelogenesis imperfecta: a classification and catalogue for the 21st century, Oral Dis. 9 (1) (2003) 19–23.

[109] I. Thesleff, Genetic basis of tooth development and dental defects, Acta Odontol. Scand. 58 (5) (2000) 191–194.

[110] M. MacDougall, Dental structural diseases mapping to human chromosome 4q21, Connect. Tissue Res. 44 (Suppl. 1) (2003) 285–291.

[111] P. Papagerakis, M. Peuchmaur, D. Hotton, L. Ferkdadji, P. Delmas, S. Sasaki, et al., Aberrant gene expression in epithelial cells of mixed odontogenic tumors, J. Dent. Res. 78 (1) (1999) 20–30.

[112] K.L. Weerheijm, M. Duggal, I. Mejara, L. Papagiannoulis, G. Koch, L.C. Martens, et al., Judgement criteria for molar incisor hypomineralisation (MIH) in epidemiologic studies: a summary of the European meeting on MIH held in Athens, 2003, Eur. J. Paed. Dent. 4 (3) (2003) 110–113.

[113] S.J. Fomon, J. Ekstrand, E.E. Ziegler, Fluoride intake and prevalence of dental fluorosis: trends in fluoride intake with special attention to infants, J. Public Health Dent. 60 (3) (2000) 131–139.

[114] S. Arte, S. Pirinen, Hypodontia. Orphanet encyclopedia, 2003. <http://www.orpha.net/data/patho/GB/uk-hypodontia.pdf>. update 2004.

[115] S.R. Line, Variation of tooth number in mammalian dentition: connecting genetics, development, and evolution, Evol. Dev. 5 (3) (2003) 295–304.
[116] I. Satokata, R. Maas, Msx1 deficient mice exhibit cleft palate and abnormalities of craniofacial and tooth development, Nat. Genet. 6 (4) (1994) 348–356.
[117] H. Peters, A. Neubuser, K. Kratochwil, R. Balling, Pax9-deficient mice lack pharyngeal pouch derivatives and teeth and exhibit craniofacial and limb abnormalities, Genes Dev. 12 (17) (1998) 2735–2747.
[118] M.M. Matzuk, T.R. Kumar, A. Bradley, Different phenotypes for mice deficient in either activins or activin receptor type II, Nature 374 (6520) (1995) 356–360.
[119] T. Sasaki, Y. Ito, X. Xu, J. Han, P. Bringas, Jr, T. Maeda, et al., LEF1 is a critical epithelial survival factor during tooth morphogenesis, Dev. Biol. 278 (1) (2005) 130–143.
[120] K. Kitamura, H. Miura, S. Miyagawa-Tomita, M. Yanazawa, Y. Katoh-Fukui, R. Suzuki, et al., Mouse Pitx2 deficiency leads to anomalies of the ventral body wall, heart, extra- and periocular mesoderm and right pulmonary isomerism, Development 126 (24) (1999) 5749–5758.
[121] C.R. Lin, C. Kioussi, S. O'Connell, P. Briata, D. Szeto, F. Liu, et al., Pitx2 regulates lung asymmetry, cardiac positioning and pituitary and tooth morphogenesis, Nature 401 (6750) (1999) 279–282.
[122] A. Mostowska, A. Kobielak, B. Biedziak, W.H. Trzeciak, Novel mutation in the paired box sequence of *PAX9* gene in a sporadic form of oligodontia, Eur. J. Oral Sci. 111 (3) (2003) 272–276.
[123] A. Mostowska, A. Kobielak, W.H. Trzeciak, Molecular basis of non-syndromic tooth agenesis: mutations of *MSX1* and *PAX9* reflect their role in patterning human dentition, Eur. J. Oral Sci. 111 (5) (2003) 365–370.
[124] P. Das, D.W. Stockton, C. Bauer, L.G. Shaffer, R.N. D'Souza, T. Wright, et al., Haploinsufficiency of *PAX9* is associated with autosomal dominant hypodontia, Hum. Genet. 110 (4) (2002) 371–376.
[125] L. Lammi, K. Halonen, S. Pirinen, I. Thesleff, S. Arte, P. Nieminen, A missense mutation in *PAX9* in a family with distinct phenotype of oligodontia, Eur. J. Hum. Genet. 11 (11) (2003) 866–871.
[126] R. Perveen, I.C. Lloyd, J. Clayton-Smith, A. Churchill, V. van Heyningen, I. Hanson, et al., Phenotypic variability and asymmetry of Rieger syndrome associated with *PITX2* mutations, Invest. Ophthalmol. Vis. Sci. 41 (9) (2000) 2456–2460.
[127] P. Nieminen, Genetic basis of tooth agenesis, J. Exp. Zool. B Mol. Dev. Evol. 312B (4) (2009) 320–342.
[128] L. Lammi, S. Arte, M. Somer, H. Jarvinen, P. Lahermo, I. Thesleff, et al., Mutations in *AXIN2* cause familial tooth agenesis and predispose to colorectal cancer, Am. J. Hum. Genet. 74 (5) (2004) 1043–1050.
[129] P.J. De Coster, L.A. Marks, L.C. Martens, A. Huysseune, Dental agenesis: genetic and clinical perspectives, J. Oral Pathol. Med. 38 (1) (2009) 1–17.
[130] J.T. Wright, The molecular etiologies and associated phenotypes of amelogenesis imperfecta, Am. J. Med. Genet. A 140 (23) (2006) 2547–2555.
[131] R.J. Gorlin, M.M. Cohen, J.R.C.M. Hennekam, Syndromes of the Head and Neck, fourth ed., University Press, Oxford, 2001.
[132] S. Pirinen, Genetic craniofacial aberrations, Acta Odontol. Scand. 56 (6) (1998) 356–359.
[133] A. Bloch-Zupan, Robert James Gorlin: Chirurgien-dentiste et généticien, Le Chirurgien Dentiste de France 1295 (2007) 22–24.
[134] K. Kurisu, M.J. Tabata, [Hereditary diseases with tooth anomalies and their causal genes], Kaibogaku Zasshi 73 (3) (1998) 201–208.

[135] G.C. Townsend, M.J. Aldred, P.M. Bartold, Genetic aspects of dental disorders, Aust. Dent. J. 43 (4) (1998) 269–286.

[136] I. Miletich, P.T. Sharpe, Normal and abnormal dental development, Hum. Mol. Genet. 12 (Spec. No. 1) (2003) 69–73.

[137] N.I. Wolf, I. Harting, A.M. Innes, S. Patzer, P. Zeitler, A. Schneider, et al., Ataxia, delayed dentition and hypomyelination: a novel leukoencephalopathy, Neuropediatrics 38 (2007) 64–70.

[138] P.J. De Coster, R.M. Verbeeck, V. Holthaus, L.C. Martens, A. Vral, Seckel syndrome associated with oligodontia, microdontia, enamel hypoplasia, delayed eruption, and dentin dysmineralization: a new variant? J. Oral Pathol. Med. 35 (10) (2006) 639–641.

[139] S. Ayme, B. Urbero, D. Oziel, E. Lecouturier, A.C. Biscarat, [Information on rare diseases: the Orphanet project], Rev. Med. Interne. 19 (Suppl. 3) (1998) 376S–377S.

[140] S. Ayme, [Orphanet, an information site on rare diseases], Soins Jan–Feb (672) (2003) 46–47.

[141] J.P. Fryns, T.J. de Ravel, London dysmorphology database, London neurogenetics database and dysmorphology photo library on CD-ROM [Version 3] 2001 R.M. Winter, M. Baraitser, Oxford University Press, ISBN 019851780, pound sterling 1595, Hum. Genet. 111 (1) (2002) 113.

[142] S.S. Guest, C.D. Evans, R.M. Winter, The online London dysmorphology database, Genet. Med. 1 (5) (1999) 207–212.

[143] R.M. Winter, M. Baraitser, The London dysmorphology database, J. Med. Genet. 24 (8) (1987) 509–510.

[144] P.N. Kantaputra, M. Paramee, A. Kaewkhampa, A. Hoshino, M. Lees, M. McEntagart, et al., Cleft lip with cleft palate, ankyloglossia, and hypodontia are associated with *TBX22* mutations, J. Dent. Res. 90 (4) (2011) 450–455.

[145] H.F. McKeown, D.L. Robinson, C. Elcock, M. al-Sharood, A.H. Brook, Tooth dimensions in hypodontia patients, their unaffected relatives and a control group measured by a new image analysis system, Eur. J. Orthod. 24 (2) (2002) 131–141.

[146] H. Vastardis, N. Karimbux, S.W. Guthua, J.G. Seidman, C.E. Seidman, A human *MSX1* homeodomain missense mutation causes selective tooth agenesis, Nat. Genet. 13 (4) (1996) 417–421.

[147] G. Hu, H. Vastardis, A.J. Bendall, Z. Wang, M. Logan, H. Zhang, et al., Haploinsufficiency of *MSX1*: a mechanism for selective tooth agenesis, Mol. Cell. Biol. 18 (10) (1998) 6044–6051.

[148] D. Jumlongras, M. Bei, J.M. Stimson, W.F. Wang, S.R. DePalma, C.E. Seidman, et al., A nonsense mutation in *MSX1* causes Witkop syndrome, Am. J. Hum. Genet. 69 (1) (2001) 67–74.

[149] M.J. van den Boogaard, M. Dorland, F.A. Beemer, H.K. van Amstel, MSX1 mutation is associated with orofacial clefting and tooth agenesis in humans, Nat. Genet. 24 (4) (2000) 342–343.

[150] M. Kamamoto, J. Machida, S. Yamaguchi, M. Kimura, T. Ono, P.A. Jezewski, et al., Clinical and functional data implicate the Arg(151)Ser variant of MSX1 in familial hypodontia, Eur. J. Hum. Genet. 19 (8) (2011) 844–850.

[151] A.C. Lidral, B.C. Reising, The role of MSX1 in human tooth agenesis, J. Dent. Res. 81 (4) (2002) 274–278.

[152] K. Xuan, F. Jin, Y.L. Liu, L.T. Yuan, L.Y. Wen, F.S. Yang, et al., Identification of a novel missense mutation of *MSX1* gene in Chinese family with autosomal-dominant oligodontia, Arch. Oral Biol. 53 (8) (2008) 773–779.

[153] S. Devadas, B. Varma, J. Mungara, T. Joseph, T.R. Saraswathi, Witkop tooth and nail syndrome: a case report, Int. J. Paediatr. Dent. 15 (5) (2005) 364–369.
[154] S.J. Hodges, K.E. Harley, Witkop tooth and nail syndrome: report of two cases in a family, Int. J. Paediatr. Dent. 9 (3) (1999) 207–211.
[155] C.D. Hudson, C.J. Witkop, Autosomal dominant hypodontia with nail dysgenesis. Report of twenty-nine cases in six families, Oral Surg. Oral Med. Oral Pathol. 39 (3) (1975) 409–423.
[156] T.H. Redpath, G.B. Winter, Autosomal dominant ectodermal dysplasia with significant dental defects, Br. Dent. J. 126 (3) (1969) 123–128.
[157] G.M. Wicomb, L.X. Stephen, P. Beighton, Dental implications of tooth–nail dysplasia (Witkop syndrome): a report of an affected family and an approach to dental management, J. Clin. Pediatr. Dent. 28 (2) (2004) 107–112.
[158] H. Engbers, J.J. van der Smagt, R. van 't Slot, J.R. Vermeesch, R. Hochstenbach, M. Poot, Wolf–Hirschhorn syndrome facial dysmorphic features in a patient with a terminal 4p16.3 deletion telomeric to the WHSCR and WHSCR 2 regions, Eur. J. Hum. Genet. 17 (1) (2009) 129–132.
[159] D. Naf, L.A. Wilson, R.A. Bergstrom, R.S. Smith, N.C. Goodwin, A. Verkerk, et al., Mouse models for the Wolf–Hirschhorn deletion syndrome, Hum. Mol. Genet. 10 (2) (2001) 91–98.
[160] J.M. Abrams, Y. Jiao, Keeping it simple: what mouse models of Wolf–Hirschhorn syndrome can tell us about large chromosomal deletions, Dis. Model. Mech. 2 (7–8) (2009) 315–316.
[161] C. Catela, D. Bilbao-Cortes, E. Slonimsky, P. Kratsios, N. Rosenthal, P. Te Welscher, Multiple congenital malformations of Wolf–Hirschhorn syndrome are recapitulated in Fgfrl1 null mice, Dis. Model. Mech. 2 (5–6) (2009) 283–294.
[162] M. Zollino, R. Lecce, R. Fischetto, M. Murdolo, F. Faravelli, A. Selicorni, et al., Mapping the Wolf–Hirschhorn syndrome phenotype outside the currently accepted WHS critical region and defining a new critical region, WHSCR-2, Am. J. Hum. Genet. 72 (3) (2003) 590–597.
[163] S.B. Babich, C. Banducci, P. Teplitsky, Dental characteristics of the Wolf–Hirschhorn syndrome: a case report, Spec. Care Dentist. 24 (4) (2004) 229–231.
[164] A. Battaglia, J.C. Carey, T.J. Wright, Wolf–Hirschhorn (4p-) syndrome, Adv. Pediatr. 48 (2001) 75–113.
[165] G.H. Breen, Taurodontism, an unreported dental finding in Wolf–Hirschhorn (4p-) syndrome, ASDC J. Dent. Child. 65 (5) (1998) 344–356.
[166] P.S. Iwanowski, S. Stengel-Rutkowski, L. Anderlik, J. Pilch, A.T. Midro, Physical and developmental phenotype analyses in a boy with Wolf–Hirschhorn syndrome, Genet. Couns. 16 (1) (2005) 31–40.
[167] N.J. Johnston, D.L. Franklin, Dental findings of a child with Wolf–Hirschhorn syndrome, Int. J. Paediatr. Dent. 16 (2) (2006) 139–142.
[168] S.G. Kant, A. Van Haeringen, E. Bakker, I. Stec, D. Donnai, P. Mollevanger, et al., Pitt-Rogers-Danks syndrome and Wolf–Hirschhorn syndrome are caused by a deletion in the same region on chromosome 4p 16.3, J. Med. Genet. 34 (7) (1997) 569–572.
[169] L.J. Lo, M.S. Noordhoff, Y.R. Chen, Cleft lip and hemangioma: a patient with Wolf–Hirschhorn syndrome, Ann. Plast. Surg. 32 (5) (1994) 539–541.
[170] P. Nieminen, J. Kotilainen, Y. Aalto, S. Knuutila, S. Pirinen, I. Thesleff, *MSX1* gene is deleted in Wolf–Hirschhorn syndrome patients with oligodontia, J. Dent. Res. 82 (12) (2003) 1013–1017.

[171] M. Sase, K. Hasegawa, R. Honda, M. Sumie, M. Nakata, N. Sugino, et al., Ultrasonographic findings of facial dysmorphism in Wolf–Hirschhorn syndrome, Am. J. Perinatol. 22 (2) (2005) 99–102.

[172] M. Zollino, C. Di Stefano, G. Zampino, P. Mastroiacovo, T.J. Wright, G. Sorge, et al., Genotype-phenotype correlations and clinical diagnostic criteria in Wolf–Hirschhorn syndrome, Am. J. Med. Genet. 94 (3) (2000) 254–261.

[173] P. Nieminen, S. Arte, D. Tanner, L. Paulin, S. Alaluusua, I. Thesleff, et al., Identification of a nonsense mutation in the *PAX9* gene in molar oligodontia, Eur. J. Hum. Genet. 9 (10) (2001) 743–746.

[174] A. Mostowska, B. Biedziak, W.H. Trzeciak, A novel mutation in *PAX9* causes familial form of molar oligodontia, Eur. J. Hum. Genet. 14 (2) (2006) 173–179.

[175] D.W. Stockton, P. Das, M. Goldenberg, R.N. D'Souza, P.I. Patel, Mutation of *PAX9* is associated with oligodontia, Nat. Genet. 24 (1) (2000) 18–19.

[176] S.A. Frazier-Bowers, D.C. Guo, A. Cavender, L. Xue, B. Evans, T. King, et al., A novel mutation in human *PAX9* causes molar oligodontia, J. Dent. Res. 81 (2) (2002) 129–133.

[177] V. Tallon-Walton, M.C. Manzanares-Cespedes, S. Arte, P. Carvalho-Lobato, I. Valdivia-Gandur, A. Garcia-Susperregui, et al., Identification of a novel mutation in the *PAX9* gene in a family affected by oligodontia and other dental anomalies, Eur. J. Oral Sci. 115 (6) (2007) 427–432.

[178] J. Zhao, Q. Hu, Y. Chen, S. Luo, L. Bao, Y. Xu, A novel missense mutation in the paired domain of human *PAX9* causes oligodontia, Am. J. Med. Genet. A 143 (21) (2007) 2592–2597.

[179] L. Hansen, S. Kreiborg, H. Jarlov, E. Niebuhr, H. Eiberg, A novel nonsense mutation in *PAX9* is associated with marked variability in number of missing teeth, Eur. J. Oral Sci. 115 (4) (2007) 330–333.

[180] Y. Wang, H. Wu, J. Wu, H. Zhao, X. Zhang, G. Mues, et al., Identification and functional analysis of two novel *PAX9* mutations, Cells Tissues Organs 189 (1–4) (2009) 80–87.

[181] Y. Wang, J.C. Groppe, J. Wu, T. Ogawa, G. Mues, R.N. D'Souza, et al., Pathogenic mechanisms of tooth agenesis linked to paired domain mutations in human *PAX9*, Hum. Mol. Genet. 18 (15) (2009) 2863–2874.

[182] B.A. Amendt, E.V. Semina, W.L. Alward, Rieger syndrome: a clinical, molecular, and biochemical analysis, Cell Mol. Life Sci. 57 (11) (2000) 1652–1666.

[183] B.A. Amendt, L.B. Sutherland, E.V. Semina, A.F. Russo, The molecular basis of Rieger syndrome. Analysis of Pitx2 homeodomain protein activities, J. Biol. Chem. 273 (32) (1998) 20066–20072.

[184] T.A. Hjalt, E.V. Semina, Current molecular understanding of Axenfeld-Rieger syndrome, Expert Rev. Mol. Med. 7 (25) (2005) 1–17.

[185] F. Idrees, A. Bloch-Zupan, S.L. Free, D. Vaideanu, P.J. Thompson, P. Ashley, et al., A novel homeobox mutation in the *PITX2* gene in a family with Axenfeld-Rieger syndrome associated with brain, ocular, and dental phenotypes, Am. J. Med. Genet. B Neuropsychiatr. Genet. 141 (2) (2006) 184–191.

[186] Y. Wang, H. Zhao, X. Zhang, H. Feng, Novel identification of a four-base-pair deletion mutation in PITX2 in a Rieger syndrome family, J. Dent. Res. 82 (12) (2003) 1008–1012.

[187] Z. Tumer, D. Bach-Holm, Axenfeld-Rieger syndrome and spectrum of PITX2 and FOXC1 mutations, Eur. J. Hum. Genet. 17 (12) (2009) 1527–1539.

[188] N. Weisschuh, E. De Baere, B. Wissinger, Z. Tumer, Clinical utility gene card for: Axenfeld-Rieger syndrome, Eur. J. Hum. Genet. 19 (2011) 3.

[189] D. Kelberman, L. Islam, S.E. Holder, T.S. Jacques, P. Calvas, R.C. Hennekam, et al., Digenic inheritance of mutations in FOXC1 and PITX2: correlating transcription factor function and Axenfeld-Rieger disease severity, Hum. Mutat. 32 (10) (2011) 1144–1152.

[190] M.F. Lu, C. Pressman, R. Dyer, R.L. Johnson, J.F. Martin, Function of Rieger syndrome gene in left-right asymmetry and craniofacial development, Nature 401 (6750) (1999) 276–278.

[191] A.M. Sclafani, J.M. Skidmore, H. Ramaprakash, A. Trumpp, P.J. Gage, D.M. Martin, Nestin-Cre mediated deletion of Pitx2 in the mouse, Genesis 44 (7) (2006) 336–344.

[192] J.K. Brooks, P.J. Coccaro, Jr., M.A. Zarbin, The Rieger anomaly concomitant with multiple dental, craniofacial, and somatic midline anomalies and short stature, Oral Surg. Oral Med. Oral Pathol. 68 (6) (1989) 717–724.

[193] A.K. Jena, O.P. Kharbanda, Axenfeld-Rieger syndrome: report on dental and craniofacial findings, J. Clin. Pediatr. Dent. 30 (1) (2005) 83–88.

[194] E.M. O'Dwyer, D.C. Jones, Dental anomalies in Axenfeld-Rieger syndrome, Int. J. Paediatr. Dent. 15 (6) (2005) 459–463.

[195] J. Singh, K. Pannu, G. Lehl, The Rieger syndrome: orofacial manifestations. Case report of a rare condition, Quintessence Int. 34 (9) (2003) 689–692.

[196] S. Dressler, P. Meyer-Marcotty, N. Weisschuh, A. Jablonski-Momeni, K. Pieper, G. Gramer, et al., Dental and craniofacial anomalies associated with Axenfeld-Rieger syndrome with PITX2 mutation, Case Report Med. 2010 (2010) 621984.

[197] S. Nawaz, J. Klar, M. Wajid, M. Aslam, M. Tariq, J. Schuster, et al., WNT10A missense mutation associated with a complete Odonto-Onycho-Dermal Dysplasia syndrome, Eur. J. Hum. Genet. 17 (12) (2009) 1600–1605.

[198] L. Adaimy, E. Chouery, H. Megarbane, S. Mroueh, V. Delague, E. Nicolas, et al., Mutation in WNT10A is associated with an autosomal recessive ectodermal dysplasia: the odonto-onycho-dermal dysplasia, Am. J. Hum. Genet. 81 (4) (2007) 821–828.

[199] P. Kantaputra, W. Sripathomsawat, WNT10A and isolated hypodontia, Am. J. Med. Genet. A 155A (5) (2011) 1119–1122.

[200] Y. Zhang, P. Tomann, T. Andl, N.M. Gallant, J. Huelsken, B. Jerchow, et al., Reciprocal requirements for EDA/EDAR/NF-kappaB and Wnt/beta-catenin signaling pathways in hair follicle induction, Dev. Cell 17 (1) (2009) 49–61.

[201] W. Liu, X. Dong, M. Mai, R.S. Seelan, K. Taniguchi, K.K. Krishnadath, et al., Mutations in AXIN2 cause colorectal cancer with defective mismatch repair by activating beta-catenin/TCF signalling, Nat. Genet. 26 (2) (2000) 146–147.

[202] H.M. Yu, B. Jerchow, T.J. Sheu, B. Liu, F. Costantini, J.E. Puzas, et al., The role of Axin2 in calvarial morphogenesis and craniosynostosis, Development 132 (8) (2005) 1995–2005.

[203] M.L. Marvin, S.M. Mazzoni, C.M. Herron, S. Edwards, S.B. Gruber, E.M. Petty, et al., AXIN2-associated autosomal dominant ectodermal dysplasia and neoplastic syndrome, Am. J. Med. Genet. A 155A (4) (2011) 898–902.

[204] N. Callahan, A. Modesto, R. Meira, F. Seymen, A. Patir, A.R. Vieira, Axis inhibition protein 2 (AXIN2) polymorphisms and tooth agenesis, Arch. Oral Biol. 54 (1) (2009) 45–49.

[205] B. Bergendal, J. Klar, C. Stecksen-Blicks, J. Norderyd, N. Dahl, Isolated oligodontia associated with mutations in EDARADD, AXIN2, MSX1, and PAX9 genes, Am. J. Med. Genet. A 155A (7) (2011) 1616–1622.

[206] A. Mostowska, B. Biedziak, P.P. Jagodzinski, Axis inhibition protein 2 (AXIN2) polymorphisms may be a risk factor for selective tooth agenesis, J. Hum. Genet. 51 (3) (2006) 262–266.

[207] A. Bohring, T. Stamm, C. Spaich, C. Haase, K. Spree, U. Hehr, et al., WNT10A mutations are a frequent cause of a broad spectrum of ectodermal dysplasias with sex-biased manifestation pattern in heterozygotes, Am. J. Hum. Genet. 85 (1) (2009) 97–105.

[208] M. van Geel, M. Gattas, Y. Kesler, P. Tong, H. Yan, K. Tran, et al., Phenotypic variability associated with WNT10A nonsense mutations, Br. J. Dermatol. 162 (6) (2010) 1403–1406.

[209] S. Song, D. Han, H. Qu, Y. Gong, H. Wu, X. Zhang, et al., *EDA* gene mutations underlie non-syndromic oligodontia, J. Dent. Res. 88 (2) (2009) 126–131.

[210] C.L. Ku, K. Yang, J. Bustamante, A. Puel, H. von Bernuth, O.F. Santos, et al., Inherited disorders of human Toll-like receptor signaling: immunological implications, Immunol. Rev. 203 (2005) 10–20.

[211] J. Laurikkala, J. Pispa, H.S. Jung, P. Nieminen, M. Mikkola, X. Wang, et al., Regulation of hair follicle development by the TNF signal ectodysplasin and its receptor Edar, Development 129 (10) (2002) 2541–2553.

[212] M.L. Mikkola, Molecular aspects of hypohidrotic ectodermal dysplasia, Am. J. Med. Genet. A 149A (9) (2009) 2031–2036.

[213] M. Priolo, C. Lagana, Ectodermal dysplasias: a new clinical-genetic classification, J. Med. Genet. 38 (9) (2001) 579–585.

[214] M. Pinheiro, N. Freire-Maia, Ectodermal dysplasias: a clinical classification and a causal review, Am. J. Med. Genet. 53 (2) (1994) 153–162.

[215] J. Lamartine, Towards a new classification of ectodermal dysplasias, Clin. Exp. Dermatol. 28 (4) (2003) 351–355.

[216] A. Yang, R. Schweitzer, D. Sun, M. Kaghad, N. Walker, R.T. Bronson, et al., p63 is essential for regenerative proliferation in limb, craniofacial and epithelial development, Nature 398 (6729) (1999) 714–718.

[217] J. Laurikkala, M.L. Mikkola, M. James, M. Tummers, A.A. Mills, I. Thesleff, p63 regulates multiple signalling pathways required for ectodermal organogenesis and differentiation, Development 133 (8) (2006) 1553–1563.

[218] M. Pinheiro, N. Freire-Maia, Christ-Siemens-Touraine syndrome – a clinical and genetic analysis of a large Brazilian kindred: II. Affected males, Am. J. Med. Genet. 4 (2) (1979) 123–128.

[219] M. Pinheiro, N. Freire-Maia, Christ-Siemens-Touraine syndrome – a clinical and genetic analysis of a large Brazilian kindred: III. Carrier detection, Am. J. Med. Genet. 4 (2) (1979) 129–134.

[220] M. Pinheiro, N. Freire-Maia, Christ-Siemens-Touraine syndrome – a clinical and genetic analysis of a large Brazilian kindred: I. Affected females, Am. J. Med. Genet. 4 (2) (1979) 113–122.

[221] A.L. Aswegan, K.D. Josephson, R. Mowbray, R.M. Pauli, R.A. Spritz, M.S. Williams, Autosomal dominant hypohidrotic ectodermal dysplasia in a large family, Am. J. Med. Genet. 72 (4) (1997) 462–467.

[222] T. Hashiguchi, S. Yotsumoto, T. Kanzaki, Mutations in the ED1 gene in Japanese families with X-linked hypohidrotic ectodermal dysplasia, Exp. Dermatol. 12 (4) (2003) 518–522.

[223] L. Ho, M.S. Williams, R.A. Spritz, A gene for autosomal dominant hypohidrotic ectodermal dysplasia (EDA3) maps to chromosome 2q11-q13, Am. J. Hum. Genet. 62 (5) (1998) 1102–1106.

[224] J. Kere, A.K. Srivastava, O. Montonen, J. Zonana, N. Thomas, B. Ferguson, et al., X-linked anhidrotic (hypohidrotic) ectodermal dysplasia is caused by mutation in a novel transmembrane protein, Nat. Genet. 13 (4) (1996) 409–416.

[225] S.A. Wisniewski, A. Kobielak, W.H. Trzeciak, K. Kobielak, Recent advances in understanding of the molecular basis of anhidrotic ectodermal dysplasia: discovery of a ligand, ectodysplasin A and its two receptors, J. Appl. Genet. 43 (1) (2002) 97–107.

[226] N. Chassaing, S. Bourthoumieu, M. Cossee, P. Calvas, M.C. Vincent, Mutations in EDAR account for one-quarter of non-ED1-related hypohidrotic ectodermal dysplasia, Hum. Mutat. 27 (3) (2006) 255–259.

[227] C. Cluzeau, S. Hadj-Rabia, M. Jambou, S. Mansour, P. Guigue, S. Masmoudi, et al., Only four genes (EDA1, EDAR, EDARADD, and WNT10A) account for 90% of hypohidrotic/anhidrotic ectodermal dysplasia cases, Hum. Mutat. 32 (1) (2011) 70–72.

[228] A.K. Srivastava, J. Pispa, A.J. Hartung, Y. Du, S. Ezer, T. Jenks, et al., The tabby phenotype is caused by mutation in a mouse homologue of the EDA gene that reveals novel mouse and human exons and encodes a protein (ectodysplasin-A) with collagenous domains, Proc. Natl. Acad. Sci. U.S.A. 94 (24) (1997) 13069–13074.

[229] R. Peterkova, P. Kristenova, H. Lesot, S. Lisi, J.L. Vonesch, J.L. Gendrault, et al., Different morphotypes of the tabby (EDA) dentition in the mouse mandible result from a defect in the mesio-distal segmentation of dental epithelium, Orthod. Craniofac. Res. 5 (4) (2002) 215–226.

[230] I. Thesleff, M.L. Mikkola, Death receptor signaling giving life to ectodermal organs, Sci. STKE 2002 (131) (2002) E22.

[231] P.J. Crawford, M.J. Aldred, A. Clarke, Clinical and radiographic dental findings in X linked hypohidrotic ectodermal dysplasia, J. Med. Genet. 28 (3) (1991) 181–185.

[232] D. Glavina, M. Majstorovic, O. Lulic-Dukic, H. Juric, Hypohidrotic ectodermal dysplasia: dental features and carriers detection, Coll. Antropol. 25 (1) (2001) 303–310.

[233] B. Kargul, T. Alcan, U. Kabalay, M. Atasu, Hypohidrotic ectodermal dysplasia: dental, clinical, genetic and dermatoglyphic findings of three cases, J. Clin. Pediatr. Dent. 26 (1) (2001) 5–12.

[234] C.I. Gros, F. Clauss, F. Obry, M.C. Maniere, M. Schmittbuhl, Quantification of taurodontism: interests in the early diagnosis of hypohidrotic ectodermal dysplasia, Oral Dis. 16 (3) (2010) 292–298.

[235] R. Madhan, S. Nayar, Prosthetic rehabilitation of individuals with ectodermal dysplasia, Indian J. Dent. Res. 16 (3) (2005) 114–118.

[236] L. Lo Muzio, P. Bucci, F. Carile, F. Riccitiello, C. Scotti, E. Coccia, et al., Prosthetic rehabilitation of a child affected from anhidrotic ectodermal dysplasia: a case report, J. Contemp. Dent. Pract. 6 (3) (2005) 120–126.

[237] I. Tarjan, K. Gabris, N. Rozsa, Early prosthetic treatment of patients with ectodermal dysplasia: a clinical report, J. Prosthet. Dent. 93 (5) (2005) 419–424.

[238] B. Bergendal, A. Ekman, P. Nilsson, Implant failure in young children with ectodermal dysplasia: a retrospective evaluation of use and outcome of dental implant treatment in children in Sweden, Int. J. Oral Maxillofac. Implants 23 (3) (2008) 520–524.

[239] B. Bergendal, Oligodontia ectodermal dysplasia – on signs, symptoms, genetics, and outcomes of dental treatment, Swed. Dent. J. Suppl. 205 (2010) 13–78, 7–8.

[240] M.L. Casal, J.R. Lewis, E.A. Mauldin, A. Tardivel, K. Ingold, M. Favre, et al., Significant correction of disease after postnatal administration of recombinant ectodysplasin A in canine X-linked ectodermal dysplasia, Am. J. Hum. Genet. 81 (5) (2007) 1050–1056.

[241] E.A. Mauldin, O. Gaide, P. Schneider, M.L. Casal, Neonatal treatment with recombinant ectodysplasin prevents respiratory disease in dogs with X-linked ectodermal dysplasia, Am. J. Med. Genet. A 149A (9) (2009) 2045–2049.

[242] H. Lesot, F. Clauss, M.C. Maniere, M. Schmittbuhl, Consequences of X-linked hypohidrotic ectodermal dysplasia for the human jaw bone, Front. Oral Biol. 13 (2009) 93–99.

[243] F. Clauss, M.C. Maniere, F. Obry, E. Waltmann, S. Hadj-Rabia, C. Bodemer, et al., Dento-craniofacial phenotypes and underlying molecular mechanisms in hypohidrotic ectodermal dysplasia (HED): a review, J. Dent. Res. 87 (12) (2008) 1089–1099.

[244] A.K. Yap, I. Klineberg, Dental implants in patients with ectodermal dysplasia and tooth agenesis: a critical review of the literature, Int. J. Prosthodont. 22 (3) (2009) 268–276.

[245] F.J. Kramer, C. Baethge, H. Tschernitschek, Implants in children with ectodermal dysplasia: a case report and literature review, Clin. Oral Implants Res. 18 (1) (2007) 140–146.

[246] A. Puel, K. Yang, C.L. Ku, H. von Bernuth, J. Bustamante, O.F. Santos, et al., Heritable defects of the human TLR signalling pathways, J. Endotoxin. Res. 11 (4) (2005) 220–224.

[247] E.D. Carrol, A.R. Gennery, T.J. Flood, G.P. Spickett, M. Abinun, Anhidrotic ectodermal dysplasia and immunodeficiency: the role of NEMO, Arch. Dis. Child. 88 (4) (2003) 340–341.

[248] G. Courtois, A. Smahi, J. Reichenbach, R. Doffinger, C. Cancrini, M. Bonnet, et al., A hypermorphic IkappaBalpha mutation is associated with autosomal dominant anhidrotic ectodermal dysplasia and T cell immunodeficiency, J. Clin. Invest. 112 (7) (2003) 1108–1115.

[249] G. Courtois, A. Smahi, NF-kappaB-related genetic diseases, Cell Death Differ. 13 (5) (2006) 843–851.

[250] R. Doffinger, A. Smahi, C. Bessia, F. Geissmann, J. Feinberg, A. Durandy, et al., X-linked anhidrotic ectodermal dysplasia with immunodeficiency is caused by impaired NF-kappaB signaling, Nat. Genet. 27 (3) (2001) 277–285.

[251] K.H. Orstavik, M. Kristiansen, G.P. Knudsen, K. Storhaug, A. Vege, K. Eiklid, et al., Novel splicing mutation in the NEMO (IKK-gamma) gene with severe immunodeficiency and heterogeneity of X-chromosome inactivation, Am. J. Med. Genet. A 140 (1) (2006) 31–39.

[252] E. Vinolo, H. Sebban, A. Chaffotte, A. Israel, G. Courtois, M. Veron, et al., A point mutation in NEMO associated with anhidrotic ectodermal dysplasia with immunodeficiency pathology results in destabilization of the oligomer and reduces lipopolysaccharide- and tumor necrosis factor-mediated NF-kappa B activation, J. Biol. Chem. 281 (10) (2006) 6334–6348.

[253] M. Schmidt-Supprian, W. Bloch, G. Courtois, K. Addicks, A. Israel, K. Rajewsky, et al., NEMO/IKK gamma-deficient mice model incontinentia pigmenti, Mol. Cell 5 (6) (2000) 981–992.

[254] A. Nenci, C. Becker, A. Wullaert, R. Gareus, G. van Loo, S. Danese, et al., Epithelial NEMO links innate immunity to chronic intestinal inflammation, Nature 446 (7135) (2007) 557–561.

[255] G. Uzel, The range of defects associated with nuclear factor kappaB essential modulator, Curr. Opin. Allergy Clin. Immunol. 5 (6) (2005) 513–518.

[256] G. Courtois, The NF-kappaB signaling pathway in human genetic diseases, Cell. Mol. Life Sci. 62 (15) (2005) 1682–1691.

[257] A. Puel, C. Picard, C.L. Ku, A. Smahi, J.L. Casanova, et al., Inherited disorders of NF-kappaB-mediated immunity in man, Curr. Opin. Immunol. 16 (1) (2004) 34–41.

[258] D.L. Nelson, NEMO, NFkappaB signaling and incontinentia pigmenti, Curr. Opin. Genet. Dev. 16 (3) (2006) 282–288.
[259] Orphanet, Orphanet Report Series – Prevalence of rare diseases: bibliographic data. <http://www.orpha.net/orphacom/cahiers/docs/GB/Prevalence_of_rare_diseases_by_alphabetical_list.pdf>, 2009.
[260] C. Makris, V.L. Godfrey, G. Krahn-Senftleben, T. Takahashi, J.L. Roberts, T. Schwarz, et al., Female mice heterozygous for IKK gamma/NEMO deficiencies develop a dermatopathy similar to the human X-linked disorder incontinentia pigmenti, Mol. Cell 5 (6) (2000) 969–979.
[261] D. Rudolph, W.C. Yeh, A. Wakeham, B. Rudolph, D. Nallainathan, J. Potter, et al., Severe liver degeneration and lack of NF-kappaB activation in NEMO/IKKgamma-deficient mice, Genes Dev. 14 (7) (2000) 854–862.
[262] A. Nenci, M. Huth, A. Funteh, M. Schmidt-Supprian, W. Bloch, D. Metzger, et al., Skin lesion development in a mouse model of incontinentia pigmenti is triggered by NEMO deficiency in epidermal keratinocytes and requires TNF signaling, Hum. Mol. Genet. 15 (4) (2006) 531–542.
[263] S.J. Landy, D. Donnai, Incontinentia pigmenti (Bloch–Sulzberger syndrome), J. Med. Genet. 30 (1) (1993) 53–59.
[264] T.A. Phan, O. Wargon, A.M. Turner, Incontinentia pigmenti case series: clinical spectrum of incontinentia pigmenti in 53 female patients and their relatives, Clin. Exp. Dermatol. 30 (5) (2005) 474–480.
[265] S. Minic, G.E. Novotny, D. Trpinac, M. Obradovic, Clinical features of incontinentia pigmenti with emphasis on oral and dental abnormalities, Clin. Oral Investig. 10 (4) (2006) 343–347.
[266] H.P. Wu, Y.L. Wang, H.H. Chang, G.F. Huang, M.K. Guo, Dental anomalies in two patients with incontinentia pigmenti, J. Formos. Med. Assoc. 104 (6) (2005) 427–430.
[267] E. Van den Steen, P. Bottenberg, M. Bonduelle, [Dental anomalies associated with incontinentia pigmenti or Bloch–Sulzberger syndrome], Rev. Belge Med. Dent. 59 (2) (2004) 94–99.
[268] A.L. Berlin, A.S. Paller, L.S. Chan, Incontinentia pigmenti: a review and update on the molecular basis of pathophysiology, J. Am. Acad. Dermatol. 47 (2) (2002) 169–187, quiz 188–190.
[269] A.L. Bruckner, Incontinentia pigmenti: a window to the role of NF-kappaB function, Semin. Cutan. Med. Surg. 23 (2) (2004) 116–124.
[270] C. Doruk, A.A. Bicakci, H. Babacan, Orthodontic and orthopedic treatment of a patient with incontinentia pigmenti, Angle Orthod. 73 (6) (2003) 763–768.
[271] S. Hadj-Rabia, D. Froidevaux, N. Bodak, D. Hamel-Teillac, A. Smahi, Y. Touil, et al., Clinical study of 40 cases of incontinentia pigmenti, Arch. Dermatol. 139 (9) (2003) 1163–1170.
[272] A. Smahi, G. Courtois, P. Vabres, S. Yamaoka, S. Heuertz, A. Munnich, et al., Genomic rearrangement in NEMO impairs NF-kappaB activation and is a cause of incontinentia pigmenti. The International Incontinentia Pigmenti (IP) Consortium, Nature 405 (6785) (2000) 466–472.
[273] T. Rinne, B. Hamel, H. van Bokhoven, H.G. Brunner, Pattern of p63 mutations and their phenotypes – update, Am. J. Med. Genet. A 140 (13) (2006) 1396–1406.
[274] A.A. Mills, B. Zheng, X.J. Wang, H. Vogel, D.R. Roop, A. Bradley, p63 is a p53 homologue required for limb and epidermal morphogenesis, Nature 398 (6729) (1999) 708–713.
[275] H.G. Brunner, B.C. Hamel, H. Bokhoven, P63 gene mutations and human developmental syndromes, Am. J. Med. Genet. 112 (3) (2002) 284–290.

[276] H. Vanbokhoven, G. Melino, E. Candi, W. Declercq, p63, a story of mice and men, J. Invest. Dermatol. 131 (6) (2011) 1196–1207.

[277] P.W. Buss, H.E. Hughes, A. Clarke, Twenty-four cases of the EEC syndrome: clinical presentation and management, J. Med. Genet. 32 (9) (1995) 716–723.

[278] D. Lacombe, F. Serville, D. Marchand, J. Battin, Split hand/split foot deformity and LADD syndrome in a family: overlap between the EEC and LADD syndromes, J. Med. Genet. 30 (8) (1993) 700–703.

[279] S.M. Maas, T.P. de Jong, P. Buss, R.C. Hennekam, EEC syndrome and genitourinary anomalies: an update, Am. J. Med. Genet. 63 (3) (1996) 472–478.

[280] J.R. O'Quinn, R.C. Hennekam, L.B. Jorde, M. Bamshad, Syndromic ectrodactyly with severe limb, ectodermal, urogenital, and palatal defects maps to chromosome 19, Am. J. Hum. Genet. 62 (1) (1998) 130–135.

[281] E.S. Rodini, A. Richieri-Costa, EEC syndrome: report on 20 new patients, clinical and genetic considerations, Am. J. Med. Genet. 37 (1) (1990) 42–53.

[282] N.M. Roelfsema, J.M. Cobben, The EEC syndrome: a literature study, Clin. Dysmorphol. 5 (2) (1996) 115–127.

[283] S.W. Scherer, P. Poorkaj, H. Massa, S. Soder, T. Allen, M. Nunes, et al., Physical mapping of the split hand/split foot locus on chromosome 7 and implication in syndromic ectrodactyly, Hum. Mol. Genet. 3 (8) (1994) 1345–1354.

[284] A.P. South, G.H. Ashton, C. Willoughby, I.H. Ellis, O. Bleck, T. Hamada, et al., EEC (ectrodactyly, ectodermal dysplasia, clefting) syndrome: heterozygous mutation in the p63 gene (R279H) and DNA-based prenatal diagnosis, Br. J. Dermatol. 146 (2) (2002) 216–220.

[285] M. Fete, H. vanBokhoven, S.E. Clements, F. McKeon, D.R. Roop, M.I. Koster, et al., International research symposium on ankyloblepharon-ectodermal defects-cleft lip/palate (AEC) syndrome, Am. J. Med. Genet. A 149A (9) (2009) 1885–1893.

[286] J.R. Avila, P.A. Jezewski, A.R. Vieira, I.M. Orioli, E.E. Castilla, K. Christensen, et al., PVRL1 variants contribute to non-syndromic cleft lip and palate in multiple populations, Am. J. Med. Genet. A 140 (23) (2006) 2562–2570.

[287] M.J. Barron, S.J. Brookes, C.E. Draper, D. Garrod, J. Kirkham, R.C. Shore, et al., The cell adhesion molecule nectin-1 is critical for normal enamel formation in mice, Hum. Mol. Genet. 17 (22) (2008) 3509–3520.

[288] T. Yoshida, J. Miyoshi, Y. Takai, I. Thesleff, Cooperation of nectin-1 and nectin-3 is required for normal ameloblast function and crown shape development in mouse teeth, Dev. Dyn. 239 (10) (2010) 2558–2569.

[289] P. Bowen, H.B. Armstrong, Ectodermal dysplasia, mental retardation, cleft lip/palate and other anomalies in three sibs, Clin. Genet. 9 (1) (1976) 35–42.

[290] T. Bustos, V. Simosa, J. Pinto-Cisternas, W. Abramovits, L. Jolay, L. Rodriguez, et al., Autosomal recessive ectodermal dysplasia: I. An undescribed dysplasia/malformation syndrome, Am. J. Med. Genet. 41 (4) (1991) 398–404.

[291] G. Ogur, M. Yuksel, Association of syndactyly, ectodermal dysplasia, and cleft lip and palate: report of two sibs from Turkey, J. Med. Genet. 25 (1) (1988) 37–40.

[292] E.S. Rodini, A. Richieri-Costa, Autosomal recessive ectodermal dysplasia, cleft lip/palate, mental retardation, and syndactyly: the Zlotogora-Ogur syndrome, Am. J. Med. Genet. 36 (4) (1990) 473–476.

[293] M.A. Sozen, K. Suzuki, M.M. Tolarova, T. Bustos, J.E. Fernandez Iglesias, R.A. Spritz, Mutation of PVRL1 is associated with sporadic, non-syndromic cleft lip/palate in northern Venezuela, Nat. Genet. 29 (2) (2001) 141–142.

[294] K. Suzuki, D. Hu, T. Bustos, J. Zlotogora, A. Richieri-Costa, J.A. Helms, et al., Mutations of PVRL1, encoding a cell-cell adhesion molecule/herpesvirus receptor, in cleft lip/palate-ectodermal dysplasia, Nat. Genet. 25 (4) (2000) 427–430.

[295] J. Zlotogora, Y. Zilberman, A. Tenenbaum, M.R. Wexler, Cleft lip and palate, pili torti, malformed ears, partial syndactyly of fingers and toes, and mental retardation: a new syndrome? J. Med. Genet. 24 (5) (1987) 291–293.

[296] J. Zlotogora, G. Ogur, Syndactyly, ectodermal dysplasia, and cleft lip and palate, J. Med. Genet. 25 (7) (1988) 503.

[297] J. Zlotogora, Syndactyly, ectodermal dysplasia, and cleft lip/palate, J. Med. Genet. 31 (12) (1994) 957–959.

[298] Orphanet, Orphanet Reports Series – Prevalence of rare diseases: a bibliographic survey. <http://www.orpha.net/orphacom/cahiers/docs/GB/Prevalence_of_rare_diseases.pdf>, 2007.

[299] V.L. Ruiz-Perez, J.A. Goodship, Ellis-van Creveld syndrome and Weyers acrodental dysostosis are caused by cilia-mediated diminished response to hedgehog ligands, Am. J. Med. Genet. C Semin. Med. Genet. 151C (4) (2009) 341–351.

[300] H.J. Blair, S. Tompson, Y.N. Liu, J. Campbell, K. MacArthur, C.P. Ponting, et al., Evc2 is a positive modulator of Hedgehog signalling that interacts with Evc at the cilia membrane and is also found in the nucleus, BMC Biol. 9 (2011) 14.

[301] V.L. Ruiz-Perez, H.J. Blair, M.E. Rodriguez-Andres, M.J. Blanco, A. Wilson, Y.N. Liu, et al., Evc is a positive mediator of Ihh-regulated bone growth that localises at the base of chondrocyte cilia, Development 134 (16) (2007) 2903–2912.

[302] M. Atasu, S. Biren, Ellis-van Creveld syndrome: dental, clinical, genetic and dermatoglyphic findings of a case, J. Clin. Pediatr. Dent. 24 (2) (2000) 141–145.

[303] S.M. Mintz, M.A. Siegel, P.J. Seider, An overview of oral frena and their association with multiple syndromic and nonsyndromic conditions, Oral Surg. Oral Med. Oral Pathol. Oral Radiol. Endod. 99 (3) (2005) 321–324.

[304] M.I. Mostafa, S.A. Temtamy, M.A. el-Gammal, I.M. Mazen, Unusual pattern of inheritance and orodental changes in the Ellis-van Creveld syndrome, Genet. Couns. 16 (1) (2005) 75–83.

[305] V.L. Ruiz-Perez, S.E. Ide, T.M. Strom, B. Lorenz, D. Wilson, K. Woods, et al., Mutations in a new gene in Ellis-van Creveld syndrome and Weyers acrodental dysostosis, Nat. Genet. 24 (3) (2000) 283–286.

[306] V.L. Ruiz-Perez, S.W. Tompson, H.J. Blair, C. Espinoza-Valdez, P. Lapunzina, E.O. Silva, et al., Mutations in two nonhomologous genes in a head-to-head configuration cause Ellis-van Creveld syndrome, Am. J. Hum. Genet. 72 (3) (2003) 728–732.

[307] S.W. Tompson, V.L. Ruiz-Perez, H.J. Blair, S. Barton, V. Navarro, J.L. Robson, et al., Sequencing EVC and EVC2 identifies mutations in two-thirds of Ellis-van Creveld syndrome patients, Hum. Genet. 120 (5) (2007) 663–670.

[308] S.W. Tompson, V.L. Ruiz-Perez, M.J. Wright, J.A. Goodship, Ellis-van Creveld syndrome resulting from segmental uniparental disomy of chromosome 4, J. Med. Genet. 38 (6) (2001) E18.

[309] A. Cahuana, C. Palma, W. Gonzales, E. Gean, Oral manifestations in Ellis-van Creveld syndrome: report of five cases, Pediatr. Dent. 26 (3) (2004) 277–282.

[310] M. Galdzicka, S. Patnala, M.G. Hirshman, J.F. Cai, H. Nitowsky, J.A. Egeland, et al., A new gene, EVC2, is mutated in Ellis-van Creveld syndrome, Mol. Genet. Metab. 77 (4) (2002) 291–295.

[311] M.L. Hunter, G.J. Roberts, Oral and dental anomalies in Ellis van Creveld syndrome (chondroectodermal dysplasia): report of a case, Int. J. Paediatr. Dent. 8 (2) (1998) 153–157.

[312] C.S. Katsouras, C. Thomadakis, L.K. Michalis, Cardiac Ellis-van Creveld syndrome, Int. J. Cardiol. 87 (2–3) (2003) 315–316.

[313] K. Kurian, S. Shanmugam, T. Harsh Vardah, S. Gupta, Chondroectodermal dysplasia (Ellis van Creveld syndrome): a report of three cases with review of literature, Indian J. Dent. Res. 18 (1) (2007) 31–34.

[314] V.A. McKusick, J.A. Egeland, R. Eldridge, D.E. Krusen, Dwarfism in the Amish I. The Ellis-van Creveld syndrome, Bull. Johns Hopkins Hosp. 115 (1964) 306–336.

[315] V.A. McKusick, Ellis-van Creveld syndrome and the Amish, Nat. Genet. 24 (3) (2000) 203–204.

[316] C.G. Sajeev, T.N. Roy, K. Venugopal, Images in cardiology: common atrium in a child with Ellis-van Creveld syndrome, Heart 88 (2) (2002) 142.

[317] J.M. van Hagen, J.A. Baart, J.J. Gille, [From gene to disease; EVC, EVC2, and Ellis-van Creveld syndrome], Ned. Tijdschr. Geneeskd. 149 (17) (2005) 929–931.

[318] M. Valencia, P. Lapunzina, D. Lim, R. Zannolli, D. Bartholdi, B. Wollnik, et al., Widening the mutation spectrum of EVC and EVC2: ectopic expression of Weyer variants in NIH 3T3 fibroblasts disrupts Hedgehog signaling, Hum. Mutat. 30 (12) (2009) 1667–1675.

[319] F.N. Hattab, O.M. Yassin, I.S. Sasa, Oral manifestations of Ellis-van Creveld syndrome: report of two siblings with unusual dental anomalies, J. Clin. Pediatr. Dent. 22 (2) (1998) 159–165.

[320] N.A. Aminabadi, A. Ebrahimi, S.G. Oskouei, Chondroectodermal dysplasia (Ellis-van Creveld syndrome): a case report, J. Oral Sci. 52 (2) (2010) 333–336.

[321] W. Shen, D. Han, J. Zhang, H. Zhao, H. Feng, Two novel heterozygous mutations of EVC2 cause a mild phenotype of Ellis-van Creveld syndrome in a Chinese family, Am. J. Med. Genet. A 155 (9) (2011) 2131–2136.

[322] G. Baujat, M. Le Merrer, Ellis-van Creveld syndrome, Orphanet. J. Rare Dis. 2 (2007) 27.

[323] C.J. Curry, B.D. Hall, Polydactyly, conical teeth, nail dysplasia, and short limbs: a new autosomal dominant malformation syndrome, Birth Defects Orig. Artic. Ser. 15 (5B) (1979) 253–263.

[324] T.D. Howard, A.E. Guttmacher, W. McKinnon, M. Sharma, V.A. McKusick, E.W. Jabs, Autosomal dominant postaxial polydactyly, nail dystrophy, and dental abnormalities map to chromosome 4p16, in the region containing the Ellis-van Creveld syndrome locus, Am. J. Hum. Genet. 61 (6) (1997) 1405–1412.

[325] M. Roubicek, J. Spranger, Weyers acrodental dysostosis in a family, Clin. Genet. 26 (6) (1984) 587–590.

[326] X. Ye, G. Song, M. Fan, L. Shi, E.W. Jabs, S. Huang, et al., A novel heterozygous deletion in the EVC2 gene causes Weyers acrofacial dysostosis, Hum. Genet. 119 (1–2) (2006) 199–205.

[327] S.D. Shapiro, R.J. Jorgenson, C.F. Salinas, Brief clinical report: Curry–Hall syndrome, Am. J. Med. Genet. 17 (3) (1984) 579–583.

[328] S. Spranger, G. Tariverdian, Symptomatic heterozygosity in the Ellis-van Creveld syndrome? Clin. Genet. 47 (4) (1995) 217–220.

[329] H. Weyers, [Hexadactyly, mandibular fissure and oligodontia, a new syndrome; dysostosis acrofacialis], Ann. Paediatr. 181 (1) (1953) 45–60.

[330] H. Weyers, [A correlated abnormality of the mandible and extremities (dysostosis acrofacialis)], Fortschr. Geb. Rontgenstr. 77 (5) (1952) 562–567.

[331] F.J. Wittig, S.A. Hickey, M. Kumar, Double epiglottis in Weyer's acrofacial dysostosis, J. Laryngol. Otol. 112 (10) (1998) 976–978.

[332] S. Kondo, B.C. Schutte, R.J. Richardson, B.C. Bjork, A.S. Knight, Y. Watanabe, et al., Mutations in IRF6 cause Van der Woude and popliteal pterygium syndromes, Nat. Genet. 32 (2) (2002) 285–289.

[333] M. Peyrard-Janvid, M. Pegelow, H. Koillinen, C. Larsson, I. Fransson, J. Rautio, et al., Novel and de novo mutations of the IRF6 gene detected in patients with Van der Woude or popliteal pterygium syndrome, Eur. J. Hum. Genet. 13 (12) (2005) 1261–1267.

[334] M.L. Marazita, A.C. Lidral, J.C. Murray, L.L. Field, B.S. Maher, T. Goldstein McHenry, et al., Genome scan, fine-mapping, and candidate gene analysis of non-syndromic cleft lip with or without cleft palate reveals phenotype-specific differences in linkage and association results, Hum. Hered. 68 (3) (2009) 151–170.

[335] K.D. Rutledge, C. Barger, J.H. Grant, N.H. Robin, IRF6 mutations in mixed isolated familial clefting, Am. J. Med. Genet. A 152A (12) (2010) 3107–3109.

[336] A.R. Vieira, A. Modesto, R. Meira, A.R. Barbosa, A.C. Lidral, J.C. Murray, Interferon regulatory factor 6 (IRF6) and fibroblast growth factor receptor 1 (FGFR1) contribute to human tooth agenesis, Am. J. Med. Genet. A 143 (6) (2007) 538–545.

[337] L. Desmyter, M. Ghassibe, N. Revencu, O. Boute, M. Lees, G. Francois, et al., IRF6 screening of syndromic and a priori non-syndromic cleft lip and palate patients: identification of a new type of minor VWS sign, Mol. Syndromol. 1 (2) (2010) 67–74.

[338] N.K. Rorick, A. Kinoshita, J.L. Weirather, M. Peyrard-Janvid, R.L. de Lima, M. Dunnwald, et al., Genomic strategy identifies a missense mutation in WD-repeat domain 65 (WDR65) in an individual with Van der Woude syndrome, Am. J. Med. Genet. A 155A (6) (2011) 1314–1321.

[339] C.R. Ingraham, A. Kinoshita, S. Kondo, B. Yang, S. Sajan, K.J. Trout, et al., Abnormal skin, limb and craniofacial morphogenesis in mice deficient for interferon regulatory factor 6 (Irf6), Nat. Genet. 38 (11) (2006) 1335–1340.

[340] R.W. Stottmann, B.C. Bjork, J.B. Doyle, D.R. Beier, Identification of a Van der Woude syndrome mutation in the cleft palate 1 mutant mouse, Genesis 48 (5) (2010) 303–308.

[341] A.B. Burdick, D. Bixler, C.L. Puckett, Genetic analysis in families with van der Woude syndrome, J. Craniofac. Genet. Dev. Biol. 5 (2) (1985) 181–208.

[342] C.B. Item, D. Turhani, D. Thurnher, K. Yerit, K. Sinko, G. Wittwer, et al., Van Der Woude syndrome: variable penetrance of a novel mutation (p.Arg 84Gly) of the IRF6 gene in a Turkish family, Int. J. Mol. Med. 15 (2) (2005) 247–251.

[343] H. Koillinen, F.K. Wong, J. Rautio, V. Ollikainen, A. Karsten, O. Larson, et al., Mapping of the second locus for the Van der Woude syndrome to chromosome 1p34, Eur. J. Hum. Genet. 9 (10) (2001) 747–752.

[344] M.M. Lees, R.M. Winter, S. Malcolm, H.M. Saal, L. Chitty, Popliteal pterygium syndrome: a clinical study of three families and report of linkage to the Van der Woude syndrome locus on 1q32, J. Med. Genet. 36 (12) (1999) 888–892.

[345] A. Mostowska, P. Wojcicki, K. Kobus, W.H. Trzeciak, Gene symbol: IRF6. Disease: Van der Woude syndrome, Hum. Genet. 116 (6) (2005) 534.

[346] A.L. Sertie, A.V. Sousa, S. Steman, R.C. Pavanello, M.R. Passos-Bueno, Linkage analysis in a large Brazilian family with van der Woude syndrome suggests the existence of a susceptibility locus for cleft palate at 17p11.2-11.1, Am. J. Hum. Genet. 65 (2) (1999) 433–440.

[347] M. Rizos, M.N. Spyropoulos, Van der Woude syndrome: a review. Cardinal signs, epidemiology, associated features, differential diagnosis, expressivity, genetic counselling and treatment, Eur. J. Orthod. 26 (1) (2004) 17–24.

[348] X. Wang, J. Liu, H. Zhang, M. Xiao, J. Li, C. Yang, et al., Novel mutations in the IRF6 gene for Van der Woude syndrome, Hum. Genet. 113 (5) (2003) 382–386.

[349] F.K. Wong, B. Gustafsson, Popliteal pterygium syndrome in a Swedish family – clinical findings and genetic analysis with the van der Woude syndrome locus at 1q32-q41, Acta Odontol. Scand. 58 (2) (2000) 85–88.

[350] P. Yeetong, C. Mahatumarat, P. Siriwan, N. Rojvachiranonda, K. Suphapeetiporn, V. Shotelersuk, Three novel mutations of the IRF6 gene with one associated with an unusual feature in Van der Woude syndrome, Am. J. Med. Genet. A 149A (11) (2009) 2489–2492.

[351] M. Macca, B. Franco, The molecular basis of oral-facial-digital syndrome, type 1, Am. J. Med. Genet. C Semin. Med. Genet 151C (4) (2009) 318–325.
[352] M.I. Ferrante, A. Zullo, A. Barra, S. Bimonte, N. Messaddeq, M. Studer, et al., Oral-facial-digital type I protein is required for primary cilia formation and left-right axis specification, Nat. Genet. 38 (1) (2006) 112–117.
[353] R.J. Gorlin, V.E. Anderson, C.R. Scott, Hypertrophied frenuli, oligophrenia, famflial trembling and anomalies of the hand. Report of four casesin one family and a forme fruste in another, N. Engl. J. Med. 264 (1961) 486–489.
[354] S.A. Feather, P.J. Winyard, S. Dodd, A.S. Woolf, Oral-facial-digital syndrome type 1 is another dominant polycystic kidney disease: clinical, radiological and histopathological features of a new kindred, Nephrol. Dial. Transplant. 12 (7) (1997) 1354–1361.
[355] S.A. Feather, A.S. Woolf, D. Donnai, S. Malcolm, R.M. Winter, The oral-facial-digital syndrome type 1 (OFD1), a cause of polycystic kidney disease and associated malformations, maps to Xp22.2-Xp22.3, Hum. Mol. Genet. 6 (7) (1997) 1163–1167.
[356] O.M. Fenton, S.R. Watt-Smith, The spectrum of the oro-facial digital syndrome, Br. J. Plast. Surg. 38 (4) (1985) 532–539.
[357] M.I. Ferrante, G. Giorgio, S.A. Feather, A. Bulfone, V. Wright, M. Ghiani, et al., Identification of the gene for oral-facial-digital type I syndrome, Am. J. Hum. Genet. 68 (3) (2001) 569–576.
[358] L. Romio, A.M. Fry, P.J. Winyard, S. Malcolm, A.S. Woolf, S.A. Feather, OFD1 is a centrosomal/basal body protein expressed during mesenchymal–epithelial transition in human nephrogenesis, J. Am. Soc. Nephrol. 15 (10) (2004) 2556–2568.
[359] M. Romero, B. Franco, J.S. del Pozo, A. Romance, Buccal anomalies, cephalometric analysis and genetic study of two sisters with orofaciodigital syndrome type I, Cleft. Palate Craniofac. J. 44 (6) (2007) 660–666.
[360] C. Thauvin-Robinet, M. Cossee, V. Cormier-Daire, L. Van Maldergem, A. Toutain, Y. Alembik, et al., Clinical, molecular, and genotype-phenotype correlation studies from 25 cases of oral-facial-digital syndrome type 1: a French and Belgian collaborative study, J. Med. Genet. 43 (1) (2006) 54–61.
[361] D.T. Whelan, W. Feldman, I. Dost, The oro-facial-digital syndrome, Clin. Genet. 8 (3) (1975) 205–212.
[362] P. Diz, V. Alvarez-Iglesias, J. Feijoo, J. Limeres, J. Seoane, I. Tomas, et al., A novel mutation in the OFD1 (Cxorf5) gene may contribute to oral phenotype in patients with oral-facial-digital syndrome type 1, Oral Dis. 17 (6) (2011) 610–614.
[363] A.V. Makeyev, D. Bayarsaihan, Molecular basis of Williams–Beuren syndrome: TFII-I regulated targets involved in craniofacial development, Cleft. Palate Craniofac. J. 48 (1) (2011) 109–116.
[364] A. Ohazama, P.T. Sharpe, TFII-I gene family during tooth development: candidate genes for tooth anomalies in Williams syndrome, Dev. Dyn. 236 (10) (2007) 2884–2888.
[365] S.J. Palmer, E.S. Tay, N. Santucci, T.T. Cuc Bach, J. Hook, F.A. Lemckert, et al., Expression of Gtf2ird1, the Williams syndrome-associated gene, during mouse development, Gene Expr. Patterns 7 (4) (2007) 396–404.
[366] B. Enkhmandakh, A.V. Makeyev, L. Erdenechimeg, F.H. Ruddle, N.O. Chimge, M.I. Tussie-Luna, et al., Essential functions of the Williams–Beuren syndrome-associated TFII-I genes in embryonic development, Proc. Natl. Acad. Sci. U.S.A. 106 (1) (2009) 181–186.
[367] S. Axelsson, T. Bjornland, I. Kjaer, A. Heiberg, K. Storhaug, Dental characteristics in Williams syndrome: a clinical and radiographic evaluation, Acta Odontol. Scand. 61 (3) (2003) 129–136.

[368] I. Tarjan, G. Balaton, P. Balaton, S. Varbiro, Z. Vajo, Facial and dental appearance of Williams syndrome, Postgrad. Med. J. 79 (930) (2003) 241.
[369] A. Lacroix, M. Pezet, A. Capel, D. Bonnet, M. Hennequin, M.P. Jacob, et al., [Williams–Beuren syndrome: a multidisciplinary approach], Arch. Pediatr. 16 (3) (2009) 273–282.
[370] C. Joseph, M.M. Landru, F. Bdeoui, B. Gogly, S.M. Dridi, Periodontal conditions in Williams Beuren syndrome: a series of 8 cases, Eur. Arch. Paediatr. Dent. 9 (3) (2008) 142–147.
[371] M. Moskovitz, D. Brener, S. Faibis, B. Peretz, Medical considerations in dental treatment of children with Williams syndrome, Oral Surg. Oral Med. Oral Pathol. Oral Radiol. Endod. 99 (5) (2005) 573–580.
[372] S. Amenta, C. Sofocleous, A. Kolialexi, L. Thomaidis, S. Giouroukos, E. Karavitakis, et al., Clinical manifestations and molecular investigation of 50 patients with Williams syndrome in the Greek population, Pediatr. Res. 57 (6) (2005) 789–795.
[373] S. Axelsson, Variability of the cranial and dental phenotype in Williams syndrome, Swed. Dent. J. Suppl. (170) (2005) 3–67.
[374] A. Oncag, S. Gunbay, A. Parlar, Williams syndrome, J. Clin. Pediatr. Dent. 19 (4) (1995) 301–304.
[375] J. Hertzberg, L. Nakisbendi, H.L. Needleman, B. Pober, Williams syndrome – oral presentation of 45 cases, Pediatr. Dent. 16 (4) (1994) 262–267.
[376] E. Mass, L. Belostoky, Craniofacial morphology of children with Williams syndrome, Cleft. Palate Craniofac. J. 30 (3) (1993) 343–349.
[377] M. Tassabehji, D. Donnai, Williams–Beuren Syndrome: more or less? Segmental duplications and deletions in the Williams–Beuren syndrome region provide new insights into language development, Eur. J. Hum. Genet. 14 (5) (2006) 507–508.
[378] M. Tassabehji, Williams–Beuren syndrome: a challenge for genotype-phenotype correlations, Hum. Mol. Genet. 12 (Spec. No. 2) (2003) R229–R237.
[379] C. Schubert, The genomic basis of the Williams–Beuren syndrome, Cell. Mol. Life Sci. 66 (7) (2009) 1178–1197.
[380] C. Schubert, F. Laccone, Williams–Beuren syndrome: determination of deletion size using quantitative real-time PCR, Int. J. Mol. Med. 18 (5) (2006) 799–806.
[381] B.R. Pober, Williams–Beuren syndrome, N. Engl. J. Med. 362 (3) (2010) 239–252.
[382] A. Noor, C. Windpassinger, I. Vitcu, M. Orlic, M.A. Rafiq, M. Khalid, et al., Oligodontia is caused by mutation in LTBP3, the gene encoding latent TGF-beta binding protein 3, Am. J. Hum. Genet. 84 (4) (2009) 519–523.
[383] K. Kitisin, T. Saha, T. Blake, N. Golestaneh, M. Deng, C. Kim, et al., Tgf-Beta signaling in development, Sci. STKE 2007 (399) (2007) cm1.
[384] I. Thesleff, E. Jarvinen, M. Suomalainen, Affecting tooth morphology and renewal by fine-tuning the signals mediating cell and tissue interactions, Novartis Found. Symp. 284 (2007) 142–153, discussion 153–163.
[385] M. Takahara, M. Harada, D. Guan, M. Otsuji, T. Naruse, M. Takagi, et al., Developmental failure of phalanges in the absence of growth/differentiation factor 5, Bone 35 (5) (2004) 1069–1076.
[386] E.E. Storm, T.V. Huynh, N.G. Copeland, N.A. Jenkins, D.M. Kingsley, S.J. Lee, Limb alterations in brachypodism mice due to mutations in a new member of the TGF beta-superfamily, Nature 368 (6472) (1994) 639–643.
[387] F.N. Hattab, T. al-Khateeb, M. Mansour, Oral manifestations of severe short-limb dwarfism resembling Grebe chondrodysplasia: report of a case, Oral Surg. Oral Med. Oral Pathol. Oral Radiol. Endod. 81 (5) (1996) 550–555.

[388] R.K. Bachman, Hereditary peripheral dysostosis (three cases), Proc. R. Soc. Med. 60 (1) (1967) 21–22.
[389] M. Faiyaz-Ul-Haque, W. Ahmad, A. Wahab, S. Haque, A.C. Azim, S.H. Zaidi, et al., Frameshift mutation in the cartilage-derived morphogenetic protein 1 (CDMP1) gene and severe acromesomelic chondrodysplasia resembling Grebe-type chondrodysplasia, Am. J. Med. Genet. 111 (1) (2002) 31–37.
[390] A. Giedion, A. Prader, C. Fliegel, N. Krasikov, L. Langer, A. Poznanski, Angel-shaped phalango-epiphyseal dysplasia (ASPED): identification of a new genetic bone marker, Am. J. Med. Genet. 47 (5) (1993) 765–771.
[391] M. Holder-Espinasse, F. Escande, E. Mayrargue, A. Dieux-Coeslier, D. Fron, A. Doual-Bisser, et al., Angel shaped phalangeal dysplasia, hip dysplasia, and positional teeth abnormalities are part of the brachydactyly C spectrum associated with CDMP-1 mutations, J. Med. Genet. 41 (6) (2004) e78.
[392] A. Polinkovsky, N.H. Robin, J.T. Thomas, M. Irons, A. Lynn, F.R. Goodman, et al., Mutations in CDMP1 cause autosomal dominant brachydactyly type C, Nat. Genet. 17 (1) (1997) 18–19.
[393] J.T. Thomas, K. Lin, M. Nandedkar, M. Camargo, J. Cervenka, F.P. Luyten, A human chondrodysplasia due to a mutation in a TGF-beta superfamily member, Nat. Genet. 12 (3) (1996) 315–317.
[394] M.M. Cohen, Jr., The hedgehog signaling network, Am. J. Med. Genet. A 123 (1) (2003) 5–28.
[395] J.E. Ming, E. Roessler, M. Muenke, Human developmental disorders and the Sonic hedgehog pathway, Mol. Med. Today 4 (8) (1998) 343–349.
[396] M. Oldak, T. Grzela, M. Lazarczyk, J. Malejczyk, P. Skopinski, Clinical aspects of disrupted Hedgehog signaling (Review), Int. J. Mol. Med. 8 (4) (2001) 445–452.
[397] L. Jacob, L. Lum, Hedgehog signaling pathway, Sci. STKE 2007 (407) (2007) cm6.
[398] L. Jacob, L. Lum, Deconstructing the hedgehog pathway in development and disease, Science 318 (5847) (2007) 66–68.
[399] E.H. Villavicencio, D.O. Walterhouse, P.M. Iannaccone, The sonic hedgehog-patched-gli pathway in human development and disease, Am. J. Hum. Genet. 67 (5) (2000) 1047–1054.
[400] J. Lertsirivorakul, R.K. Hall, Solitary median maxillary central incisor syndrome occurring together with oromandibular-limb hypogenesis syndrome type 1: a case report of this previously unreported combination of syndromes, Int. J. Paediatr. Dent. 18 (4) (2008) 306–311.
[401] M. Seppala, M.J. Depew, D.C. Martinelli, C.M. Fan, P.T. Sharpe, M.T. Cobourne, Gas1 is a modifier for holoprosencephaly and genetically interacts with sonic hedgehog, J. Clin. Invest. 117 (6) (2007) 1575–1584.
[402] E. Lana-Elola, P. Tylzanowski, M. Takatalo, K. Alakurtti, L. Veistinen, T.A. Mitsiadis, et al., Noggin null allele mice exhibit a microform of holoprosencephaly, Hum. Mol. Genet. 20 (20) (2011) 4005–4015.
[403] R.K. Hall, Solitary median maxillary central incisor (SMMCI) syndrome, Orphanet. J. Rare Dis. 1 (2006) 12.
[404] R.K. Hall, A. Bankier, M.J. Aldred, K. Kan, J.O. Lucas, A.G. Perks, Solitary median maxillary central incisor, short stature, choanal atresia/midnasal stenosis (SMMCI) syndrome, Oral Surg. Oral Med. Oral Pathol. Oral Radiol. Endod. 84 (6) (1997) 651–662.
[405] J.P. Fryns, H. Van den Berghe, Single central maxillary incisor and holoprosencephaly, Am. J. Med. Genet. 30 (4) (1988) 943–944.
[406] L. Garavelli, C. Zanacca, G. Caselli, G. Banchini, C. Dubourg, V. David, et al., Solitary median maxillary central incisor syndrome: clinical case with a novel mutation of sonic hedgehog, Am. J. Med. Genet. A 127 (1) (2004) 93–95.

[407] U. Hehr, C. Gross, U. Diebold, D. Wahl, U. Beudt, P. Heidemann, et al., Wide phenotypic variability in families with holoprosencephaly and a sonic hedgehog mutation, Eur. J. Pediatr. 163 (7) (2004) 347–352.
[408] L. Nanni, J.E. Ming, Y. Du, R.K. Hall, M. Aldred, A. Bankier, et al., SHH mutation is associated with solitary median maxillary central incisor: a study of 13 patients and review of the literature, Am. J. Med. Genet. 102 (1) (2001) 1–10.
[409] S. Oberoi, K. Vargervik, Velocardiofacial syndrome with single central incisor, Am. J. Med. Genet. A 132 (2) (2005) 194–197.
[410] A. Verloes, S. Lesenfants, New syndrome: clavicle hypoplasia, facial dysmorphism, severe myopia, single central incisor and peripheral neuropathy, Clin. Dysmorphol. 10 (1) (2001) 29–31.
[411] K.B. Becktor, L. Sverrild, C. Pallisgaard, J. Burhoj, I. Kjaer, Eruption of the central incisor, the intermaxillary suture, and maxillary growth in patients with a single median maxillary central incisor, Acta Odontol. Scand. 59 (6) (2001) 361–366.
[412] H.G. Artman, E. Boyden, Microphthalmia with single central incisor and hypopituitarism, J. Med. Genet. 27 (3) (1990) 192–193.
[413] S.A. Berry, M.E. Pierpont, R.J. Gorlin, Single central incisor in familial holoprosencephaly, J. Pediatr. 104 (6) (1984) 877–880.
[414] R.M. Liberfarb, O.P. Abdo, R.C. Pruett, Ocular coloboma associated with a solitary maxillary central incisor and growth failure: manifestations of holoprosencephaly, Ann. Ophthalmol. 19 (6) (1987) 226–227.
[415] E. Mass, H. Sarnat, Single maxillary central incisors in the midline, ASDC J. Dent. Child. 58 (5) (1991) 413–416.
[416] E. Tavin, E. Stecker, R. Marion, Nasal pyriform aperture stenosis and the holoprosencephaly spectrum, Int. J. Pediatr. Otorhinolaryngol. 28 (2–3) (1994) 199–204.
[417] J. Miertus, W. Borozdin, V. Frecer, G. Tonini, S. Bertok, A. Amoroso, et al., A SALL4 zinc finger missense mutation predicted to result in increased DNA binding affinity is associated with cranial midline defects and mild features of Okihiro syndrome, Hum. Genet. 119 (1–2) (2006) 154–161.
[418] E. Roessler, M. Muenke, Holoprosencephaly: a paradigm for the complex genetics of brain development, J. Inherit. Metab. Dis. 21 (5) (1998) 481–497.
[419] D. Wallis, M. Muenke, Mutations in holoprosencephaly, Hum. Mutat. 16 (2) (2000) 99–108.
[420] K.B. El-Jaick, R.F. Fonseca, M.A. Moreira, M.G. Ribeiro, A.M. Bolognese, S.O. Dias, et al., Single median maxillary central incisor: New data and mutation review, Birth Defects Res. A Clin. Mol. Teratol. 79 (8) (2007) 573–580.
[421] M.M. Cohen, Jr., Holoprosencephaly: clinical, anatomic, and molecular dimensions, Birth Defects Res. A Clin. Mol. Teratol. 76 (9) (2006) 658–673.
[422] F. Tabatabaie, L. Sonnesen, I. Kjaer, The neurocranial and craniofacial morphology in children with solitary median maxillary central incisor (SMMCI), Orthod. Craniofac. Res. 11 (2) (2008) 96–104.
[423] C. Dubourg, C. Bendavid, L. Pasquier, C. Henry, S. Odent, V. David, Holoprosencephaly, Orphanet. J. Rare Dis. 2 (2007) 8.
[424] L.E. Vissers, C.M. van Ravenswaaij, R. Admiraal, J.A. Hurst, B.B. de Vries, I.M. Janssen, et al., Mutations in a new member of the chromodomain gene family cause CHARGE syndrome, Nat. Genet. 36 (9) (2004) 955–957.
[425] S. Balci, C. Tumer, C. Karaca, O. Bartsch, Familial ring (18) mosaicism in a 23-year-old young adult with 46,XY,r(18) (::p11-- > q21::)/46,XY karyotype, intellectual disability, motor retardation and single maxillary incisor and in his phenotypically normal

mother, karyotype 47,XX, + r(18)(::p11-- > q21::)/46,XX, Am. J. Med. Genet. A 155A (5) (2011) 1129–1135.

[426] M.E. Pownall, H.V. Isaacs, FGF Signalling in Vertebrate Development, Morgan & Claypool Life Sciences, San Rafael, CA, 2010.

[427] A.O. Wilkie, E.G. Bochukova, R.M. Hansen, I.B. Taylor, S.V. Rannan-Eliya, J.C. Byren, et al., Clinical dividends from the molecular genetic diagnosis of craniosynostosis, Am. J. Med. Genet. A 140 (23) (2006) 2631–2639.

[428] M.M. Cohen, Jr., The new bone biology: pathologic, molecular, and clinical correlates, Am. J. Med. Genet. A 140 (23) (2006) 2646–2706.

[429] O.D. Klein, G. Minowada, R. Peterkova, A. Kangas, B.D. Yu, H. Lesot, et al., Sprouty genes control diastema tooth development via bidirectional antagonism of epithelial–mesenchymal FGF signaling, Dev. Cell 11 (2) (2006) 181–190.

[430] C. Charles, M. Hovorakova, Y. Ahn, D.B. Lyons, P. Marangoni, S. Churava, et al., Regulation of tooth number by fine-tuning levels of receptor-tyrosine kinase signaling, Development 138 (18) (2011) 4063–4073.

[431] H. Harada, T. Toyono, K. Toyoshima, H. Ohuchi, FGF10 maintains stem cell population during mouse incisor development, Connect. Tissue Res. 43 (2–3) (2002) 201–204.

[432] T. Jaskoll, G. Abichaker, D. Witcher, F.G. Sala, S. Bellusci, M.K. Hajihosseini, et al., FGF10/FGFR2b signaling plays essential roles during in vivo embryonic submandibular salivary gland morphogenesis, BMC Dev. Biol. 5 (2005) 11.

[433] Y. Guven, R.O. Rosti, E.B. Tuna, H. Kayserili, O. Aktoren, Orodental findings of a family with lacrimo-auriculo-dento digital (LADD) syndrome, Oral Surg. Oral Med. Oral Pathol. Oral Radiol. Endod. 106 (6) (2008) e33–e44.

[434] E. Thompson, M. Pembrey, J.M. Graham, Phenotypic variation in LADD syndrome, J. Med. Genet. 22 (5) (1985) 382–385.

[435] H.R. Wiedemann, J. Drescher, LADD syndrome: report of new cases and review of the clinical spectrum, Eur. J. Pediatr. 144 (6) (1986) 579–582.

[436] S. Meuschel-Wehner, R. Klingebiel, M. Werbs, Inner ear dysplasia in sporadic lacrimo-auriculo-dento-digital syndrome. A case report and review of the literature, ORL J. Otorhinolaryngol. Relat. Spec. 64 (5) (2002) 352–354.

[437] J.M. Milunsky, G. Zhao, T.A. Maher, R. Colby, D.B. Everman, LADD syndrome is caused by FGF10 mutations, Clin. Genet. 69 (4) (2006) 349–354.

[438] E. Onrat, D. Kaya, S.T. Onrat, Lacrimo-auriculo-dento-digital syndrome with QT prolongation, Acta Cardiol. 58 (6) (2003) 567–570.

[439] P.A. Ostuni, M. Modolo, P. Revelli, A. Secchi, C. Battista, A. Tregnaghi, et al., Lacrimo-auricolo-dento-digital syndrome mimicking primary juvenile Sjogren's syndrome, Scand. J. Rheumatol. 24 (1) (1995) 55–57.

[440] D. Ramirez, E.J. Lammer, Lacrimoauriculodentodigital syndrome with cleft lip/palate and renal manifestations, Cleft Palate Craniofac. J. 41 (5) (2004) 501–506.

[441] E. Rohmann, H.G. Brunner, H. Kayserili, O. Uyguner, G. Nurnberg, E.D. Lew, et al., Mutations in different components of FGF signaling in LADD syndrome, Nat. Genet. 38 (4) (2006) 414–417.

[442] A.M. Roodhooft, C.C. Brussaard, E. Elst, K.J. van Acker, Lacrimo-auriculo-dento-digital (LADD) syndrome with renal and foot anomalies, Clin. Genet. 38 (3) (1990) 228–232.

[443] A. Calabro, M.S. Lungarotti, P. Mastroiacovo, Lacrimo-auriculo-dento-digital (LADD) syndrome, Eur. J. Pediatr. 146 (5) (1987) 536–537.

[444] M. Cortes, A. Lambiase, M. Sacchetti, S. Aronni, S. Bonini, Limbal stem cell deficiency associated with LADD syndrome, Arch. Ophthalmol. 123 (5) (2005) 691–694.

[445] M. Entesarian, J. Dahlqvist, V. Shashi, C.S. Stanley, B. Falahat, W. Reardon, et al., FGF10 missense mutations in aplasia of lacrimal and salivary glands (ALSG), Eur. J. Hum. Genet. 15 (3) (2007) 379–382.
[446] C. Francannet, P. Vanlieferinghen, P. Dechelotte, M.F. Urbain, D. Campagne, G. Malpuech, LADD syndrome in five members of a three-generation family and prenatal diagnosis, Genet. Couns. 5 (1) (1994) 85–91.
[447] A. Haktanir, B. Degirmenci, M. Acar, R. Albayrak, A. Yucel, CT findings of head and neck anomalies in lacrimo-auriculo-dento-digital (LADD) syndrome, Dentomaxillofac. Radiol. 34 (2) (2005) 102–105.
[448] R.C. Hennekam, LADD syndrome: a distinct entity? Eur. J. Pediatr. 146 (1) (1987) 94–95.
[449] D.W. Hollister, S.H. Klein, H.J. de Jager, R.S. Lachman, D.L. Rimoin, Lacrimo-auriculo-dento-digital (LADD) syndrome, Birth Defects Orig. Artic. Ser. 10 (5) (1974) 153–166.
[450] D. Horn, R. Witkowski, Phenotype and counseling in lacrimo-auriculo-dento-digital (LADD) syndrome, Genet. Couns. 4 (4) (1993) 305–309.
[451] U.U. Inan, M.D. Yilmaz, Y. Demir, B. Degirmenci, S.S. Ermis, F. Ozturk, Characteristics of lacrimo-auriculo-dento-digital (LADD) syndrome: case report of a family and literature review, Int. J. Pediatr. Otorhinolaryngol. 70 (7) (2006) 1307–1314.
[452] M. Lehotay, M. Kunkel, H. Wehrbein, Lacrimo-auriculo-dento-digital syndrome. Case report, review of the literature, and clinical spectrum, J. Orofac. Orthop. 65 (5) (2004) 425–432.
[453] M.M. Lemmerling, B.D. Vanzieleghem, I.J. Dhooge, P.B. Van Cauwenberge, M.F. Kunnen, et al., The lacrimo-auriculo-dento-digital (LADD) syndrome: temporal bone CT findings, J. Comput. Assist. Tomogr. 23 (3) (1999) 362–364.
[454] T. Azar, J.A. Scott, J.E. Arnold, N.H. Robin, Epiglottic hypoplasia associated with lacrimo-auriculo-dental-digital syndrome, Ann. Otol. Rhinol. Laryngol. 109 (8 Pt 1) (2000) 779–781.
[455] J.S. Bamforth, P. Kaurah, Lacrimo-auriculo-dento-digital syndrome: evidence for lower limb involvement and severe congenital renal anomalies, Am. J. Med. Genet. 43 (6) (1992) 932–937.
[456] R.J. Ensink, C.W. Cremers, H.G. Brunner, Congenital conductive hearing loss in the lacrimoauriculodentodigital syndrome, Arch. Otolaryngol. Head Neck Surg. 123 (1) (1997) 97–99.
[457] H.G. Kim, L.C. Layman, The role of CHD7 and the newly identified WDR11 gene in patients with idiopathic hypogonadotropic hypogonadism and Kallmann syndrome, Mol. Cell. Endocrinol. 346 (1–2) (2011) 74–83.
[458] N. Trokovic, R. Trokovic, P. Mai, J. Partanen, Fgfr1 regulates patterning of the pharyngeal region, Genes Dev. 17 (1) (2003) 141–153.
[459] K. Takamori, R. Hosokawa, X. Xu, X. Deng, P. Bringas, Jr., Y. Chai, Epithelial fibroblast growth factor receptor 1 regulates enamel formation, J. Dent. Res. 87 (3) (2008) 238–243.
[460] I. Bailleul-Forestier, C. Gros, D. Zenaty, S. Bennaceur, J. Leger, N. de Roux, Dental agenesis in Kallmann syndrome individuals with FGFR1 mutations, Int. J. Paediatr. Dent. 20 (4) (2010) 305–312.
[461] F. de Zegher, L. Lagae, D. Declerck, F. Vinckier, Kallmann syndrome and delayed puberty associated with agenesis of lateral maxillary incisors, J. Craniofac. Genet. Dev. Biol. 15 (2) (1995) 87–89.
[462] K. Molsted, I. Kjaer, A. Giwercman, S. Vesterhauge, N.E. Skakkebaek, Craniofacial morphology in patients with Kallmann's syndrome with and without cleft lip and palate, Cleft Palate Craniofac. J. 34 (5) (1997) 417–424.

[463] N. Pitteloud, A. Meysing, R. Quinton, J.S. Acierno, Jr., A.A. Dwyer, L. Plummer, et al., Mutations in fibroblast growth factor receptor 1 cause Kallmann syndrome with a wide spectrum of reproductive phenotypes, Mol. Cell. Endocrinol. 254–255 (2006) 60–69.

[464] N. Pitteloud, J.S. Acierno, Jr., A. Meysing, A.V. Eliseenkova, J. Ma, O.A. Ibrahimi, et al., Mutations in fibroblast growth factor receptor 1 cause both Kallmann syndrome and normosmic idiopathic hypogonadotropic hypogonadism, Proc. Natl. Acad. Sci. U.S.A. 103 (16) (2006) 6281–6286.

[465] J.C. Pallais, M. Au, N. Pitteloud, S. Seminara, W.F. Crowley, Kallmann syndrome, 1993.

[466] L. Chalabreysse, F. Senni, P. Bruyere, B. Aime, C. Ollagnier, A. Bozio, et al., A new hypo/oligodontia syndrome: Carvajal/Naxos syndrome secondary to desmoplakin-dominant mutations, J. Dent. Res. 90 (1) (2011) 58–64.

[467] N. Chester, H. Babbe, J. Pinkas, C. Manning, P. Leder, Mutation of the murine Bloom's syndrome gene produces global genome destabilization, Mol. Cell. Biol. 26 (17) (2006) 6713–6726.

[468] Y. Wang, J.A. Heddle, Spontaneous and induced chromosomal damage and mutations in Bloom Syndrome mice, Mutat. Res. 554 (1–2) (2004) 131–137.

[469] N. Chester, F. Kuo, C. Kozak, C.D. O'Hara, P. Leder, Stage-specific apoptosis, developmental delay, and embryonic lethality in mice homozygous for a targeted disruption in the murine Bloom's syndrome gene, Genes Dev. 12 (21) (1998) 3382–3393.

[470] N.A. Ellis, J. German, Molecular genetics of Bloom's syndrome, Hum. Mol. Genet. (Spec. No. 5) (1996) 1457–1463.

[471] N.A. Ellis, J. Groden, T.Z. Ye, J. Straughen, D.J. Lennon, S. Ciocci, et al., The Bloom's syndrome gene product is homologous to RecQ helicases, Cell 83 (4) (1995) 655–666.

[472] K. Hanada, I.D. Hickson, Molecular genetics of RecQ helicase disorders, Cell. Mol. Life Sci. 64 (17) (2007) 2306–2322.

[473] J. German, M.M. Sanz, S. Ciocci, T.Z. Ye, N.A. Ellis, Syndrome-causing mutations of the BLM gene in persons in the Bloom's syndrome registry, Hum. Mutat. 28 (8) (2007) 743–753.

[474] C. Oddoux, C.M. Clayton, H.R. Nelson, H. Ostrer, Prevalence of Bloom syndrome heterozygotes among Ashkenazi Jews, Am. J. Hum. Genet. 64 (4) (1999) 1241–1243.

[475] B.B. Roa, C.V. Savino, C.S. Richards, Ashkenazi Jewish population frequency of the Bloom syndrome gene 2281 delta 6ins7 mutation, Genet. Test. 3 (2) (1999) 219–221.

[476] L.B. Weinstein, Selected genetic disorders affecting Ashkenazi Jewish families, Fam. Community Health 30 (1) (2007) 50–62.

[477] M.B. Mann, C.A. Hodges, E. Barnes, H. Vogel, T.J. Hassold, G. Luo, et al., Defective sister-chromatid cohesion, aneuploidy and cancer predisposition in a mouse model of type II Rothmund–Thomson syndrome, Hum. Mol. Genet. 14 (6) (2005) 813–825.

[478] P. Kumar, P.K. Sharma, R.K. Gautam, R.K. Jain, H.K. Kar, Late-onset Rothmund–Thomson syndrome, Int. J. Dermatol. 46 (5) (2007) 492–493.

[479] T.D. Roinioti, P.K. Stefanopoulos, Short root anomaly associated with Rothmund–Thomson syndrome, Oral Surg. Oral Med. Oral Pathol. Oral Radiol. Endod. 103 (1) (2007) e19–e22.

[480] M.C. Haytac, H. Oztunc, U.O. Mete, M. Kaya, Rothmund–Thomson syndrome: a case report, Oral Surg. Oral Med. Oral Pathol. Oral Radiol. Endod. 94 (4) (2002) 479–484.

[481] J. Piquero-Casals, A.Y. Okubo, M.M. Nico, Rothmund–Thomson syndrome in three siblings and development of cutaneous squamous cell carcinoma, Pediatr. Dermatol. 19 (4) (2002) 312–316.

[482] K.K. Anbari, L.A. Ierardi-Curto, J.S. Silber, N. Asada, N. Spinner, E.H. Zackai, et al., Two primary osteosarcomas in a patient with Rothmund–Thomson syndrome, Clin. Orthop. Relat. Res. 378 (2000) 213–223.
[483] D.G. Snels, J.N. Bavinck, H. Muller, Vermeer B.J., A female patient with the Rothmund–Thomson syndrome associated with anhidrosis and severe infections of the respiratory tract, Dermatology 196 (2) (1998) 260–263.
[484] B. Kerr, G.S. Ashcroft, D. Scott, M.A. Horan, M.W. Ferguson, D. Donnai, Rothmund–Thomson syndrome: two case reports show heterogeneous cutaneous abnormalities, an association with genetically programmed ageing changes, and increased chromosomal radiosensitivity, J. Med. Genet. 33 (11) (1996) 928–934.
[485] J.P. Leusink, J.J. Tolboom, C.M. Weemaes, R.J. Koopman, [An infant with short stature and red cheeks (Rothmund–Thomson syndrome)], Tijdschr. Kindergeneeskd. 59 (6) (1991) 219–223.
[486] A. Nanda, A.J. Kanwar, M.M. Kapoor, D.M. Thappa, B.D. Radotra, C. Vaishnavi, et al., Rothmund–Thomson syndrome in two siblings, Pediatr. Dermatol. 6 (4) (1989) 325–328.
[487] V. Zinke, M. Lesche, H. Schuler, [An interesting case: the Rothmund syndrome], Stomatol. DDR 35 (8) (1985) 467–470.
[488] J.G. Hall, R.A. Pagon, K.M. Wilson, Rothmund–Thomson syndrome with severe dwarfism, Am. J. Dis. Child. 134 (2) (1980) 165–169.
[489] B.S. Kraus, M.A. Gottlieb, H.R. Meliton, The dentition in Rothmund's syndrome, J. Am. Dent. Assoc. 81 (4) (1970) 895–915.
[490] W.K. Bottomley, J.M. Box, Dental anomalies in the Rothmund–Thomson syndrome. Report of a case, Oral Surg. Oral Med. Oral Pathol. 41 (3) (1976) 321–326.
[491] J. Yin, Y.T. Kwon, A. Varshavsky, W. Wang, RECQL4, mutated in the Rothmund–Thomson and RAPADILINO syndromes, interacts with ubiquitin ligases UBR1 and UBR2 of the N-end rule pathway, Hum. Mol. Genet. 13 (20) (2004) 2421–2430.
[492] T. Simon, J. Kohlhase, C. Wilhelm, M. Kochanek, B. De Carolis, F. Berthold, Multiple malignant diseases in a patient with Rothmund–Thomson syndrome with RECQL4 mutations: case report and literature review, Am. J. Med. Genet. A 152A (6) (2010) 1575–1579.
[493] L.L. Galante, J.E. Schwarzbauer, Requirements for sulfate transport and the diastrophic dysplasia sulfate transporter in fibronectin matrix assembly, J. Cell Biol. 179 (5) (2007) 999–1009.
[494] A. Forlino, B. Gualeni, F. Pecora, S.D. Torre, R. Piazza, C. Tiveron, et al., Insights from a transgenic mouse model on the role of SLC26A2 in health and disease, Novartis Found. Symp. 273 (2006) 193–206, discussion 206–212, 261–264.
[495] A. Forlino, R. Piazza, C. Tiveron, S. Della Torre, L. Tatangelo, L. Bonafe, et al., A diastrophic dysplasia sulfate transporter (SLC26A2) mutant mouse: morphological and biochemical characterization of the resulting chondrodysplasia phenotype, Hum. Mol. Genet. 14 (6) (2005) 859–871.
[496] P.A. Dawson, D. Markovich, Pathogenetics of the human SLC26 transporters, Curr. Med. Chem. 12 (4) (2005) 385–396.
[497] E. Karlstedt, E. Isotalo, M.L. Haapanen, M. Kalland, S. Pirinen, I. Kaitila, Correlation between speech outcome and cephalometric dimensions in patients with diastrophic dysplasia, J. Craniofac. Genet. Dev. Biol. 18 (1) (1998) 38–43.
[498] E. Karlstedt, I. Kaitila, S. Pirinen, Phenotypic features of dentition in diastrophic dysplasia, J. Craniofac. Genet. Dev. Biol. 16 (3) (1996) 164–173.
[499] J. Kere, Overview of the SLC26 family and associated diseases, Novartis Found. Symp. 273 (2006) 2–11, discussion 11–18, 261–264.

[500] A. Rintala, E. Marttinen, S.L. Rantala, I. Kaitila, Cleft palate in diastrophic dysplasia. Morphology, results of treatment and complications, Scand. J. Plast. Reconstr. Surg. 20 (1) (1986) 45–49.

[501] A. Rossi, A. Superti-Furga, Mutations in the diastrophic dysplasia sulfate transporter (DTDST) gene (SLC26A2): 22 novel mutations, mutation review, associated skeletal phenotypes, and diagnostic relevance, Hum. Mutat. 17 (3) (2001) 159–171.

[502] Y. Yokoyama, [Diastrophic dysplasia], Ryoikibetsu Shokogun Shirizu 33 (2001) 560–561.

[503] P.A. Dawson, Sulfate in fetal development, Semin. Cell Dev. Biol. 22 (2011) 653–659.

[504] M. Zenker, J. Mayerle, M.M. Lerch, A. Tagariello, K. Zerres, P.R. Durie, et al., Deficiency of UBR1, a ubiquitin ligase of the N-end rule pathway, causes pancreatic dysfunction, malformations and mental retardation (Johanson–Blizzard syndrome), Nat. Genet. 37 (12) (2005) 1345–1350.

[505] R. Auslander, O. Nevo, R. Diukman, E. Morrad, M. Bardicef, H. Abramovici, Johanson–Blizzard syndrome: a prenatal ultrasonographic diagnosis, Ultrasound Obstet. Gynecol. 13 (6) (1999) 450–452.

[506] M. Baraitser, S.V. Hodgson, The Johanson–Blizzard syndrome, J. Med. Genet. 19 (4) (1982) 302–303.

[507] J. Braun, A. Lerner, R. Gershoni-Baruch, The temporal bone in the Johanson–Blizzard syndrome. A CT study, Pediatr. Radiol. 21 (8) (1991) 580–583.

[508] J.L. Bresson, J. Schmitz, J.M. Saudubray, G. Lesec, J.A. Hummel, J. Rey, [Johanson–Blizzard's syndrome: another cause of pancreatic lipomatosis (author's transl)], Arch. Fr. Pediatr. 37 (1) (1980) 21–24.

[509] D.L. Daentl, J.L. Frias, E.F. Gilbert, J.M. Opitz, The Johanson–Blizzard syndrome: case report and autopsy findings, Am. J. Med. Genet. 3 (2) (1979) 129–135.

[510] M. Dumic, J. Ille, G. Bobonj, R. Kordic, S. Batinica, [The Johanson–Blizzard syndrome], Lijec. Vjesn. 120 (5) (1998) 114–116.

[511] P.R. Durie, Inherited and congenital disorders of the exocrine pancreas, Gastroenterologist 4 (3) (1996) 169–187.

[512] P.R. Durie, Inherited causes of exocrine pancreatic dysfunction, Can. J. Gastroenterol. 11 (2) (1997) 145–152.

[513] C.R. Fichter, G.A. Johnson, S.R. Braddock, J.D. Tobias, Perioperative care of the child with the Johanson–Blizzard syndrome, Paediatr. Anaesth. 13 (1) (2003) 72–75.

[514] J.W. Fox, G.T. Golden, M.T. Edgerton, Surgical correction of the absent nasal alae of the Johanson–Blizzard syndrome, Plast. Reconstr. Surg. 57 (4) (1976) 484–486.

[515] R. Gershoni-Baruch, A. Lerner, J. Braun, Y. Katzir, T.C. Iancu, A. Benderly, Johanson–Blizzard syndrome: clinical spectrum and further delineation of the syndrome, Am. J. Med. Genet. 35 (4) (1990) 546–551.

[516] N.S. Gould, J.B. Paton, A.R. Bennett, Johanson–Blizzard syndrome: clinical and pathological findings in 2 sibs, Am. J. Med. Genet. 33 (2) (1989) 194–199.

[517] C. Guzman, A. Carranza, Two siblings with exocrine pancreatic hypoplasia and orofacial malformations (Donlan syndrome and Johanson–Blizzard syndrome), J. Pediatr. Gastroenterol. Nutr. 25 (3) (1997) 350–353.

[518] J.A. Hurst, M. Baraitser, Johanson–Blizzard syndrome, J. Med. Genet. 26 (1) (1989) 45–48.

[519] K. Iwasaki, [Johanson–Blizzard syndrome], Ryoikibetsu Shokogun Shirizu (34 Pt 2) (2001) 24–25.

[520] N.L. Jones, P.M. Hofley, P.R. Durie, Pathophysiology of the pancreatic defect in Johanson–Blizzard syndrome: a disorder of acinar development, J. Pediatr. 125 (3) (1994) 406–408.

[521] M. Kaeriyama, N. Sasaki, A. Okuno, [Johanson–Blizzard syndrome], Nippon Rinsho. 44 (7) (1986) 1677–1682.
[522] S. Kobayashi, K. Ohmori, J. Sekiguchi, Johanson–Blizzard syndrome facial anomaly and its correction using a microsurgical bone graft and tripartite osteotomy, J. Craniofac. Surg. 6 (5) (1995) 382–385.
[523] T. Koizumi, [Johanson–Blizzard syndrome], Ryoikibetsu Shokogun Shirizu 1 (1993) 162–164.
[524] T. Koizumi, [Johanson–Blizzard syndrome], Ryoikibetsu Shokogun Shirizu 30 (Pt 5) (2000) 198–200.
[525] K. Kristjansson, W.H. Hoffman, D.B. Flannery, M.J. Cohen, Johanson–Blizzard syndrome and hypopituitarism, J. Pediatr. 113 (5) (1988) 851–853.
[526] M.L. Kulkarni, S.K. Shetty, K.S. Kallambella, P.M. Kulkarni, Johanson – blizzard syndrome, Indian J. Pediatr. 71 (12) (2004) 1127–1129.
[527] A. Lerner, T. Iancu, [Johanson–Blizzard syndrome], Harefuah 115 (11) (1988) 342–344.
[528] M.K. Mardini, M. Ghandour, N.A. Sakati, W.L. Nyhan, Johanson–Blizzard syndrome in a large inbred kindred with three involved members, Clin. Genet. 14 (5) (1978) 247–250.
[529] V. Maunoury, S. Nieuwarts, J. Ferri, O. Ernst, [Pancreatic lipomatosis revealing Johanson–Blizzard syndrome], Gastroenterol. Clin. Biol. 23 (10) (1999) 1099–1101.
[530] J.N. McHeik, L. Hendiri, P. Vabres, M. Berthier, J. Cardona, D. Bonneau, et al., [Johanson-Blizzard syndrome: a case report], Arch. Pediatr. 9 (11) (2002) 1163–1165.
[531] J.B. Moeschler, M.S. Lubinsky, Johanson–Blizzard syndrome with normal intelligence, Am. J. Med. Genet. 22 (1) (1985) 69–73.
[532] J.B. Moeschler, M.J. Polak, J.J. Jenkins, 3rd, R.S. Amato, The Johanson–Blizzard syndrome: a second report of full autopsy findings, Am. J. Med. Genet. 26 (1) (1987) 133–138.
[533] N. Motohashi, S. Pruzansky, D. Day, Roentgencephalometric analysis of craniofacial growth in the Johanson–Blizzard syndrome, J. Craniofac. Genet. Dev. Biol. 1 (1) (1981) 57–72.
[534] K. Nagashima, H. Yagi, T. Kuroume, A case of Johanson–Blizzard syndrome complicated by diabetes mellitus, Clin. Genet. 43 (2) (1993) 98–100.
[535] K. Ono, T. Ogawa, N. Matsuda, Oral findings in Johanson–Blizzard syndrome, J. Oral Med. 42 (1) (1987) 14–16, 66.
[536] J.F. Prater, K. D'Addio, Johanson–Blizzard syndrome – a case study, behavioral manifestations, and successful treatment strategies, Biol. Psychiatry 51 (6) (2002) 515–517.
[537] P. Reichart, S. Flatz, M. Burdelski, [Ectodermal dysplasia and exocrine pancreatic insufficiency – a familial syndrome], Dtsch. Zahnarztl. Z. 34 (3) (1979) 263–265.
[538] F. Rosanowski, U. Hoppe, T. Hies, U. Eysholdt, [Johanson–Blizzard syndrome. A complex dysplasia syndrome with aplasia of the nasal alae and inner ear deafness], Hno 46 (10) (1998) 876–878.
[539] S. Rudnik-Schoneborn, B. Keller, F.A. Beemer, K. Pistor, H.F. Swanenburg de Veye, K. Zerres, [Johanson–Blizzard syndrome], Klin. Padiatr. 203 (1) (1991) 33–38.
[540] B.K. Sandhu, M.J. Brueton, Concurrent pancreatic and growth hormone insufficiency in Johanson–Blizzard syndrome, J. Pediatr. Gastroenterol. Nutr. 9 (4) (1989) 535–538.
[541] J. Sarles, [The Johanson–Blizzard syndrome], Arch. Pediatr. 10 (6) (2003) 553–554; author reply 554.
[542] W.J. Steinbach, R.L. Hintz, Diabetes mellitus and profound insulin resistance in Johanson–Blizzard syndrome, J. Pediatr. Endocrinol. Metab. 13 (9) (2000) 1633–1636.

[543] H.F. Swanenburg de Veye, J.A. Heineman-de-Boer, F.A. Beemer, A child of high intelligence with the Johanson–Blizzard syndrome, Genet. Couns. 2 (1) (1991) 21–25.

[544] P.G. Szilagyi, J. Corsetti, C.M. Callahan, K. McCormick, L.A. Metlay, Pancreatic exocrine aplasia, clinical features of leprechaunism, and abnormal gonadotropin regulation, Pediatr. Pathol. 7 (1) (1987) 51–61.

[545] T. Takahashi, M. Fujishima, S. Tsuchida, M. Enoki, G. Takada, Johanson–Blizzard syndrome: loss of glucagon secretion response to insulin-induced hypoglycemia, J. Pediatr. Endocrinol. Metab. 17 (8) (2004) 1141–1144.

[546] D.R. Trellis, R.E. Clouse, Johanson–Blizzard syndrome. Progression of pancreatic involvement in adulthood, Dig. Dis. Sci. 36 (3) (1991) 365–369.

[547] P.H. Vanlieferinghen, C. Borderon, C.H. Francannet, P. Gembara, P. Dechelotte, Johanson–Blizzard syndrome. a new case with autopsy findings, Genet. Couns. 12 (3) (2001) 245–250.

[548] P. Vanlieferinghen, D. Gallot, C. Francannet, F. Meyer, P. Dechelotte, Prenatal ultrasonographic diagnosis of a recurrent case of Johanson–Blizzard syndrome, Genet. Couns. 14 (1) (2003) 105–107.

[549] D.D. Vaughn, A.A. Jabra, E.K. Fishman, Pancreatic disease in children and young adults: evaluation with CT, Radiographics 18 (5) (1998) 1171–1187.

[550] M.W. Vieira, V.L. Lopes, H. Teruya, L. Guimaraes-Lamonato, L.C. Oliveira, C.D. Costa, [Johanson–Blizzard syndrome: the importance of differential diagnostic in pediatrics], J. Pediatr. (Rio. J.) 78 (5) (2002) 433–436.

[551] M. Zenker, J. Mayerle, M.M. Lerch, A. Tagariello, K. Zerres, P.R. Durie, et al., Corrigendum: Deficiency of UBR1, a ubiquitin ligase of the N-end rule pathway, causes pancreatic dysfunction, malformations and mental retardation (Johanson–Blizzard syndrome), Nat. Genet. 38 (2) (2006) 265.

[552] K. Zerres, E.A. Holtgrave, The Johanson–Blizzard syndrome: report of a new case with special reference to the dentition and review of the literature, Clin. Genet. 30 (3) (1986) 177–183.

[553] M.M. Lerch, M. Zenker, S. Turi, J. Mayerle, Developmental and metabolic disorders of the pancreas, Endocrinol. Metab. Clin. North Am. 35 (2) (2006) 219–241, vii.

[554] N. Rezaei, M. Sabbaghian, Z. Liu, M. Zenker, Eponym: Johanson–Blizzard syndrome, Eur. J. Pediatr. 170 (2) (2011) 179–183.

[555] S. Glaser, J. Schaft, S. Lubitz, K. Vintersten, F. van der Hoeven, K.R. Tufteland, et al., Multiple epigenetic maintenance factors implicated by the loss of Mll2 in mouse development, Development 133 (8) (2006) 1423–1432.

[556] D. Cogulu, O. Oncag, E. Celen, F. Ozkinay, Kabuki syndrome with additional dental findings: a case report, J. Dent. Child. (Chic.) 75 (2) (2008) 185–187.

[557] W. Courtens, A. Rassart, J.J. Stene, E. Vamos, Further evidence for autosomal dominant inheritance and ectodermal abnormalities in Kabuki syndrome, Am. J. Med. Genet. 93 (3) (2000) 244–249.

[558] N. Matsumoto, N. Niikawa, Kabuki make-up syndrome: a review, Am. J. Med. Genet. 117C (1) (2003) 57–65.

[559] D. Petzold, E. Kratzsch, C. Opitz, S. Tinschert, The Kabuki syndrome: four patients with oral abnormalities, Eur. J. Orthod. 25 (1) (2003) 13–19.

[560] G.M. Abdel-Salam, H.H. Afifi, M.M. Eid, T.H. El-Badry, N. Kholoussi, Ectodermal abnormalities in patients with Kabuki syndrome, Pediatr. Dermatol. 28 (5) (2011) 507–511.

[561] M.C. Hannibal, K.J. Buckingham, S.B. Ng, J.E. Ming, A.E. Beck, M.J. McMillin, et al., Spectrum of MLL2 (ALR) mutations in 110 cases of Kabuki syndrome, Am. J. Med. Genet. A 155A (7) (2011) 1511–1516.

[562] L. Micale, B. Augello, C. Fusco, A. Selicorni, M.N. Loviglio, M.C. Silengo, et al., Mutation spectrum of MLL2 in a cohort of Kabuki syndrome patients, Orphanet. J. Rare Dis. 6 (2011) 38.

[563] S.B. Ng, A.W. Bigham, K.J. Buckingham, M.C. Hannibal, M.J. McMillin, H.I. Gildersleeve, et al., Exome sequencing identifies MLL2 mutations as a cause of Kabuki syndrome, Nat. Genet. 42 (9) (2010) 790–793.

[564] A.D. Paulussen, A.P. Stegmann, M.J. Blok, D. Tserpelis, C. Posma-Velter, Y. Detisch, et al., MLL2 mutation spectrum in 45 patients with Kabuki syndrome, Hum. Mutat. 32 (2) (2011) E2018–E2025.

[565] Y. Li, N. Bogershausen, Y. Alanay, P.O. Simsek Kiper, N. Plume, K. Keupp, et al., A mutation screen in patients with Kabuki syndrome, Hum. Genet. 130 (6) (2011) 715–724.

[566] C.S. Teixeira, C.R. Silva, R.S. Honjo, D.R. Bertola, L.M. Albano, C.A. Kim, Dental evaluation of Kabuki syndrome patients, Cleft Palate Craniofac. J. 46 (6) (2009) 668–673.

[567] G. Spano, G. Campus, A. Bortone, V. Lai, P.F. Luglie, Oral features in Kabuki make-up syndrome, Eur. J. Paediatr. Dent. 9 (3) (2008) 149–152.

[568] C.T. Rocha, I.T. Peixoto, P.M. Fernandes, C.P. Torres, A.M. de Queiroz, Dental findings in Kabuki make-up syndrome: a case report, Spec. Care Dentist. 28 (2) (2008) 53–57.

[569] B.M. dos Santos, R.R. Ribeiro, A.S. Stuani, F.W. de Paula e Silva, A.M. de Queiroz, Kabuki make-up (Niikawa-Kuroki) syndrome: dental and craniofacial findings in a Brazilian child, Braz. Dent. J. 17 (3) (2006) 249–254.

[570] K. Matsune, T. Shimizu, T. Tohma, Y. Asada, H. Ohashi, T. Maeda, Craniofacial and dental characteristics of Kabuki syndrome, Am. J. Med. Genet. 98 (2) (2001) 185–190.

[571] A.A. Mhanni, H.G. Cross, A.E. Chudley, Kabuki syndrome: description of dental findings in 8 patients, Clin. Genet. 56 (2) (1999) 154–157.

[572] A.A. Mhanni, A.E. Chudley, Genetic landmarks through philately – Kabuki theater and Kabuki syndrome, Clin. Genet. 56 (2) (1999) 116–117.

[573] N.M. Maas, T. Van de Putte, C. Melotte, A. Francis, C.T. Schrander-Stumpel, D. Sanlaville, et al., The C20orf133 gene is disrupted in a patient with Kabuki syndrome, J. Med. Genet. 44 (9) (2007) 562–569.

[574] C. Bacino, ROR2-Related Robinow Syndrome, in: R.A. Pagon, T.D. Bird, C.R. Dolan, K. Stephens, (Eds.), GeneReviews [Internet], University of Washington, Seattle (WA), 1993.

[575] N. Eronat, D. Cogulu, F. Ozkinay, A case report on autosomal recessive Robinow syndrome, Eur. J. Paediatr. Dent. 10 (3) (2009) 147–150.

[576] J.F. Mazzeu, E. Pardono, A.M. Vianna-Morgante, A. Richieri-Costa, C. Ae Kim, D. Brunoni, et al., Clinical characterization of autosomal dominant and recessive variants of Robinow syndrome, Am. J. Med. Genet. A 143 (4) (2007) 320–325.

[577] A. Shaw, C. Longman, M. Irving, M. Splitt, Neonatal teeth in X-linked Opitz (G/BBB) syndrome, Clin. Dysmorphol. 15 (3) (2006) 185–186.

[578] G. da Silva Dalben, A. Richieri-Costa, L.A. de Assis Taveira, Tooth abnormalities and soft tissue alterations in patients with G/BBB syndrome, Oral Dis. 14 (8) (2008) 747–753.

[579] P. Nieminen, N.V. Morgan, A.L. Fenwick, S. Parmanen, L. Veistinen, M.L. Mikkola, et al., Inactivation of IL11 signaling causes craniosynostosis, delayed tooth eruption, and supernumerary teeth, Am. J. Hum. Genet. 89 (1) (2011) 67–81.

[580] R. Peterkova, H. Lesot, M. Peterka, Phylogenetic memory of developing mammalian dentition, J. Exp. Zool. B Mol. Dev. Evol. 306 (3) (2006) 234–250.

[581] S.J. Zou, R.N. D'Souza, T. Ahlberg, A.L. Bronckers, Tooth eruption and cementum formation in the Runx2/Cbfa1 heterozygous mouse, Arch. Oral Biol. 48 (9) (2003) 673–677.

[582] C.R. Chung, K. Tsuji, A. Nifuji, T. Komori, K. Soma, M. Noda, Micro-CT evaluation of tooth, calvaria and mechanical stress-induced tooth movement in adult Runx2/Cbfa1 heterozygous knock-out mice, J. Med. Dent. Sci. 51 (1) (2004) 105–113.

[583] S. Mundlos, L.F. Huang, P. Selby, B.R. Olsen, Cleidocranial dysplasia in mice, Ann. N.Y. Acad. Sci. 785 (1996) 301–302.

[584] F. Otto, A.P. Thornell, T. Crompton, A. Denzel, K.C. Gilmour, I.R. Rosewell, et al., Cbfa1, a candidate gene for cleidocranial dysplasia syndrome, is essential for osteoblast differentiation and bone development, Cell 89 (5) (1997) 765–771.

[585] D.O. Sillence, H.E. Ritchie, P.B. Selby, Animal model: skeletal anomalies in mice with cleidocranial dysplasia, Am. J. Med. Genet. 27 (1) (1987) 75–85.

[586] A.D. Angle, J. Rebellato, Dental team management for a patient with cleidocranial dysostosis, Am. J. Orthod. Dentofacial. Orthop. 128 (1) (2005) 110–117.

[587] S.C. Cooper, C.M. Flaitz, D.A. Johnston, B. Lee, J.T. Hecht, A natural history of cleidocranial dysplasia, Am. J. Med. Genet. 104 (1) (2001) 1–6.

[588] A.L. Counts, M.D. Rohrer, H. Prasad, P. Bolen, An assessment of root cementum in cleidocranial dysplasia, Angle Orthod. 71 (4) (2001) 293–298.

[589] G.J. Feldman, N.H. Robin, L.A. Brueton, E. Robertson, E.M. Thompson, J. Siegel-Bartelt, et al., A gene for cleidocranial dysplasia maps to the short arm of chromosome 6, Am. J. Hum. Genet. 56 (4) (1995) 938–943.

[590] I. Golan, A. Waldeck, U. Baumert, J. Strutz, D. Mussig, [Anomalies of the skull in cleidocranial dysplasia], Hno 52 (12) (2004) 1061–1066.

[591] T. Komori, Regulation of bone development and maintenance by Runx2, Front. Biosci. 13 (2008) 898–903.

[592] B. Lee, K. Thirunavukkarasu, L. Zhou, L. Pastore, A. Baldini, J. Hecht, et al., Missense mutations abolishing DNA binding of the osteoblast-specific transcription factor OSF2/CBFA1 in cleidocranial dysplasia, Nat. Genet. 16 (3) (1997) 307–310.

[593] S. Mundlos, Cleidocranial dysplasia: clinical and molecular genetics, J. Med. Genet. 36 (3) (1999) 177–182.

[594] S. Mundlos, F. Otto, C. Mundlos, J.B. Mulliken, A.S. Aylsworth, S. Albright, et al., Mutations involving the transcription factor CBFA1 cause cleidocranial dysplasia, Cell 89 (5) (1997) 773–779.

[595] T. Pal, D. Napierala, T.A. Becker, M. Loscalzo, D. Baldridge, B. Lee, et al., The presence of germ line mosaicism in cleidocranial dysplasia, Clin. Genet. 71 (6) (2007) 589–591.

[596] V.C. Petropoulos, T.J. Balshi, S.F. Balshi, G.J. Wolfinger, Treatment of a patient with cleidocranial dysplasia using osseointegrated implants: a patient report, Int. J. Oral Maxillofac. Implants 19 (2) (2004) 282–287.

[597] Z. Suba, G. Balaton, S. Gyulai-Gaal, P. Balaton, J. Barabas, I. Tarjan, Cleidocranial dysplasia: diagnostic criteria and combined treatment, J. Craniofac. Surg. 16 (6) (2005) 1122–1126.

[598] A. Verstrynge, C. Carels, A. Verdonck, W. Mollemans, G. Willems, J. Schoenaers, [Dentomaxillary and -facial problems in cleidocranial dysplasia], Ned. Tijdschr. Tandheelkd. 113 (2) (2006) 69–74.

[599] M.M. Cohen, Jr., Perspectives on *RUNX* genes: an update, Am. J. Med. Genet. A 149A (12) (2009) 2629–2646.

[600] N. Suda, M. Hattori, K. Kosaki, A. Banshodani, K. Kozai, K. Tanimoto, et al., Correlation between genotype and supernumerary tooth formation in cleidocranial dysplasia, Orthod. Craniofac. Res. 13 (4) (2010) 197–202.

[601] E. Half, D. Bercovich, P. Rozen, Familial adenomatous polyposis, Orphanet. J. Rare Dis. 4 (2009) 22.

[602] A.R. Clarke, Cancer genetics: mouse models of intestinal cancer, Biochem. Soc. Trans. 35 (Pt 5) (2007) 1338–1341.

[603] R. Fodde, W. Edelmann, K. Yang, C. van Leeuwen, C. Carlson, B. Renault, et al., A targeted chain-termination mutation in the mouse Apc gene results in multiple intestinal tumors, Proc. Natl. Acad. Sci. U.S.A. 91 (19) (1994) 8969–8973.

[604] S. Hasegawa, T. Sato, H. Akazawa, H. Okada, A. Maeno, M. Ito, et al., Apoptosis in neural crest cells by functional loss of APC tumor suppressor gene, Proc. Natl. Acad. Sci. U.S.A. 99 (1) (2002) 297–302.

[605] L.K. Su, K.W. Kinzler, B. Vogelstein, A.C. Preisinger, A.R. Moser, C. Luongo, et al., Multiple intestinal neoplasia caused by a mutation in the murine homolog of the APC gene, Science 256 (5057) (1992) 668–670.

[606] X.P. Wang, D.J. O'Connell, J.J. Lund, I. Saadi, M. Kuraguchi, A. Turbe-Doan, et al., Apc inhibition of Wnt signaling regulates supernumerary tooth formation during embryogenesis and throughout adulthood, Development 136 (11) (2009) 1939–1949.

[607] E.J. Gardner, Follow-up study of a family group exhibiting dominant inheritance for a syndrome including intestinal polyps, osteomas, fibromas and epidermal cysts, Am. J. Hum. Genet. 14 (1962) 376–390.

[608] W. Carl, M.A. Sullivan, Dental abnormalities and bone lesions associated with familial adenomatous polyposis: report of cases, J. Am. Dent. Assoc. 119 (1) (1989) 137–139.

[609] M. Karazivan, K. Manoukian, B. Lalonde, [Familial adenomatous polyposis or Gardner syndrome – review of the literature and presentation of 2 clinical cases], J. Can. Dent. Assoc. 66 (1) (2000) 26–30.

[610] J.O. Sondergaard, S. Bulow, H. Jarvinen, J. Wolf, I.N. Witt, G. Tetens, Dental anomalies in familial adenomatous polyposis coli, Acta Odontol. Scand. 45 (1) (1987) 61–63.

[611] M.A. Wijn, J.J. Keller, F.M. Giardiello, H.S. Brand, Oral and maxillofacial manifestations of familial adenomatous polyposis, Oral Dis. 13 (4) (2007) 360–365.

[612] R.J. Woods, R.G. Sarre, G.C. Ctercteko, D.J. Jagelman, J.W. Smith, P.M. Duchesneau, et al., Occult radiologic changes in the skull and jaw in familial adenomatous polyposis coli, Dis. Colon Rectum 32 (4) (1989) 304–306.

[613] F.J. Hes, M. Nielsen, E.C. Bik, D. Konvalinka, J.T. Wijnen, E. Bakker, et al., Somatic APC mosaicism: an underestimated cause of polyposis coli, Gut 57 (1) (2008) 71–76.

[614] M. Nielsen, E. Bik, F.J. Hes, M.H. Breuning, H.F. Vasen, E. Bakker, et al., Genotype-phenotype correlations in 19 Dutch cases with APC gene deletions and a literature review, Eur. J. Hum. Genet. 15 (10) (2007) 1034–1042.

[615] M. Nielsen, F.J. Hes, F.M. Nagengast, M.M. Weiss, E.M. Mathus-Vliegen, H. Morreau, et al., Germline mutations in APC and MUTYH are responsible for the majority of families with attenuated familial adenomatous polyposis, Clin. Genet. 71 (5) (2007) 427–433.

[616] R.W. Burt, K.W. Jasperson, APC-Associated Polyposis Conditions, in: R.A. Pagon, T.D. Bird, C.R. Dolan, K. Stephens, (Eds.), GeneReviews [Internet], University of Washington, Seattle (WA), 1993.

[617] Y.L. Wallis, D.G. Morton, C.M. McKeown, F. Macdonald, Molecular analysis of the APC gene in 205 families: extended genotype-phenotype correlations in FAP and evidence for the role of APC amino acid changes in colorectal cancer predisposition, J. Med. Genet. 36 (1) (1999) 14–20.

[618] X.P. Wang, J. Fan, Molecular genetics of supernumerary tooth formation, Genesis 49 (4) (2011) 261–277.

[619] M. Payne, J.A. Anderson, J. Cook, Gardner's syndrome – a case report, Br. Dent. J. 193 (7) (2002) 383–384.

[620] M. Coccia, S.P. Brooks, T.R. Webb, K. Christodoulou, I.O. Wozniak, V. Murday, et al., X-linked cataract and Nance-Horan syndrome are allelic disorders, Hum. Mol. Genet. 18 (14) (2009) 2643–2655.

[621] D. Stambolian, J. Favor, W. Silvers, P. Avner, V. Chapman, E. Zhou, Mapping of the X-linked cataract (Xcat) mutation, the gene implicated in the Nance Horan syndrome, on the mouse X chromosome, Genomics 22 (2) (1994) 377–380.

[622] K.M. Huang, J. Wu, M.K. Duncan, C. Moy, A. Dutra, J. Favor, et al., Xcat, a novel mouse model for Nance-Horan syndrome inhibits expression of the cytoplasmic-targeted Nhs1 isoform, Hum. Mol. Genet. 15 (2) (2006) 319–327.

[623] A. Toutain, A.D. Ayrault, C. Moraine, Mental retardation in Nance-Horan syndrome: clinical and neuropsychological assessment in four families, Am. J. Med. Genet. 71 (3) (1997) 305–314.

[624] A. Reches, Y. Yaron, K. Burdon, O. Crystal-Shalit, D. Kidron, M. Malcov, et al., Prenatal detection of congenital bilateral cataract leading to the diagnosis of Nance-Horan syndrome in the extended family, Prenat. Diagn. 27 (7) (2007) 662–664.

[625] K.M. Huang, J. Wu, S.P. Brooks, A.J. Hardcastle, R.A. Lewis, D. Stambolian, Identification of three novel NHS mutations in families with Nance-Horan syndrome, Mol. Vis. 13 (2007) 470–474.

[626] R.J. Florijn, W. Loves, L.J. Maillette de Buy Wenniger-Prick, M.M. Mannens, N. Tijmes, S.P. Brooks, et al., New mutations in the NHS gene in Nance-Horan syndrome families from the Netherlands, Eur. J. Hum. Genet. 14 (9) (2006) 986–990.

[627] S. Sharma, S.L. Ang, M. Shaw, D.A. Mackey, J. Gecz, J.W. McAvoy, et al., Nance-Horan syndrome protein, NHS, associates with epithelial cell junctions, Hum. Mol. Genet. 15 (12) (2006) 1972–1983.

[628] V.L. Ramprasad, A. Thool, S. Murugan, D. Nancarrow, P. Vyas, S.K. Rao, et al., Truncating mutation in the NHS gene: phenotypic heterogeneity of Nance-Horan syndrome in an Asian Indian family, Invest. Ophthalmol. Vis. Sci. 46 (1) (2005) 17–23.

[629] S.P. Brooks, N.D. Ebenezer, S. Poopalasundaram, O.J. Lehmann, A.T. Moore, A.J. Hardcastle, Identification of the gene for Nance-Horan syndrome (NHS), J. Med. Genet. 41 (10) (2004) 768–771.

[630] S. Brooks, N. Ebenezer, S. Poopalasundaram, E. Maher, P. Francis, A. Moore, et al., Refinement of the X-linked cataract locus (CXN) and gene analysis for CXN and Nance-Horan syndrome (NHS), Ophthalmic. Genet. 25 (2) (2004) 121–131.

[631] A. Toutain, N. Ronce, B. Dessay, L. Robb, C. Francannet, M. Le Merrer, et al., Nance-Horan syndrome: linkage analysis in 4 families refines localization in Xp22.31-p22.13 region, Hum. Genet. 99 (2) (1997) 256–261.

[632] D. Stambolian, R.A. Lewis, K. Buetow, A. Bond, R. Nussbaum, Nance-Horan syndrome: localization within the region Xp21.1-Xp22.3 by linkage analysis, Am. J. Hum. Genet. 47 (1) (1990) 13–19.

[633] R.A. Lewis, R.L. Nussbaum, D. Stambolian, Mapping X-linked ophthalmic diseases. IV. Provisional assignment of the locus for X-linked congenital cataracts and microcornea (the Nance-Horan syndrome) to Xp22.2-p22.3, Ophthalmology 97 (1) (1990) 110–120; discussion 111–120.

[634] A. Toutain, B. Dessay, N. Ronce, M.I. Ferrante, J. Tranchemontagne, R. Newbury-Ecob, et al., Refinement of the NHS locus on chromosome Xp22.13 and analysis of five candidate genes, Eur. J. Hum. Genet. 10 (9) (2002) 516–520.

[635] W.E. Nance, M. Warburg, D. Bixler, E.M. Helveston, Congenital X-linked cataract, dental anomalies and brachymetacarpalia, Birth Defects Orig. Artic. Ser. 10 (4) (1974) 285–291.

[636] S. Sharma, K.S. Koh, C. Collin, A. Dave, A. McMellon, Y. Sugiyama, et al., NHS-A isoform of the NHS gene is a novel interactor of ZO-1, Exp. Cell Res. 315 (14) (2009) 2358–2372.

[637] S.P. Brooks, M. Coccia, H.R. Tang, N. Kanuga, L.M. Machesky, M. Bailly, et al., The Nance-Horan syndrome protein encodes a functional WAVE homology domain (WHD) and is important for co-ordinating actin remodelling and maintaining cell morphology, Hum. Mol. Genet. 19 (12) (2010) 2421–2432.

[638] P. Kantaputra, I. Miletich, H.J. Ludecke, E.Y. Suzuki, V. Praphanphoj, R. Shivdasani, et al., Tricho-rhino-phalangeal syndrome with supernumerary teeth, J. Dent. Res. 87 (11) (2008) 1027–1031.

[639] Z. Gai, T. Gui, Y. Muragaki, The function of TRPS1 in the development and differentiation of bone, kidney, and hair follicles, Histol. Histopathol. 26 (7) (2011) 915–921.

[640] D.M. Piscopo, E.B. Johansen, R. Derynck, Identification of the GATA factor TRPS1 as a repressor of the osteocalcin promoter, J. Biol. Chem. 284 (46) (2009) 31690–31703.

[641] A. Paterson, P.S. Thomas, Abnormal modelling of the humeral head in the tricho-rhino-phalangeal syndrome: a new radiological observation, Australas Radiol. 44 (3) (2000) 325–327.

[642] M. Dumic, J. Ille, M. Mikecin, M. Cvitkovic, V. Hitrec, K. Potocki, [Trichorhinophalangeal syndrome], Lijec. Vjesn. 115 (5–6) (1993) 163–165.

[643] Y. Ohta, T. Ohura, H. Iwamoto, T. Nobata, T. Kawamoto, Z. Kinoshita, et al., [A case of tricho-rhino-phalangeal syndrome], Nihon Kyosei Shika Gakkai Zasshi 46 (2) (1987) 427–437.

[644] J.I. Sommermater, C. Stoll, F. Obry, W. Bacon, R. Haag, [Dento-maxillo-facial abnormalities and tricho-rhino-phalangeal syndrome: apropos of a case], Rev. Odontostomatol. (Paris) 7 (3) (1978) 195–200.

[645] S.S. Gellis, M. Feingold, Picture of the month. Tricho-rhino-phalangeal syndrome, Am. J. Dis. Child. 124 (1) (1972) 89–90.

[646] R.D. Parkhurst, G.S. Light, G.E. Bacon, J.C. Gall, Jr., Tricho-rhino-phalangeal syndrome with hypoglycemia: case report with endocrine function studies, South Med. J. 65 (4) (1972) 457–459.

[647] R.J. Gorlin, M.M. Cohen, Jr., J. Wolfson, Tricho-rhino-phalangeal syndrome, Am. J. Dis. Child. 118 (4) (1969) 595–599.

[648] G. Machuca, F. Martinez, C. Machuca, P. Bullon, Craniofacial and oral manifestations of trichorhinophalangeal syndrome type I (Giedion's syndrome): a case report, Oral Surg. Oral Med. Oral Pathol. Oral Radiol. Endod. 84 (1) (1997) 35–39.

[649] D. Braga, A.M. Manganoni, R. Gavazzoni, G. Pasolini, G. De Panfilis, A case of trichorhinophalangeal syndrome, type I, Cutis 53 (2) (1994) 92–94.

[650] C.G. Bennett, C.J. Hill, J.L. Frias, Facial and oral findings in trichorhinophalangeal syndrome type 1 (characteristics of TRPS 1), Pediatr. Dent. 3 (4) (1981) 348–352.

[651] I.E. Hussels, Trichorhinophalangeal syndrome in two sibs, Birth Defects Orig. Artic. Ser. 7 (7) (1971) 301–303.

[652] R.M. Goodman, R. Trilling, M. Hertz, H. Horoszowski, P. Merlob, S. Reisner, New clinical observations in the trichorhinophalangeal syndrome, J. Craniofac. Genet. Dev. Biol. 1 (1) (1981) 15–29.

[653] J. Ferrando, J.A. Del Olmo, J. Bassas, E. Fernandez, R. Fontarnau, [Trichorhinophalangeal syndrome (Giedion)], Med. Cutan Ibero. Lat. Am. 9 (5) (1981) 351–360.

[654] P. Scheffer, M. Verdier, G. Finidori, [Trichorhinophalangeal syndrome: analysis of craniofacial architecture in six cases (author's transl)], Rev. Stomatol. Chir. Maxillofac. 82 (4) (1981) 230–233.

[655] A. Giedion, [Tricho-rhino-phalangeal syndrome], Helv. Paediatr. Acta 21 (5) (1966) 475–485.

[656] T.H. Malik, D. Von Stechow, R.T. Bronson, R.A. Shivdasani, Deletion of the GATA domain of TRPS1 causes an absence of facial hair and provides new insights into the bone disorder in inherited tricho-rhino-phalangeal syndromes, Mol. Cell. Biol. 22 (24) (2002) 8592–8600.

[657] D. Napierala, X. Garcia-Rojas, K. Sam, K. Wakui, C. Chen, R. Mendoza-Londono, et al., Mutations and promoter SNPs in RUNX2, a transcriptional regulator of bone formation, Mol. Genet. Metab. 86 (1–2) (2005) 257–268.

[658] P. Kantaputra, P. Tanpaiboon, T. Porntaveetus, A. Ohazama, P. Sharpe, A. Rauch, et al., The smallest teeth in the world are caused by mutations in the PCNT gene, Am. J. Med. Genet. A 155A (6) (2011) 1398–1403.

[659] A. Rauch, The shortest of the short: pericentrin mutations and beyond, Best Pract. Res. Clin. Endocrinol. Metab. 25 (1) (2011) 125–130.

[660] R. Ramsebner, M. Ludwig, T. Parzefall, T. Lucas, W.D. Baumgartner, O. Bodamer, et al., A FGF3 mutation associated with differential inner ear malformation, microtia, and microdontia, Laryngoscope 120 (2) (2010) 359–364.

[661] S. Riazuddin, Z.M. Ahmed, R.S. Hegde, S.N. Khan, I. Nasir, U. Shaukat, et al., Variable expressivity of FGF3 mutations associated with deafness and LAMM syndrome, BMC Med. Genet. 12 (2011) 21.

[662] A. Sensi, S. Ceruti, P. Trevisi, F. Gualandi, M. Busi, I. Donati, et al., LAMM syndrome with middle ear dysplasia associated with compound heterozygosity for FGF3 mutations, Am. J. Med. Genet. A 155A (5) (2011) 1096–1101.

[663] M. Tekin, B.O. Hismi, S. Fitoz, H. Ozdag, F.B. Cengiz, A. Sirmaci, et al., Homozygous mutations in fibroblast growth factor 3 are associated with a new form of syndromic deafness characterized by inner ear agenesis, microtia, and microdontia, Am. J. Hum. Genet. 80 (2) (2007) 338–344.

[664] M. Tekin, H. Ozturkmen Akay, S. Fitoz, S. Birnbaum, F.B. Cengiz, L. Sennaroglu, et al., Homozygous FGF3 mutations result in congenital deafness with inner ear agenesis, microtia, and microdontia, Clin. Genet. 73 (6) (2008) 554–565.

[665] O. Alsmadi, B.F. Meyer, F. Alkuraya, S. Wakil, F. Alkayal, H. Al-Saud, et al., Syndromic congenital sensorineural deafness, microtia and microdontia resulting from a novel homoallelic mutation in fibroblast growth factor 3 (FGF3), Eur. J. Hum. Genet. 17 (1) (2009) 14–21.

[666] K.B. Hunter, T. Lucke, J. Spranger, S.F. Smithson, H. Alpay, J.L. Andre, et al., Schimke immunoosseous dysplasia: defining skeletal features, Eur. J. Pediatr. 169 (7) (2010) 801–811.

[667] J. Visootsak, J.M. Graham, Jr., Klinefelter syndrome and other sex chromosomal aneuploidies, Orphanet. J. Rare Dis. 1 (2006) 42.

[668] M.R. Benjamin, F.S. Rodrigo, R.J. Gorlin, Multiple macrodontic multituberculism, Am. J. Med. Genet. 120A (2) (2003) 283–285.

[669] T. Yoda, Y. Ishii, Y. Honma, E. Sakai, S. Enomoto, Multiple macrodonts with odontoma in a mother and son – a variant of Ekman-Westborg–Julin syndrome. Report of a case, Oral Surg. Oral Med. Oral Pathol. Oral Radiol. Endod. 85 (3) (1998) 301–303.

[670] J.A. Nemes, M. Alberth, The Ekman-Westborg and Julin trait: report of a case, Oral Surg. Oral Med. Oral Pathol. Oral Radiol. Endod. 102 (5) (2006) 659–662.

[671] B. Ekman-Westborg, P. Julin, Multiple anomalies in dental morphology: macrodontia, multituberculism, central cusps, and pulp invaginations. Report of a case, Oral Surg. Oral Med. Oral Pathol. 38 (2) (1974) 217–222.

[672] J.V. Ruch, Patterned distribution of differentiating dental cells: facts and hypotheses, J. Biol. Buccale 18 (2) (1990) 91–98.
[673] J.V. Ruch, Tooth crown morphogenesis and cytodifferentiations: candid questions and critical comments, Connect. Tissue Res. 32 (1–4) (1995) 1–8.
[674] J.V. Ruch, H. Lesot, C. Begue-Kirn, Odontoblast differentiation, Int. J. Dev. Biol. 39 (1) (1995) 51–68.
[675] Y. Tanaka, I. Naruse, T. Maekawa, H. Masuya, T. Shiroishi, S. Ishii, et al., Abnormal skeletal patterning in embryos lacking a single Cbp allele: a partial similarity with Rubinstein–Taybi syndrome, Proc. Natl. Acad. Sci. U.S.A. 94 (19) (1997) 10215–10220.
[676] M.A. Wood, M.P. Kaplan, A. Park, E.J. Blanchard, A.M. Oliveira, T.L. Lombardi, et al., Transgenic mice expressing a truncated form of CREB-binding protein (CBP) exhibit deficits in hippocampal synaptic plasticity and memory storage, Learn. Mem. 12 (2) (2005) 111–119.
[677] S.A. Josselyn, What's right with my mouse model? New insights into the molecular and cellular basis of cognition from mouse models of Rubinstein–Taybi Syndrome, Learn. Mem. 12 (2) (2005) 80–83.
[678] Y. Oike, N. Takakura, A. Hata, T. Kaname, M. Akizuki, Y. Yamaguchi, et al., Mice homozygous for a truncated form of CREB-binding protein exhibit defects in hematopoiesis and vasculo-angiogenesis, Blood 93 (9) (1999) 2771–2779.
[679] Z. Zhang, C. Hofmann, E. Casanova, G. Schutz, B. Lutz, Generation of a conditional allele of the CBP gene in mouse, Genesis 40 (2) (2004) 82–89.
[680] O. Bartsch, J. Labonte, B. Albrecht, D. Wieczorek, S. Lechno, U. Zechner, et al., Two patients with EP300 mutations and facial dysmorphism different from the classic Rubinstein–Taybi syndrome, Am. J. Med. Genet. A 152A (1) (2010) 181–184.
[681] A. Bloch-Zupan, J. Stachtou, D. Emmanouil, B. Arveiler, D. Griffiths, D. Lacombe, Oro-dental features as useful diagnostic tool in Rubinstein–Taybi syndrome, Am. J. Med. Genet. A 143 (6) (2007) 570–573.
[682] M.A. Baker, Dental and oral manifestations of Rubinstein–Taybi syndrome: report of case, ASDC J. Dent. Child. 54 (5) (1987) 369–371.
[683] S. Ikuno, K. Shinoda, T. Koizumi, A. Fujii, Y. Ito, E. Sobue, et al., [Dental findings on the Rubinstein–Taybi syndrome; a case report], Shoni Shikagaku Zasshi 25 (1) (1987) 148–155.
[684] M.J. Kinirons, Oral aspects of Rubenstein – Taybi syndrome, Br. Dent. J. 154 (2) (1983) 46–47.
[685] D.G. Gardner, S.S. Girgis, Talon cusps: a dental anomaly in the Rubinstein–Taybi syndrome, Oral Surg. Oral Med. Oral Pathol. 47 (6) (1979) 519–521.
[686] B. Rohlfing, K. Lewis, E.B. Singleton, Rubinstein–Taybi syndrome. Report of an unusual case, Am. J. Dis. Child. 121 (1) (1971) 71–74.
[687] R.C. Hennekam, J.M. Van Doorne, Oral aspects of Rubinstein–Taybi syndrome, Am. J. Med. Genet. Suppl. 6 (1990) 42–47.
[688] O. Bartsch, S. Schmidt, M. Richter, S. Morlot, E. Seemanova, G. Wiebe, et al., DNA sequencing of CREBBP demonstrates mutations in 56% of patients with Rubinstein–Taybi syndrome (RSTS) and in another patient with incomplete RSTS, Hum. Genet. 117 (5) (2005) 485–493.
[689] J.H. Roelfsema, S.J. White, Y. Ariyurek, D. Bartholdi, D. Niedrist, F. Papadia, et al., Genetic heterogeneity in Rubinstein–Taybi syndrome: mutations in both the CBP and EP300 genes cause disease, Am. J. Hum. Genet. 76 (4) (2005) 572–580.
[690] I. Coupry, C. Roudaut, M. Stef, M.A. Delrue, M. Marche, I. Burgelin, et al., Molecular analysis of the CBP gene in 60 patients with Rubinstein–Taybi syndrome, J. Med. Genet. 39 (6) (2002) 415–421.

[691] R.C. Hennekam, M.J. Van Den Boogaard, B.J. Sibbles, H.G. Van Spijker, Rubinstein–Taybi syndrome in The Netherlands, Am. J. Med. Genet. Suppl. 6 (1990) 17–29.

[692] H. Taybi, J.H. Rubinstein, Broad thumbs and toes, and unusual facial features; a probable mental retardation syndrome, Am. J. Roentgenol. Radium Ther. Nucl. Med. 93 (1965) 362–366.

[693] J.H. Rubinstein, H. Taybi, Broad thumbs and toes and facial abnormalities. A possible mental retardation syndrome, Am. J. Dis. Child. 105 (1963) 588–608.

[694] J.H. Rubinstein, Broad thumb-hallux (Rubinstein–Taybi) syndrome 1957-1988, Am. J. Med. Genet. Suppl. 6 (1990) 3–16.

[695] C.Y. Gregory-Evans, M. Moosajee, M.D. Hodges, D.S. Mackay, L. Game, N. Vargesson, et al., SNP genome scanning localizes oto-dental syndrome to chromosome 11q13 and microdeletions at this locus implicate FGF3 in dental and inner-ear disease and FADD in ocular coloboma, Hum. Mol. Genet. 16 (20) (2007) 3482–3493.

[696] Y. Alvarez, M.T. Alonso, V. Vendrell, L.C. Zelarayan, P. Chamero, T. Theil, et al., Requirements for FGF3 and FGF10 during inner ear formation, Development 130 (25) (2003) 6329–6338.

[697] S.L. Mansour, J.M. Goddard, M.R. Capecchi, Mice homozygous for a targeted disruption of the proto-oncogene int-2 have developmental defects in the tail and inner ear, Development 117 (1) (1993) 13–28.

[698] E.P. Hatch, C.A. Noyes, X. Wang, T.J. Wright, S.L. Mansour, Fgf3 is required for dorsal patterning and morphogenesis of the inner ear epithelium, Development 134 (20) (2007) 3615–3625.

[699] A. Bloch-Zupan, J.R. Goodman, Otodental syndrome, Orphanet J. Rare Dis. 1 (2006) 5.

[700] R.J. Chen, H.S. Chen, L.M. Lin, C.C. Lin, R.J. Jorgenson, et al., 'Otodental' dysplasia, Oral Surg. Oral Med. Oral Pathol. 66 (3) (1988) 353–358.

[701] J.D. Colter, H.O. Sedano, Otodental syndrome: a case report, Pediatr. Dent. 27 (6) (2005) 482–485.

[702] R.A. Cook, J.R. Cox, R.J. Jorgenson, Otodental dysplasia: a five year study, Ear Hear 2 (2) (1981) 90–94.

[703] L.S. Levin, R.J. Jorgenson, Otodental dysplasia: a previously undescribed syndrome, Birth Defects 10 (1974) 310–312.

[704] L.S. Levin, R.J. Jorgenson, R.A. Cook, Otodental dysplasia: a 'new' ectodermal dysplasia, Clin. Genet. 8 (2) (1975) 136–144.

[705] A.J. Mesaros, Jr., J.W. Basden, Otodental syndrome, Gen. Dent. 44 (5) (1996) 427–429.

[706] L. Santos-Pinto, M.P. Oviedo, A. Santos-Pinto, H.I. Iost, N.S. Seale, A.K. Reddy, Otodental syndrome: three familial case reports, Pediatr. Dent. 20 (3) (1998) 208–211.

[707] H.O. Sedano, L.C. Moreira, R.A. de Souza, A.B. Moleri, Otodental syndrome: a case report and genetic considerations, Oral Surg. Oral Med. Oral Pathol. Oral Radiol. Endod. 92 (3) (2001) 312–317.

[708] C.J. Witkop, Jr., K.K. Gundlach, W.J. Streed, J.J. Sauk, Jr., Globodontia in the otodental syndrome, Oral Surg. Oral Med. Oral. Pathol. 41 (4) (1976) 472–483.

[709] Z. Fan, T. Yamaza, J.S. Lee, J. Yu, S. Wang, G. Fan, et al., BCOR regulates mesenchymal stem cell function by epigenetic mechanisms, Nat. Cell Biol. 11 (8) (2009) 1002–1009.

[710] J.A. Blake, C.J. Bult, J.T. Eppig, J.A. Kadin, J.E. Richardson, The mouse genome database genotypes:phenotypes, Nucleic Acids Res. 37 (Database issue) (2009) D712–D719.

[711] J. Cai, S. Kwak, J.M. Lee, E.J. Kim, M.J. Lee, G.H. Park, et al., Function analysis of mesenchymal Bcor in tooth development by using RNA interference, Cell Tissue Res. 341 (2) (2010) 251–258.

[712] P. Hedera, J.L. Gorski, Oculo-facio-cardio-dental syndrome: skewed X chromosome inactivation in mother and daughter suggest X-linked dominant Inheritance, Am. J. Med. Genet. 123A (3) (2003) 261–266.
[713] H. Numabe, Y. Numabe, [Oculo-facio-cardio-dental syndrome], Ryoikibetsu Shokogun Shirizu 34 (Pt 2) (2001) 350–351.
[714] E. McGovern, M. Al-Mudaffer, C. McMahon, D. Brosnahan, P. Fleming, W. Reardon, Oculo-facio-cardio-dental syndrome in a mother and daughter, Int. J. Oral Maxillofac. Surg. 35 (11) (2006) 1060–1062.
[715] H. Turkkahraman, M. Sarioglu, Oculo-facio-cardio-dental syndrome: report of a rare case, Angle Orthod. 76 (1) (2006) 184–186.
[716] H. Tsukawaki, M. Tsuji, T. Kawamoto, K. Ohyama, Three cases of oculo-facio-cardio-dental (OFCD) syndrome, Cleft Palate Craniofac. J. 42 (5) (2005) 467–476.
[717] D. Horn, M. Chyrek, S. Kleier, S. Luttgen, H. Bolz, et al., Novel mutations in BCOR in three patients with oculo-facio-cardio-dental syndrome, but none in Lenz microphthalmia syndrome, Eur. J. Hum. Genet. 13 (5) (2005) 563–569.
[718] T. Kawamoto, N. Motohashi, K. Ohyama, A case of oculo-facio-cardio-dental syndrome with integrated orthodontic-prosthodontic treatment, Cleft Palate Craniofac. J. 41 (1) (2004) 84–94.
[719] I. Barthelemy, L. Samuels, D.M. Kahn, S.A. Schendel, Oculo-facio-cardio-dental syndrome: two new cases, J. Oral Maxillofac. Surg. 59 (8) (2001) 921–925.
[720] B.R. Schulze, Rare dental abnormalities seen in oculo-facio-cardio-dental (OFCD) syndrome: three new cases and review of nine patients, Am. J. Med. Genet. 82 (5) (1999) 429–435.
[721] C. Opitz, D. Horn, R. Lehmann, T. Dimitrova, K. Fasmers-Henke, Oculo-facio-cardio-dental (OFCD) syndrome, J. Orofac. Orthop. 59 (3) (1998) 178–185.
[722] H.L. Obwegeser, R.J. Gorlin, Oculo-facio-cardio-dental (OFCD) syndrome, Clin. Dysmorphol. 6 (3) (1997) 281–283.
[723] R.J. Gorlin, A.H. Marashi, H.L. Obwegeser, Oculo-facio-cardio-dental (OFCD) syndrome, Am. J. Med. Genet. 63 (1) (1996) 290–292.
[724] D. Ng, N. Thakker, C.M. Corcoran, D. Donnai, R. Perveen, A. Schneider, et al., Oculofaciocardiodental and Lenz microphthalmia syndromes result from distinct classes of mutations in BCOR, Nat. Genet. 36 (4) (2004) 411–416.
[725] S. Oberoi, A.E. Winder, J. Johnston, K. Vargervik, A.M. Slavotinek, Case reports of oculofaciocardiodental syndrome with unusual dental findings, Am. J. Med. Genet. A 136 (3) (2005) 275–277.
[726] P.J. Francis, V. Berry, A.J. Hardcastle, E.R. Maher, A.T. Moore, S.S. Bhattacharya, A locus for isolated cataract on human Xp, J. Med. Genet. 39 (2) (2002) 105–109.
[727] D. Cogulu, F. Ertugrul, Dental management of a patient with oculo-facio-cardio-dental syndrome, J. Dent. Child. (Chic.) 75 (3) (2008) 306–308.
[728] F. Brancati, A. Sarkozy, B. Dallapiccola, KBG syndrome, Orphanet J. Rare Dis. 1 (2006) 50.
[729] J. Herrmann, P.D. Pallister, W. Tiddy, J.M. Opitz, The KBG syndrome – a syndrome of short stature, characteristic facies, mental retardation, macrodontia and skeletal anomalies, Birth Defects Orig. Artic. Ser. 11 (5) (1975) 7–18.
[730] K.L. Skjei, M.M. Martin, A.M. Slavotinek, KBG syndrome: report of twins, neurological characteristics, and delineation of diagnostic criteria, Am. J. Med. Genet. A 143 (3) (2007) 292–300.
[731] A.M. Davanzo, G. Rosalia, M. Biondi, D. De Brasi, A.R. Colucci, A. Panetta, et al., Eight isolated cases of KBG syndrome: a new hypothesis of study, Eur. Rev. Med. Pharmacol. Sci. 9 (1) (2005) 49–52.

[732] F. Brancati, M.G. D'Avanzo, M.C. Digilio, A. Sarkozy, M. Biondi, D. De Brasi, et al., KBG syndrome in a cohort of Italian patients, Am. J. Med. Genet. A 131 (2) (2004) 144–149.

[733] G.H. Maegawa, J.C. Leite, T.M. Felix, H.L. da Silveira, H.E. da Silveira, Clinical variability in KBG syndrome: report of three unrelated families, Am. J. Med. Genet. A 131 (2) (2004) 150–154.

[734] M. Tekin, A. Kavaz, M. Berberoglu, S. Fitoz, M. Ekim, G. Ocal, et al., The KBG syndrome: confirmation of autosomal dominant inheritance and further delineation of the phenotype, Am. J. Med. Genet. A 130 (3) (2004) 284–287.

[735] P.A. Dowling, P. Fleming, R.J. Gorlin, M. King, N.C. Nevin, M. McEntagart, The KBG syndrome, characteristic dental findings: a case report, Int. J. Paediatr. Dent. 11 (2) (2001) 131–134.

[736] S.F. Smithson, E.M. Thompson, A.G. McKinnon, I.S. Smith, R.M. Winter, The KBG syndrome, Clin. Dysmorphol. 9 (2) (2000) 87–91.

[737] M. Mathieu, M. Helou, G. Morin, P. Dolhem, B. Devauchelle, C. Piussan, The KBG syndrome: an additional sporadic case, Genet. Couns. 11 (1) (2000) 33–35.

[738] K. Devriendt, M. Holvoet, J.P. Fryns, Further delineation of the KBG syndrome, Genet. Couns. 9 (3) (1998) 191–194.

[739] M.R. Rivera-Vega, N. Leyva Juarez, S.A. Cuevas-Covarrubias, S.H. Kofman-Alfaro, Congenital heart defect and conductive hypoacusia in a patient with the KBG syndrome, Clin. Genet. 50 (4) (1996) 278–279.

[740] D. Soekarman, P. Volcke, J.P. Fryns, The KBG syndrome: follow-up data on three affected brothers, Clin. Genet. 46 (4) (1994) 283–286.

[741] M. Zollino, A. Battaglia, M.G. D'Avanzo, M.M. Della Bruna, R. Marini, G. Scarano, et al., Six additional cases of the KBG syndrome: clinical reports and outline of the diagnostic criteria, Am. J. Med. Genet. 52 (3) (1994) 302–307.

[742] J.P. Fryns, M. Haspeslagh, Mental retardation, short stature, minor skeletal anomalies, craniofacial dysmorphism and macrodontia in two sisters and their mother. Another variant example of the KBG syndrome? Clin. Genet. 26 (1) (1984) 69–72.

[743] I. Tollaro, B. v, C. Calzolari, F. Franchini, M.L. Giovannucci Uzielli, P.L. Vieri, [Dento-maxillo-facial anomalies in the KBG syndrome], Minerva. Stomatol. 33 (3) (1984) 437–446.

[744] H. Kumar, N. Prabhu, A. Cameron, KBG syndrome: review of the literature and findings of 5 affected patients, Oral Surg. Oral Med. Oral Pathol. Oral Radiol. Endod. 108 (3) (2009) e72–e79.

[745] A. Sirmaci, M. Spiliopoulos, F. Brancati, E. Powell, D. Duman, A. Abrams, et al., Mutations in ANKRD11 cause KBG syndrome, characterized by intellectual disability, skeletal malformations, and macrodontia, Am. J. Hum. Genet. 89 (2) (2011) 289–294.

[746] P.J. De Coster, M. Cornelissen, A. De Paepe, L.C. Martens, A. Vral, Abnormal dentin structure in two novel gene mutations [COL1A1, Arg134Cys] and [ADAMTS2, Trp795-to-ter] causing rare type I collagen disorders, Arch. Oral Biol. 52 (2) (2007) 101–109.

[747] S. Unger, F. Antoniazzi, M. Brugnara, Y. Alanay, A. Caglayan, K. Lachlan, et al., Clinical and radiographic delineation of odontochondrodysplasia, Am. J. Med. Genet. A 146A (6) (2008) 770–778.

[748] J. Goldblatt, P. Carman, P. Sprague, Unique dwarfing, spondylometaphyseal skeletal dysplasia, with joint laxity and dentinogenesis imperfecta, Am. J. Med. Genet. 39 (2) (1991) 170–172.

[749] M. Castori, P. Cascone, M. Valiante, L. Laino, G. Iannetti, R.C. Hennekam, et al., Elsahy–Waters syndrome: evidence for autosomal recessive inheritance, Am. J. Med. Genet. A 152A (11) (2010) 2810–2815.
[750] T. Sreenath, T. Thyagarajan, B. Hall, G. Longenecker, R. D'Souza, S. Hong, et al., Dentin sialophosphoprotein knockout mouse teeth display widened predentin zone and develop defective dentin mineralization similar to human dentinogenesis imperfecta type III, J. Biol. Chem. 278 (27) (2003) 24874–24880.
[751] S. Suzuki, T. Sreenath, N. Haruyama, C. Honeycutt, A. Terse, A. Cho, et al., Dentin sialoprotein and dentin phosphoprotein have distinct roles in dentin mineralization, Matrix Biol. 28 (4) (2009) 221–229.
[752] S. Xiao, C. Yu, X. Chou, W. Yuan, Y. Wang, L. Bu, et al., Dentinogenesis imperfecta 1 with or without progressive hearing loss is associated with distinct mutations in DSPP, Nat. Genet. 27 (2) (2001) 201–204.
[753] M.J. Barron, S.T. McDonnell, I. Mackie, M.J. Dixon, Hereditary dentine disorders: dentinogenesis imperfecta and dentine dysplasia, Orphanet J. Rare Dis. 3 (2008) 31.
[754] Guideline on oral health care/dental management of heritable dental development anomalies, Pediatr. Dent. 30 (Suppl. 7) (2008) 196–201.
[755] P.S. Hart, T.C. Hart, Disorders of human dentin, Cells Tissues Organs 186 (1) (2007) 70–77.
[756] J.W. Kim, J.P. Simmer, Hereditary dentin defects, J. Dent. Res. 86 (5) (2007) 392–399.
[757] B. Malmgren, S. Lindskog, A. Elgadi, S. Norgren, Clinical, histopathologic, and genetic investigation in two large families with dentinogenesis imperfecta type II, Hum. Genet. 114 (5) (2004) 491–498.
[758] D.A. McKnight, P. Suzanne Hart, T.C. Hart, J.K. Hartsfield, A. Wilson, J.T. Wright, et al., A comprehensive analysis of normal variation and disease-causing mutations in the human DSPP gene, Hum. Mutat. 29 (12) (2008) 1392–1404.
[759] E.D. Shields, D. Bixler, A.M. el-Kafrawy, A proposed classification for heritable human dentine defects with a description of a new entity, Arch. Oral Biol. 18 (4) (1973) 543–553.
[760] Y. Song, C. Wang, B. Peng, X. Ye, G. Zhao, M. Fan, et al., Phenotypes and genotypes in 2 DGI families with different DSPP mutations, Oral Surg. Oral Med. Oral Pathol. Oral Radiol. Endod. 102 (3) (2006) 360–374.
[761] Y. Yamakoshi, J.C. Hu, M. Fukae, H. Zhang, J.P. Simmer, Dentin glycoprotein: the protein in the middle of the dentin sialophosphoprotein chimera, J. Biol. Chem. 280 (17) (2005) 17472–17479.
[762] X. Zhang, J. Zhao, C. Li, S. Gao, C. Qiu, P. Liu, et al., DSPP mutation in dentinogenesis imperfecta Shields type II, Nat. Genet. 27 (2) (2001) 151–152.
[763] T.C. Hart, P.S. Hart, Genetic studies of craniofacial anomalies: clinical implications and applications, Orthod. Craniofac. Res. 12 (3) (2009) 212–220.
[764] S.K. Lee, K.E. Lee, Y.H. Hwang, M. Kida, T. Tsutsumi, T. Ariga, et al., Identification of the DSPP mutation in a new kindred and phenotype–genotype correlation, Oral Dis. 17 (3) (2011) 314–319.
[765] P. Nieminen, L. Papagiannoulis-Lascarides, J. Waltimo-Siren, P. Ollila, S. Karjalainen, S. Arte, et al., Frameshift mutations in dentin phosphoprotein and dependence of dentin disease phenotype on mutation location, J. Bone Miner. Res. 26 (4) (2011) 873–880.
[766] M. MacDougall, J. Dong, Dentinogenesis imperfecta type IIINORD, (Ed.), NORD Guide to Rare Disorders, Lippincott Williams & Wilkins, Philadelphia, PA, 2003, pp. 177–178.

[767] J.W. Kim, J.C. Hu, J.I. Lee, S.K. Moon, Y.J. Kim, K.T. Jang, et al., Mutational hot spot in the DSPP gene causing dentinogenesis imperfecta type II, Hum. Genet. 116 (3) (2005) 186–191.

[768] J. Dong, T. Gu, L. Jeffords, M. MacDougall, Dentin phosphoprotein compound mutation in dentin sialophosphoprotein causes dentinogenesis imperfecta type III, Am. J. Med. Genet. A 132 (3) (2005) 305–309.

[769] M. MacDougall, L.G. Jeffords, T.T. Gu, C.B. Knight, G. Frei, B.E. Reus, et al., Genetic linkage of the dentinogenesis imperfecta type III locus to chromosome 4q, J. Dent. Res. 78 (6) (1999) 1277–1282.

[770] L.S. Levin, S.H. Leaf, R.J. Jelmini, J.J. Rose, K.N. Rosenbaum, Dentinogenesis imperfecta in the Brandywine isolate (DI type III): clinical, radiologic, and scanning electron microscopic studies of the dentition, Oral Surg. Oral Med. Oral Pathol. 56 (3) (1983) 267–274.

[771] C.J. Witkop, Jr., C.J. MacLean, P.J. Schmidt, J.L. Henry, Medical and dental findings in the Brandywine isolate, Ala. J. Med. Sci. 3 (4) (1966) 382–403.

[772] M.L. Beattie, J.W. Kim, S.G. Gong, C.A. Murdoch-Kinch, J.P. Simmer, J.C. Hu, Phenotypic variation in dentinogenesis imperfecta/dentin dysplasia linked to 4q21, J. Dent. Res. 85 (4) (2006) 329–333.

[773] S.K. Lee, J.C. Hu, K.E. Lee, J.P. Simmer, J.W. Kim, A dentin sialophosphoprotein mutation that partially disrupts a splice acceptor site causes type II dentin dysplasia, J. Endod. 34 (12) (2008) 1470–1473.

[774] D.A. McKnight, J.P. Simmer, P.S. Hart, T.C. Hart, L.W. Fisher, Overlapping DSPP mutations cause dentin dysplasia and dentinogenesis imperfecta, J. Dent. Res. 87 (12) (2008) 1108–1111.

[775] T. Thyagarajan, T. Sreenath, A. Cho, J.T. Wright, A.B. Kulkarni, Reduced expression of dentin sialophosphoprotein is associated with dysplastic dentin in mice overexpressing TGF-β 1 in teeth, J. Biol. Chem. 79 (2000) 11.

[776] T.L. Comer, T.G. Gound, Hereditary pattern for dentinal dysplasia type Id: a case report, Oral Surg. Oral Med. Oral Pathol. Oral Radiol. Endod. 94 (1) (2002) 51–53.

[777] P.E. Shankly, I.C. Mackie, P. Sloan, Dentinal dysplasia type I: report of a case, Int. J. Paediatr. Dent. 9 (1) (1999) 37–42.

[778] A.R. Vieira, A. Modesto, M.G. Cabral, Dentinal dysplasia type I: report of an atypical case in the primary dentition, ASDC J. Dent. Child. 65 (2) (1998) 141–144.

[779] G. Ansari, J.S. Reid, Dentinal dysplasia type I: review of the literature and report of a family, ASDC J. Dent. Child. 64 (6) (1997) 429–434.

[780] M.K. O'Carroll, The diagnosis of dentinal dysplasia Type I, Dentomaxillofac. Radiol. 23 (1) (1994) 52–53.

[781] C.V. Brenneise, K.R. Conway, Dentin dysplasia, type II: report of 2 new families and review of the literature, Oral Surg. Oral Med. Oral Pathol. Oral Radiol. Endod. 87 (6) (1999) 752–755.

[782] M. Melnick, J.R. Eastman, L.I. Goldblatt, M. Michaud, D. Bixler, Dentin dysplasia, type II: a rare autosomal dominant disorder, Oral Surg. Oral Med. Oral Pathol. 44 (4) (1977) 592–599.

[783] J.C. Hu, J.P. Simmer, Developmental biology and genetics of dental malformations, Orthod. Craniofac. Res 10 (2) (2007) 45–52.

[784] S. Parekh, A. Kyriazidou, A. Bloch-Zupan, G. Roberts, Multiple pulp stones and shortened roots of unknown etiology, Oral Surg. Oral Med. Oral Pathol. Oral Radiol. Endod. 101 (6) (2006) e139–e142.

[785] C.J. Witkop, Jr., Hereditary defects of dentin, Dent. Clin. North Am. 19 (1) (1975) 25–45.

[786] C.J. Witkop, Jr., Amelogenesis imperfecta, dentinogenesis imperfecta and dentin dysplasia revisited: problems in classification, J. Oral Pathol. 17 (9–10) (1988) 547–553.

[787] P.N. Kantaputra, Dentinogenesis imperfecta-associated syndromes, Am. J. Med. Genet. 104 (1) (2001) 75–78.

[788] J. Bonaventure, R. Stanescu, V. Stanescu, J.C. Allain, M.P. Muriel, D. Ginisty, et al., Type II collagen defect in two sibs with the Goldblatt syndrome, a chondrodysplasia with dentinogenesis imperfecta, and joint laxity, Am. J. Med. Genet. 44 (6) (1992) 738–753.

[789] M.A. da Fonseca, Dental findings in the Schimke immuno-osseous dysplasia, Am. J. Med. Genet. 93 (2) (2000) 158–160.

[790] D.L. Wedgwood, J.B. Curran, C.L. Lavelle, J.R. Trott, Cranio-facial and dental anomalies in the Branchio-Skeleto-Genital (BSG) syndrome with suggestions for more appropriate nomenclature, Br. J. Oral Surg. 21 (2) (1983) 94–102.

[791] P.N. Kantaputra, A newly recognized syndrome of skeletal dysplasia with opalescent and rootless teeth, Oral Surg. Oral Med. Oral Pathol. Oral Radiol. Endod. 92 (3) (2001) 303–307.

[792] P.N. Kantaputra, Apparently new osteodysplastic and primordial short stature with severe microdontia, opalescent teeth, and rootless molars in two siblings, Am. J. Med. Genet. 111 (4) (2002) 420–428.

[793] H.E. Christiansen, U. Schwarze, S.M. Pyott, A. Alswaid, M. Al Balwi, S. Alrasheed, et al., Homozygosity for a missense mutation in SERPINH1, which encodes the collagen chaperone protein HSP47, results in severe recessive osteogenesis imperfecta, Am. J. Hum. Genet. 86 (3) (2010) 389–398.

[794] I. Aubin, C.P. Adams, S. Opsahl, D. Septier, C.E. Bishop, N. Auge, et al., A deletion in the gene encoding sphingomyelin phosphodiesterase 3 (Smpd3) results in osteogenesis and dentinogenesis imperfecta in the mouse, Nat. Genet. 37 (8) (2005) 803–805.

[795] M. Goldberg, S. Opsahl, I. Aubin, D. Septier, C. Chaussain-Miller, A. Boskey, et al., Sphingomyelin degradation is a key factor in dentin and bone mineralization: lessons from the fro/fro mouse. The chemistry and histochemistry of dentin lipids, J. Dent. Res. 87 (1) (2008) 9–13.

[796] G.E. Lopez Franco, A. Huang, N. Pleshko Camacho, R.D. Blank, Dental phenotype of the col1a2(oim) mutation: DI is present in both homozygotes and heterozygotes, Bone 36 (6) (2005) 1039–1046.

[797] J.R. Shapiro, D.J. McBride, Jr., N.S. Fedarko, OIM and related animal models of osteogenesis imperfecta, Connect. Tissue Res. 31 (4) (1995) 265–268.

[798] A.S. Kamoun-Goldrat, M.F. Le Merrer, Animal models of osteogenesis imperfecta and related syndromes, J. Bone Miner. Metab. 25 (4) (2007) 211–218.

[799] T. Binger, M. Rucker, W.J. Spitzer, Dentofacial rehabilitation by osteodistraction, augmentation and implantation despite osteogenesis imperfecta, Int. J. Oral Maxillofac. Surg. 35 (6) (2006) 559–562.

[800] R.K. Hall, M.C. Maniere, J. Palamara, J. Hemmerle, Odontoblast dysfunction in osteogenesis imperfecta: an LM, SEM, and ultrastructural study, Connect. Tissue Res. 43 (2–3) (2002) 401–405.

[801] J. Kindelan, M. Tobin, D. Roberts-Harry, R.A. Loukota, Orthodontic and orthognathic management of a patient with osteogenesis imperfecta and dentinogenesis imperfecta: a case report, J. Orthod. 30 (4) (2003) 291–296.

[802] M. Koreeda-Miura, T. Onishi, T. Ooshima, Significance of histopathologic examination in the diagnosis of dentin defects associated with type IV osteogenesis imperfecta: two case reports, Oral Surg. Oral Med. Oral Pathol. Oral Radiol. Endod. 95 (1) (2003) 85–89.

[803] C.Y. Lee, S.K. Ertel, Bone graft augmentation and dental implant treatment in a patient with osteogenesis imperfecta: review of the literature with a case report, Implant Dent. 12 (4) (2003) 291–295.

[804] E. Madenci, K. Yilmaz, M. Yilmaz, Y. Coskun, Alendronate treatment in osteogenesis imperfecta, J. Clin. Rheumatol. 12 (2) (2006) 53–56.

[805] B. Malmgren, S. Norgren, Dental aberrations in children and adolescents with osteogenesis imperfecta, Acta Odontol. Scand. 60 (2) (2002) 65–71.

[806] B. Malmgren, S. Lindskog, Assessment of dysplastic dentin in osteogenesis imperfecta and dentinogenesis imperfecta, Acta Odontol. Scand. 61 (2) (2003) 72–80.

[807] D. Pallos, P.S. Hart, J.R. Cortelli, S. Vian, J.T. Wright, J. Korkko, et al., Novel COL1A1 mutation (G559C) [correction of G599C] associated with mild osteogenesis imperfecta and dentinogenesis imperfecta, Arch. Oral Biol. 46 (5) (2001) 459–470.

[808] D. Rios, A.L. Vieira, L.M. Tenuta, M.A. Machado, Osteogenesis imperfecta and dentinogenesis imperfecta: associated disorders, Quintessence Int. 36 (9) (2005) 695–701.

[809] P.J. Roughley, F. Rauch, F.H. Glorieux, Osteogenesis imperfecta – clinical and molecular diversity, Eur. Cell Mater. 5 (2003) 41–47; discussion 47.

[810] K. Sanches, A.M. de Queiroz, A.C. de Freitas, K.V. Serrano, Clinical features, dental findings and dental care management in osteogenesis imperfecta. J. Clin, Pediatr. Dent. 30 (1) (2005) 77–82.

[811] J. Waltimo-Siren, M. Kolkka, S. Pynnonen, K. Kuurila, I. Kaitila, O. Kovero, Craniofacial features in osteogenesis imperfecta: a cephalometric study, Am. J. Med. Genet. A 133A (2) (2005) 142–150.

[812] J.C. Marini, A. Forlino, W.A. Cabral, A.M. Barnes, J.D. San Antonio, S. Milgrom, et al., Consortium for osteogenesis imperfecta mutations in the helical domain of type I collagen: regions rich in lethal mutations align with collagen binding sites for integrins and proteoglycans, Hum. Mutat. 28 (3) (2007) 209–221.

[813] D. Basel, R.D. Steiner, Osteogenesis imperfecta: recent findings shed new light on this once well-understood condition, Genet. Med. 11 (6) (2009) 375–385.

[814] G. Baujat, A.S. Lebre, V. Cormier-Daire, M. Le Merrer, [Osteogenesis imperfecta, diagnosis information (clinical and genetic classification)], Arch. Pediatr. 15 (5) (2008) 789–791.

[815] A. Kamoun-Goldrat, D. Ginisty, M. Le Merrer, Effects of bisphosphonates on tooth eruption in children with osteogenesis imperfecta, Eur. J. Oral Sci. 116 (3) (2008) 195–198.

[816] A.S. Kamoun-Goldrat, M.F. Le Merrer, [Osteogenesis imperfecta and dentinogenesis imperfecta: diagnostic frontiers and importance in dentofacial orthopedics], Orthod. Fr. 78 (2) (2007) 89–99.

[817] M.S. Cheung, F.H. Glorieux, Osteogenesis Imperfecta: update on presentation and management, Rev. Endocr. Metab. Disord. 9 (2) (2008) 153–160.

[818] F.H. Glorieux, Osteogenesis imperfecta, Best Pract. Res. Clin. Rheumatol. 22 (1) (2008) 85–100.

[819] F.S. Van Dijk, I.M. Nesbitt, P.G. Nikkels, A. Dalton, E.M. Bongers, J.M. van de Kamp, et al., CRTAP mutations in lethal and severe osteogenesis imperfecta: the importance of combining biochemical and molecular genetic analysis, Eur. J. Hum. Genet. 17 (12) (2009) 1560–1569.

[820] F.S. van Dijk, I.M. Nesbitt, P.G. Nikkels, A. Dalton, E.M. Bongers, J.M. van de Kamp, et al., PPIB mutations cause severe osteogenesis imperfecta, Am. J. Hum. Genet. 85 (4) (2009) 521–527.

[821] F. Rauch, L. Lalic, P. Roughley, F.H. Glorieux, Genotype-phenotype correlations in nonlethal Osteogenesis imperfecta caused by mutations in the helical domain of collagen type I, Eur. J. Hum. Genet. 18 (6) (2010) 642–647.
[822] E.P. Homan, F. Rauch, I. Grafe, C. Lietman, J.A. Doll, B. Dawson, et al., Mutations in SERPINF1 cause Osteogenesis imperfecta Type VI, J. Bone Miner. Res. 26 (12) (2011) 2798–2803.
[823] P. Lapunzina, M. Aglan, S. Temtamy, J.A. Caparros-Martin, M. Valencia, R. Leton, et al., Identification of a frameshift mutation in Osterix in a patient with recessive osteogenesis imperfecta, Am J Hum Genet 87 (1) (2010) 110–114.
[824] J. O'Sullivan, C.C. Bitu, S.B. Daly, J.E. Urquhart, M.J. Barron, S.S. Bhaskar, et al., Whole-exome sequencing identifies FAM20A mutations as a cause of amelogenesis imperfecta and gingival hyperplasia syndrome, Am. J. Hum. Genet. 88 (5) (2011) 616–620.
[825] G.B. Winter, K.W. Lee, N.W. Johnson, Hereditary amelogenesis imperfecta. A rare autosomal dominant type, Br. Dent. J. 127 (4) (1969) 157–164.
[826] S. Poulsen, H. Gjorup, D. Haubek, G. Haukali, H. Hintze, H. Lovschall, et al., Amelogenesis imperfecta – a systematic literature review of associated dental and orofacial abnormalities and their impact on patients, Acta Odontol. Scand. 66 (4) (2008) 193–199.
[827] J.T. Wright, B. Daly, D. Simmons, S. Hong, S.P. Hart, T.C. Hart, et al., Human enamel phenotype associated with amelogenesis imperfecta and a kallikrein-4 (g.2142 G > A) proteinase mutation, Eur. J. Oral Sci. 114 (Suppl. 1) (2006) 13–17, discussion 39–41, 379
[828] S. Hart, T. Hart, C. Gibson, J.T. Wright, Mutational analysis of X-linked amelogenesis imperfecta in multiple families, Arch. Oral Biol. 45 (1) (2000) 79–86.
[829] P.S. Hart, T.C. Hart, J.P. Simmer, J.T. Wright, A nomenclature for X-linked amelogenesis imperfecta, Arch. Oral Biol. 47 (4) (2002) 255–260.
[830] P.S. Hart, M.J. Aldred, P.J. Crawford, N.J. Wright, T.C. Hart, J.T. Wright, Amelogenesis imperfecta phenotype-genotype correlations with two amelogenin gene mutations, Arch. Oral Biol. 47 (4) (2002) 261–265.
[831] P.S. Hart, M.D. Michalec, W.K. Seow, T.C. Hart, J.T. Wright, Identification of the enamelin (g.8344delG) mutation in a new kindred and presentation of a standardized ENAM nomenclature, Arch. Oral Biol. 48 (8) (2003) 589–596.
[832] T.C. Hart, P.S. Hart, M.C. Gorry, M.D. Michalec, O.H. Ryu, C. Uygur, et al., Novel ENAM mutation responsible for autosomal recessive amelogenesis imperfecta and localised enamel defects, J. Med. Genet. 40 (12) (2003) 900–906.
[833] P.S. Hart, T.C. Hart, M.D. Michalec, O.H. Ryu, D. Simmons, S. Hong, et al., Mutation in kallikrein 4 causes autosomal recessive hypomaturation amelogenesis imperfecta, J. Med. Genet. 41 (7) (2004) 545–549.
[834] P.J. Crawford, M. Aldred, A. Bloch-Zupan, Amelogenesis imperfecta, Orphanet J. Rare Dis. 2 (2007) 17.
[835] M.H. Rajpar, K. Harley, C. Laing, R.M. Davies, M.J. Dixon, Mutation of the gene encoding the enamel-specific protein, enamelin, causes autosomal-dominant amelogenesis imperfecta, Hum. Mol. Genet. 10 (16) (2001) 1673–1677.
[836] J.W. Kim, F. Seymen, B.P. Lin, B. Kiziltan, K. Gencay, J.P. Simmer, et al., ENAM mutations in autosomal-dominant amelogenesis imperfecta, J. Dent. Res. 84 (3) (2005) 278–282.
[837] J.W. Kim, J.P. Simmer, T.C. Hart, P.S. Hart, M.D. Ramaswami, J.D. Bartlett, et al., MMP-20 mutation in autosomal recessive pigmented hypomaturation amelogenesis imperfecta, J. Med. Genet. 42 (3) (2005) 271–275.

[838] J. Dong, D. Amor, M.J. Aldred, T. Gu, M. Escamilla, M. MacDougall, et al., DLX3 mutation associated with autosomal dominant amelogenesis imperfecta with taurodontism, Am. J. Med. Genet. A 133 (2) (2005) 138–141.

[839] G. Mendoza, T.J. Pemberton, K. Lee, R. Scarel-Caminaga, R. Mehrian-Shai, C. Gonzalez-Quevedo, et al., A new locus for autosomal dominant amelogenesis imperfecta on chromosome 8q24.3, Hum. Genet. 120 (5) (2007) 653–662.

[840] J.W. Kim, S.K. Lee, Z.H. Lee, J.C. Park, K.E. Lee, M.H. Lee, et al., FAM83H mutations in families with autosomal-dominant hypocalcified amelogenesis imperfecta, Am. J. Hum. Genet. 82 (2) (2008) 489–494.

[841] S.K. Lee, Mutational spectrum of FAM83H: the C-terminal portion is required for tooth enamel calcification, Hum. Mutat. 29 (8) (2008) E95–E99.

[842] W. El-Sayed, J.C. Hu, J.D. Bartlett, K.E. Lee, B.P. Lin, J.P. Simmer, et al., Mutations in the beta propeller WDR72 cause autosomal-recessive hypomaturation amelogenesis imperfecta, Am. J. Hum. Genet. 85 (5) (2009) 699–705.

[843] E.C. Lau, T.K. Mohandas, L.J. Shapiro, H.C. Slavkin, M.L. Snead, Human and mouse amelogenin gene loci are on the sex chromosomes, Genomics 4 (2) (1989) 162–168.

[844] M. MacDougall, D. Simmons, T.T. Gu, K. Forsman-Semb, C.K. Mardh, M. Mesbah, et al., Cloning, characterization and immunolocalization of human ameloblastin, Eur. J. Oral Sci. 108 (4) (2000) 303–310.

[845] D. Deutsch, E. Fermon, J. Lustmann, L. Dafni, Z. Mao, V. Leytin, Tuftelin mRNA is expressed in a human ameloblastoma tumor, Connect. Tissue Res. 39 (1–3) (1998) 177–184; discussion 187–194.

[846] J.Y. Sire, T. Davit-Beal, S. Delgado, X. Gu, The origin and evolution of enamel mineralization genes, Cells Tissues Organs 186 (1) (2007) 25–48.

[847] C.W. Gibson, Z.A. Yuan, B. Hall, G. Longenecker, E. Chen, T. Thyagarajan, et al., Amelogenin-deficient mice display an amelogenesis imperfecta phenotype, J. Biol. Chem. 276 (34) (2001) 31871–31875.

[848] S.K. Prakash, C.W. Gibson, J.T. Wright, C. Boyd, T. Cormier, R. Sierra, et al., Tooth enamel defects in mice with a deletion at the Arhgap 6/Amel X locus, Calcif. Tissue Int. 77 (1) (2005) 23–29.

[849] Y. Li, Z.A. Yuan, M.A. Aragon, A.B. Kulkarni, C.W. Gibson, Comparison of body weight and gene expression in amelogenin null and wild-type mice, Eur. J. Oral Sci. 114 (Suppl. 1) (2006) 190–193; discussion 201–202, 381.

[850] J.C. Hu, Y.H. Chun, T. Al Hazzazzi, J.P. Simmer, Enamel formation and amelogenesis imperfecta, Cells Tissues Organs 186 (1) (2007) 78–85.

[851] D.B. Ravassipour, P.S. Hart, T.C. Hart, A.V. Ritter, M. Yamauchi, C. Gibson, et al., Unique enamel phenotype associated with amelogenin gene (AMELX) codon 41 point mutation, J. Dent. Res. 79 (7) (2000) 1476–1481.

[852] B. Richard, S. Delgado, P. Gorry, J.Y. Sire, A study of polymorphism in human AMELX, Arch. Oral Biol. 52 (11) (2007) 1026–1031.

[853] J.T. Wright, T.C. Hart, P.S. Hart, D. Simmons, C. Suggs, B. Daley, et al., Human and mouse enamel phenotypes resulting from mutation or altered expression of AMEL, ENAM, MMP20 and KLK4, Cells Tissues Organs 189 (1–4) (2009) 224–229.

[854] I. Bailleul-Forestier, M. Molla, A. Verloes, A. Berdal, The genetic basis of inherited anomalies of the teeth. Part 1: clinical and molecular aspects of non-syndromic dental disorders, Eur. J. Med. Genet. 51 (4) (2008) 273–291.

[855] S.A. Kindelan, A.H. Brook, L. Gangemi, N. Lench, F.S. Wong, J. Fearne, et al., Detection of a novel mutation in X-linked amelogenesis imperfecta, J. Dent. Res. 79 (12) (2000) 1978–1982.

[856] P.J. Crawford, M.J. Aldred, X-linked amelogenesis imperfecta. Presentation of two kindreds and a review of the literature, Oral Surg. Oral Med. Oral Pathol. 73 (4) (1992) 449–455.

[857] M. Lagerstrom, N. Dahl, Y. Nakahori, Y. Nakagome, B. Backman, U. Landegren, et al., A deletion in the amelogenin gene (AMG) causes X-linked amelogenesis imperfecta (AIH1), Genomics 10 (4) (1991) 971–975.

[858] M. Lagerstrom-Fermer, M. Nilsson, B. Backman, E. Salido, L. Shapiro, U. Pettersson, et al., Amelogenin signal peptide mutation: correlation between mutations in the amelogenin gene (AMGX) and manifestations of X-linked amelogenesis imperfecta, Genomics 26 (1) (1995) 159–162.

[859] M. Lagerstrom-Fermer, U. Landegren, Understanding enamel formation from mutations causing X-linked amelogenesis imperfecta, Connect. Tissue Res. 32 (1–4) (1995) 241–246.

[860] N.J. Lench, G.B. Winter, Characterisation of molecular defects in X-linked amelogenesis imperfecta (AIH1), Hum. Mutat. 5 (3) (1995) 251–259.

[861] H. Seedorf, M. Klaften, F. Eke, H. Fuchs, U. Seedorf, M. Hrabe de Angelis, A mutation in the enamelin gene in a mouse model, J. Dent. Res. 86 (8) (2007) 764–768.

[862] H. Masuya, K. Shimizu, H. Sezutsu, Y. Sakuraba, J. Nagano, A. Shimizu, et al., Enamelin (Enam) is essential for amelogenesis: ENU-induced mouse mutants as models for different clinical subtypes of human amelogenesis imperfecta (AI), Hum. Mol. Genet. 14 (5) (2005) 575–583.

[863] H. Seedorf, I.N. Springer, E. Grundner-Culemann, H.K. Albers, A. Reis, H. Fuchs, et al., Amelogenesis imperfecta in a new animal model – a mutation in chromosome 5 (human 4q21), J. Dent. Res. 83 (8) (2004) 608–612.

[864] J. Dong, T.T. Gu, D. Simmons, M. MacDougall, Enamelin maps to human chromosome 4q21 within the autosomal dominant amelogenesis imperfecta locus, Eur. J. Oral Sci. 108 (5) (2000) 353–358.

[865] M. Kida, T. Ariga, T. Shirakawa, H. Oguchi, Y. Sakiyama, Autosomal-dominant hypoplastic form of amelogenesis imperfecta caused by an enamelin gene mutation at the exon–intron boundary, J. Dent. Res. 81 (11) (2002) 738–742.

[866] C.K. Mardh, B. Backman, G. Holmgren, J.C. Hu, J.P. Simmer, K. Forsman-Semb, A nonsense mutation in the enamelin gene causes local hypoplastic autosomal dominant amelogenesis imperfecta (AIH2), Hum. Mol. Genet. 11 (9) (2002) 1069–1074.

[867] S.J. Gutierrez, M. Chaves, D.M. Torres, I. Briceno, Identification of a novel mutation in the enamelin gene in a family with autosomal-dominant amelogenesis imperfecta, Arch. Oral Biol. 52 (5) (2007) 503–506.

[868] V.K. Gopinath, T.P. Yoong, C.Y. Yean, M. Ravichandran, Identifying polymorphism in enamelin gene in amelogenesis imperfecta (AI), Arch. Oral Biol. 53 (10) (2008) 937–940.

[869] D. Ozdemir, P.S. Hart, E. Firatli, G. Aren, O.H. Ryu, T.C. Hart, Phenotype of ENAM mutations is dosage-dependent, J. Dent. Res. 84 (11) (2005) 1036–1041.

[870] C.C. Hu, T.C. Hart, B.R. Dupont, J.J. Chen, X. Sun, Q. Qian, et al., Cloning human enamelin cDNA, chromosomal localization, and analysis of expression during tooth development, J. Dent. Res. 79 (4) (2000) 912–919.

[871] A. Pavlic, M. Petelin, T. Battelino, Phenotype and enamel ultrastructure characteristics in patients with ENAM gene mutations g.13185-13186insAG and 8344delG, Arch. Oral Biol. 52 (3) (2007) 209–217.

[872] Y. Ding, M.R. Estrella, Y.Y. Hu, H.L. Chan, H.D. Zhang, J.W. Kim, et al., Fam83h is associated with intracellular vesicles and ADHCAI, J. Dent. Res. 88 (11) (2009) 991–996.

[873] J.T. Wright, S. Frazier-Bowers, D. Simmons, K. Alexander, P. Crawford, S.T. Han, et al., Phenotypic variation in FAM83H-associated amelogenesis imperfecta, J. Dent. Res. 88 (4) (2009) 356–360.

[874] P.S. Hart, S. Becerik, D. Cogulu, G. Emingil, D. Ozdemir-Ozenen, S.T. Han, et al., Novel FAM83H mutations in Turkish families with autosomal dominant hypocalcified amelogenesis imperfecta, Clin. Genet. 75 (4) (2009) 401–404.

[875] H.K. Hyun, S.K. Lee, K.E. Lee, H.Y. Kang, E.J. Kim, P.H. Choung, et al., Identification of a novel FAM83H mutation and microhardness of an affected molar in autosomal dominant hypocalcified amelogenesis imperfecta, Int. Endod. J. 42 (11) (2009) 1039–1043.

[876] J.D. Bartlett, D.H. Skobe, J.T. Lee, Y. Wright, A.B. Li, C.W. Kulkarni, et al., A developmental comparison of matrix metalloproteinase-20 and amelogenin null mouse enamel, Eur. J. Oral Sci. 114 (Suppl. 1) (2006) 18–23; discussion 39–41, 379.

[877] D. Bouvier, J.P. Duprez, D. Bois, Rehabilitation of young patients with amelogenesis imperfecta: a report of two cases, ASDC J. Dent. Child. 63 (6) (1996) 443–447.

[878] D. Bouvier, J.P. Duprez, C. Pirel, B. Vincent, Amelogenesis imperfecta – a prosthetic rehabilitation: A clinical report, J. Prosthet Dent. 82 (2) (1999) 130–131.

[879] I. Kostoulas, S. Kourtis, D. Andritsakis, A. Doukoudakis, Functional and esthetic rehabilitation in amelogenesis imperfecta with all-ceramic restorations: a case report, Quintessence Int. 36 (5) (2005) 329–338.

[880] C. Sabatini, S. Guzman-Armstrong, A conservative treatment for amelogenesis imperfecta with direct resin composite restorations: a case report, J. Esthet. Restor. Dent. 21 (3) (2009) 161–169; discussion 170.

[881] H. Gjorup, D. Haubek, H. Hintze, G. Haukali, H. Lovschall, J.M. Hertz, et al., Hypocalcified type of amelogenesis imperfecta in a large family: clinical, radiographic, and histological findings, associated dento-facial anomalies, and resulting treatment load, Acta Odontol. Scand. (2009) 1–8.

[882] S.M. Stines, Treatment of hypomaturation-type amelogenesis imperfecta with indirect no-preparation resin veneers fabricated with CAD/CAM Cerec 3D, version 3.03, Int. J. Comput. Dent. 11 (1) (2008) 41–50.

[883] A.P. Pires Dos Santos, C.M. Cabral, L.F. Moliterno, B.H. Oliveira, Amelogenesis imperfecta: report of a successful transitional treatment in the mixed dentition, J. Dent. Child. (Chic) 75 (2) (2008) 201–206.

[884] D. Gemalmaz, F. Isik, A. Keles, D. Kuker, Use of adhesively inserted full-ceramic restorations in the conservative treatment of amelogenesis imperfecta: a case report, J. Adhes. Dent. 5 (3) (2003) 235–242.

[885] F.G. Robinson, J.E. Haubenreich, Oral rehabilitation of a young adult with hypoplastic amelogenesis imperfecta: a clinical report, J. Prosthet. Dent. 95 (1) (2006) 10–13.

[886] L.S. Turkun, Conservative restoration with resin composites of a case of amelogenesis imperfecta, Int. Dent. J. 55 (1) (2005) 38–41.

[887] B.E. Turk, D.H. Lee, Y. Yamakoshi, A. Klingenhoff, E. Reichenberger, J.T. Wright, et al., MMP-20 is predominately a tooth-specific enzyme with a deep catalytic pocket that hydrolyzes type V collagen, Biochemistry 45 (12) (2006) 3863–3874.

[888] P. Papagerakis, H.K. Lin, K.Y. Lee, Y. Hu, J.P. Simmer, J.D. Bartlett, J.C. Hu, et al., Premature stop codon in MMP20 causing amelogenesis imperfecta, J. Dent. Res. 87 (1) (2008) 56–59.

[889] D. Ozdemir, P.S. Hart, O.H. Ryu, S.J. Choi, M. Ozdemir-Karatas, E. Firatli, et al., MMP20 active-site mutation in hypomaturation amelogenesis imperfecta, J. Dent. Res. 84 (11) (2005) 1031–1035.

[890] G. Stephanopoulos, M.E. Garefalaki, K. Lyroudia, Genes and related proteins involved in amelogenesis imperfecta, J. Dent. Res. 84 (12) (2005) 1117–1126.

[891] M.J. Aldred, P.J. Crawford, E. Roberts, C.M. Gillespie, N.S. Thomas, I. Fenton, et al., Genetic heterogeneity in X-linked amelogenesis imperfecta, Genomics 14 (3) (1992) 567–573.

[892] S.K. Lee, Z.H. Lee, S.J. Lee, B.D. Ahn, Y.J. Kim, S.H. Lee, et al., DLX3 mutation in a new family and its phenotypic variations, J. Dent. Res. 87 (4) (2008) 354–357.

[893] M.I. Morasso, A. Grinberg, G. Robinson, T.D. Sargent, K.A. Mahon, Placental failure in mice lacking the homeobox gene Dlx3, Proc. Natl. Acad. Sci. U.S.A. 96 (1) (1999) 162–167.

[894] J. Hwang, T. Mehrani, S.E. Millar, M.I. Morasso, Dlx3 is a crucial regulator of hair follicle differentiation and cycling, Development 135 (18) (2008) 3149–3159.

[895] P.J. Crawford, R.D. Evans, M.J. Aldred, Amelogenesis imperfecta: autosomal dominant hypomaturation-hypoplasia type with taurodontism, Br. Dent. J. 164 (3) (1988) 71–73.

[896] S.J. Choi, G.D. Roodman, J.Q. Feng, I.S. Song, K. Amin, P.S. Hart, et al., In vivo impact of a 4 bp deletion mutation in the *DLX3* gene on bone development, Dev. Biol. 325 (1) (2009) 129–137.

[897] J.T. Wright, S.P. Hong, D. Simmons, B. Daly, D. Uebelhart, H.U. Luder, DLX3 c.561_562delCT mutation causes attenuated phenotype of tricho-dento-osseous syndrome, Am. J. Med. Genet. A 146 (3) (2008) 343–349.

[898] F. Lezot, B. Thomas, S.R. Greene, D. Hotton, Z.A. Yuan, B. Castaneda, et al., Physiological implications of DLX homeoproteins in enamel formation, J. Cell. Physiol. 216 (3) (2008) 688–697.

[899] M. Islam, A.G. Lurie, E. Reichenberger, Clinical features of tricho-dento-osseous syndrome and presentation of three new cases: an addition to clinical heterogeneity, Oral Surg. Oral Med. Oral Pathol. Oral Radiol. Endod. 100 (6) (2005) 736–742.

[900] R.J. Haldeman, L.F. Cooper, T.C. Hart, C. Phillips, C. Boyd, G.E. Lester, et al., Increased bone density associated with DLX3 mutation in the tricho-dento-osseous syndrome, Bone 35 (4) (2004) 988–997.

[901] A.P. Dodds, S.A. Cox, C.A. Suggs, C. Boyd, R. Ruiz, T.C. Hart, et al., Characterization and mRNA expression in an unusual odontogenic lesion in a patient with tricho-dento-osseous syndrome, Histol. Histopathol. 18 (3) (2003) 849–854.

[902] J.A. Price, J.T. Wright, K. Kula, D.W. Bowden, T.C. Hart, A common *DLX3* gene mutation is responsible for tricho-dento-osseous syndrome in Virginia and North Carolina families, J. Med. Genet. 35 (10) (1998) 825–828.

[903] S. Ghoul-Mazgar, D. Hotton, F. Lezot, C. Blin-Wakkach, A. Asselin, J.M. Sautier, et al., Expression pattern of Dlx3 during cell differentiation in mineralized tissues, Bone 37 (6) (2005) 799–809.

[904] J.A. Price, D.W. Bowden, J.T. Wright, M.J. Pettenati, T.C. Hart, Identification of a mutation in DLX3 associated with tricho-dento-osseous (TDO) syndrome, Hum. Mol. Genet. 7 (3) (1998) 563–569.

[905] J.A. Price, J.T. Wright, S.J. Walker, P.J. Crawford, M.J. Aldred, T.C. Hart, et al., Tricho-dento-osseous syndrome and amelogenesis imperfecta with taurodontism are genetically distinct conditions, Clin. Genet. 56 (1) (1999) 35–40.

[906] T.C. Hart, D.W. Bowden, J. Bolyard, K. Kula, K. Hall, J.T. Wright, et al., Genetic linkage of the tricho-dento-osseous syndrome to chromosome 17q21, Hum. Mol. Genet. 6 (13) (1997) 2279–2284.

[907] J.T. Wright, K. Kula, K. Hall, J.H. Simmons, T.C. Hart, Analysis of the tricho-dento-osseous syndrome genotype and phenotype, Am. J. Med. Genet. 72 (2) (1997) 197–204.

[908] G.R. Ogden, Tricho-dento-osseous syndrome, Ann. Dent. 46 (2) (1987) 12–14.

[909] R. Lolli, G. Addessi, C. Valgiusti, [A clinical case of tricho-dento-osseous syndrome], Riv. Odontostomatol. Implantoprotesi. (2) (1984) 29–32.

[910] S.D. Shapiro, F.L. Quattromani, R.J. Jorgenson, R.S. Young, Tricho-dento-osseous syndrome: heterogeneity or clinical variability, Am. J. Med. Genet. 16 (2) (1983) 225–236.

[911] F. Quattromani, S.D. Shapiro, R.S. Young, R.J. Jorgenson, J.W. Parker, R. Blumhardt, et al., Clinical heterogeneity in the tricho-dento-osseous syndrome, Hum. Genet. 64 (2) (1983) 116–121.

[912] M. Melnick, E.D. Shields, A.H. El-Kafrawy, Tricho-dento-osseous syndrome: a scanning electron microscopic analysis, Clin. Genet. 12 (1) (1977) 17–27.

[913] S. Gulmen, P.A. Pullon, L.W. O'Brien, Tricho-dento-osseous syndrome, J. Endod. 2 (4) (1976) 117–120.

[914] R.J. Jorgenson, R.W. Warson, Dental abnormalities in the tricho-dento-osseous syndrome, Oral Surg. Oral Med. Oral Pathol. 36 (5) (1973) 693–700.

[915] J. Lichtenstein, R. Warson, R. Jorgenson, J.P. Dorst, V.A. McKusick, The tricho-dento-osseous (TDO) syndrome, Am. J. Hum. Genet. 24 (5) (1972) 569–582.

[916] K. Kula, K. Hall, T. Hart, J.T. Wright, Craniofacial morphology of the tricho-dento-osseous syndrome, Clin. Genet. 50 (6) (1996) 446–454.

[917] P. Nieminen, P.L. Lukinmaa, H. Alapulli, M. Methuen, T. Suojarvi, S. Kivirikko, et al., DLX3 homeodomain mutations cause tricho-dento-osseous syndrome with novel phenotypes, Cells Tissues Organs 194 (1) (2011) 49–59.

[918] H. Koshiba, O. Kimura, M. Nakata, C.J. Witkop, Jr., Clinical, genetic, and histologic features of the trichoonychodental (TOD) syndrome, Oral Surg. Oral Med. Oral Pathol. 46 (3) (1978) 376–385.

[919] I. Normand de la Tranchade, H. Bonarek, J.M. Marteau, M.J. Boileau, J. Nancy, Amelogenesis imperfecta and nephrocalcinosis: a new case of this rare syndrome, J. Clin. Pediatr. Dent. 27 (2) (2003) 171–175.

[920] L.M. Paula, N.S. Melo, E.N. Silva Guerra, D.H. Mestrinho, A.C. Acevedo, Case report of a rare syndrome associating amelogenesis imperfecta and nephrocalcinosis in a consanguineous family, Arch. Oral Biol. 50 (2) (2005) 237–242.

[921] E.L. Dellow, K.E. Harley, R.J. Unwin, O. Wrong, G.B. Winter, B.J. Parkins, et al., Amelogenesis imperfecta, nephrocalcinosis, and hypocalciuria syndrome in two siblings from a large family with consanguineous parents, Nephrol. Dial. Transplant 13 (12) (1998) 3193–3196.

[922] R.K. Hall, P. Phakey, J. Palamara, D.A. McCredie, Amelogenesis imperfecta and nephrocalcinosis syndrome. Case studies of clinical features and ultrastructure of tooth enamel in two siblings, Oral Surg. Oral Med. Oral Pathol. Oral Radiol. Endod. 79 (5) (1995) 583–592.

[923] P. Phakey, J. Palamara, R.K. Hall, D.A. McCredie, Ultrastructural study of tooth enamel with amelogenesis imperfecta in AI-nephrocalcinosis syndrome, Connect. Tissue Res. 32 (1–4) (1995) 253–259.

[924] M. Lubinsky, C. Angle, P.W. Marsh, C.J. Witkop, Jr., Syndrome of amelogenesis imperfecta, nephrocalcinosis, impaired renal concentration, and possible abnormality of calcium metabolism, Am. J. Med. Genet. 20 (2) (1985) 233–243.

[925] N. Suda, Y. Kitahara, K. Ohyama, A case of amelogenesis imperfecta, cleft lip and palate and polycystic kidney disease, Orthod. Craniofac. Res. 9 (1) (2006) 52–56.

[926] D. Kessel, C.M. Hall, D.G. Shaw, Two unusual cases of nephrocalcinosis in infancy, Pediatr. Radiol. 22 (6) (1992) 470–471.

[927] D. MacGibbon, Generalized enamel hypoplasia and renal dysfunction, Aust. Dent. J. 17 (1) (1972) 61–63.

[928] Z. Kirzioglu, K.G. Ulu, M.T. Sezer, S. Yuksel, The relationship of amelogenesis imperfecta and nephrocalcinosis syndrome, Med. Oral Patol. Oral Cir. Bucal. 14 (11) (2009) e579–e582.

[929] L. Hunter, L.D. Addy, J. Knox, N. Drage, Is amelogenesis imperfecta an indication for renal examination? Int. J. Paediatr. Dent. 17 (1) (2007) 62–65.

[930] I.K. Jalili, N.J. Smith, A progressive cone-rod dystrophy and amelogenesis imperfecta: a new syndrome, J. Med. Genet. 25 (11) (1988) 738–740.

[931] M. Michaelides, A. Bloch-Zupan, G.E. Holder, D.M. Hunt, A.T. Moore, An autosomal recessive cone-rod dystrophy associated with amelogenesis imperfecta, J. Med. Genet. 41 (6) (2004) 468–473.

[932] D.A. Parry, A.J. Mighell, W. El-Sayed, R.C. Shore, I.K. Jalili, H. Dollfus, et al., Mutations in CNNM4 cause Jalili syndrome, consisting of autosomal-recessive cone-rod dystrophy and amelogenesis imperfecta, Am. J. Hum. Genet. 84 (2) (2009) 266–273.

[933] L.M. Downey, T.J. Keen, I.K. Jalili, J. McHale, M.J. Aldred, S.P. Robertson, et al., Identification of a locus on chromosome 2q11 at which recessive amelogenesis imperfecta and cone-rod dystrophy cosegregate, Eur. J. Hum. Genet. 10 (12) (2002) 865–869.

[934] B. Polok, P. Escher, A. Ambresin, E. Chouery, S. Bolay, I. Meunier, et al., Mutations in CNNM4 cause recessive cone-rod dystrophy with amelogenesis imperfecta, Am. J. Hum. Genet. 84 (2) (2009) 259–265.

[935] D.R. Bertola, R. Antequera, M.J. Rodovalho, R.S. Honjo, L.M. Albano, I.M. Furquim, et al., Brachyolmia with amelogenesis imperfecta: further evidence of a distinct entity, Am. J. Med. Genet. A 149A (3) (2009) 532–534.

[936] R.S. Houlston, G.B. Winter, P.M. Speight, J. Fairhurst, I.K. Temple, Taurodontism and disproportionate short stature, Clin. Dysmorphol. 3 (3) (1994) 251–254.

[937] V. Rao, R.E. Morton, I.D. Young, Spondyloepimetaphyseal dysplasia and abnormal dentition in siblings: a new autosomal recessive syndrome, Clin. Dysmorphol. 6 (1) (1997) 3–12.

[938] A. Verloes, P. Jamblin, L. Koulischer, J.P. Bourguignon, A new form of skeletal dysplasia with amelogenesis imperfecta and platyspondyly, Clin. Genet. 49 (1) (1996) 2–5.

[939] G. Guazzi, S. Palmeri, A. Malandrini, G. Ciacci, R. Di Perri, G. Mancini, et al., Ataxia, mental deterioration, epilepsy in a family with dominant enamel hypoplasia: a variant of Kohlschutter-Tonz syndrome? Am. J. Med. Genet. 50 (1) (1994) 79–83.

[940] M. Petermoller, J. Kunze, G. Gross-Selbeck, Kohlschutter syndrome: syndrome of epilepsy – dementia – amelogenesis imperfecta, Neuropediatrics 24 (6) (1993) 337–338.

[941] A. Kohlschutter, D. Chappuis, C. Meier, O. Tonz, F. Vassella, N. Herschkowitz, et al., Familial epilepsy and yellow teeth – a disease of the CNS associated with enamel hypoplasia, Helv. Paediatr. Acta 29 (4) (1974) 283–294.

[942] J. Christodoulou, R.K. Hall, S. Menahem, I.J. Hopkins, J.G. Rogers, A syndrome of epilepsy, dementia, and amelogenesis imperfecta: genetic and clinical features, J. Med. Genet. 25 (12) (1988) 827–830.

[943] E. Haberlandt, C. Svejda, S. Felber, S. Baumgartner, B. Gunther, G. Utermann, et al., Yellow teeth, seizures, and mental retardation: a less severe case of Kohlschutter-Tonz syndrome, Am. J. Med. Genet. A 140 (3) (2006) 281–283.

[944] J. Zlotogora, A. Fuks, Z. Borochowitz, Y. Tal, Kohlschutter-Tonz syndrome: epilepsy, dementia, and amelogenesis imperfecta, Am. J. Med. Genet. 46 (4) (1993) 453–454.

[945] D. Donnai, P.I. Tomlin, R.M. Winter, Kohlschutter syndrome in siblings, Clin. Dysmorphol. 14 (3) (2005) 123–126.

[946] T. Wygold, G. Kurlemann, G. Schuierer, Kohlschutter syndrome--an example of a rare progressive neuroectodermal disease. Case report and review of the literature, Klin. Padiatr. 208 (5) (1996) 271–275.

[947] S.A. Musumeci, M. Elia, R. Ferri, C. Romano, C. Scuderi, S. Del Gracco, A further family with epilepsy, dementia and yellow teeth: the Kohlschutter syndrome, Brain Dev. 17 (2) (1995) 133–138, discussion 142–133.

[948] C.J. Witkop, Jr., L.J. Brearley, W.C. Gentry, Jr., Hypoplastic enamel, onycholysis, and hypohidrosis inherited as an autosomal dominant trait. A review of ectodermal dysplasia syndromes, Oral Surg. Oral Med. Oral Pathol. 39 (1) (1975) 71–86.

[949] C.J. Witkop, Genetics, SSO Schweiz Monatsschr. Zahnheilkd. 82 (9) (1972) 917–941.

[950] H.G. Jackson, O.P. Leyva, Síndrome amelo-ónico-hipohidrótico asociado con taurodontismo. Reporte de un caso, ADM 44 (1987) 102–104.

[951] K.A. Sastry, A. Ruprecht, A.M. Suliman, Hypoplastic enamel, onycholysis and hypohydrosis: a report of two cases, J. Oral Med. 38 (1) (1983) 21–23.

[952] K.R. Ong, S. Visram, S. McKaig, L.A. Brueton, Sensorineural deafness, enamel abnormalities and nail abnormalities: a case report of Heimler syndrome in identical twin girls, Eur. J. Med. Genet. 49 (2) (2006) 187–193.

[953] A. Heimler, J.E. Fox, J.E. Hershey, P. Crespi, Sensorineural hearing loss, enamel hypoplasia, and nail abnormalities in sibs, Am. J. Med. Genet. 39 (2) (1991) 192–195.

[954] C. Pollak, M. Floy, B. Say, Sensorineural hearing loss and enamel hypoplasia with subtle nail findings: another family with Heimler's syndrome, Clin. Dysmorphol. 12 (1) (2003) 55–58.

[955] M. Tischkowitz, C. Clenaghan, S. Davies, L. Hunter, J. Potts, S. Verhoef, et al., Amelogenesis imperfecta, sensorineural hearing loss, and Beau's lines, a second case report of Heimler's syndrome, J. Med. Genet. 36 (12) (1999) 941–943.

[956] T. Asaka, M. Akiyama, T. Domon, W. Nishie, K. Natsuga, Y. Fujita, et al., Type XVII collagen is a key player in tooth enamel formation, Am. J. Pathol. 174 (1) (2009) 91–100.

[957] Q.J. Jiang, J. Uitto, Animal models of epidermolysis bullosa – targets for gene therapy, J. Invest. Dermatol. 124 (3) (2005) xi–xiii.

[958] J.W. Bauer, M. Laimer, Gene therapy of epidermolysis bullosa, Expert Opin. Biol. Ther. 4 (9) (2004) 1435–1443.

[959] P. Vijayaraj, G. Sohl, T.M. Magin, Keratin transgenic and knockout mice: functional analysis and validation of disease-causing mutations, Methods Mol. Biol. 360 (2007) 203–251.

[960] A. Fritsch, S. Loeckermann, J.S. Kern, A. Braun, M.R. Bosl, T.A. Bley, et al., A hypomorphic mouse model of dystrophic epidermolysis bullosa reveals mechanisms of disease and response to fibroblast therapy, J. Clin. Invest. 118 (5) (2008) 1669–1679.

[961] C. Larrazabal-Moron, A. Boronat-Lopez, M. Penarrocha-Diago, M. Penarrocha-Diago, Oral rehabilitation with bone graft and simultaneous dental implants in a patient with epidermolysis bullosa: a clinical case report, J. Oral Maxillofac. Surg. 67 (7) (2009) 1499–1502.

[962] A.K. Louloudiadis, K.A. Louloudiadis, Case report: Dystrophic Epidermolysis Bullosa: dental management and oral health promotion, Eur. Arch. Paediatr. Dent. 10 (1) (2009) 42–45.

[963] J.K. Brooks, L.C. Bare, J. Davidson, L.S. Taylor, J.T. Wright, Junctional epidermolysis bullosa associated with hypoplastic enamel and pervasive failure of tooth eruption: Oral rehabilitation with use of an overdenture, Oral Surg. Oral Med. Oral Pathol. Oral Radiol. Endod. 105 (4) (2008) e24–e28.

[964] F. Stavropoulos, S. Abramowicz, Management of the oral surgery patient diagnosed with epidermolysis bullosa: report of 3 cases and review of the literature, J. Oral Maxillofac. Surg. 66 (3) (2008) 554–559.

[965] M. Penarrocha, J. Rambla, J. Balaguer, C. Serrano, J. Silvestre, J.V. Bagan, Complete fixed prostheses over implants in patients with oral epidermolysis bullosa, J. Oral Maxillofac. Surg. 65 (7 Suppl 1) (2007) 103–106.

[966] E. Mabuchi, N. Umegaki, H. Murota, T. Nakamura, K. Tamai, I. Katayama, Oral steroid improves bullous pemphigoid-like clinical manifestations in non-Herlitz junctional epidermolysis bullosa with COL17A1 mutation, Br. J. Dermatol. 157 (3) (2007) 596–598.

[967] H. Lee, M. Al Mardini, C. Ercoli, M.N. Smith, Oral rehabilitation of a completely edentulous epidermolysis bullosa patient with an implant-supported prosthesis: a clinical report, J. Prosthet. Dent. 97 (2) (2007) 65–69.

[968] B. Azrak, K. Kaevel, L. Hofmann, C. Gleissner, B. Willershausen, Dystrophic epidermolysis bullosa: oral findings and problems, Spec. Care Dentist. 26 (3) (2006) 111–115.

[969] F.N. Pekiner, D. Yucelten, S. Ozbayrak, E.C. Sezen, Oral-clinical findings and management of epidermolysis bullosa, J. Clin. Pediatr. Dent. 30 (1) (2005) 59–65.

[970] M.C. Serrano-Martinez, J.V. Bagan, F.J. Silvestre, M.T. Viguer, Oral lesions in recessive dystrophic epidermolysis bullosa, Oral Dis. 9 (5) (2003) 264–268.

[971] E. Calikoglu, R. Anadolu, Management of generalized pruritus in dominant dystrophic epidermolysis bullosa using low-dose oral cyclosporin, Acta Derm. Venereol. 82 (5) (2002) 380–382.

[972] I. Marini, F. Vecchiet, Sucralfate: a help during oral management in patients with epidermolysis bullosa, J. Periodontol. 72 (5) (2001) 691–695.

[973] M. Penarrocha-Diago, C. Serrano, J.M. Sanchis, F.J. Silvestre, J.V. Bagan, Placement of endosseous implants in patients with oral epidermolysis bullosa, Oral Surg. Oral Med. Oral Pathol. Oral Radiol. Endod. 90 (5) (2000) 587–590.

[974] H. Yamasaki, J. Tada, T. Yoshioka, J. Arata, Epidermolysis bullosa pruriginosa (McGrath) successfully controlled by oral cyclosporin, Br. J. Dermatol. 137 (2) (1997) 308–310.

[975] M.M. Trotter, Tube feeding vs oral intake: nutrition care for patients with epidermolysis bullosa, J. Am. Diet. Assoc. 95 (12) (1995) 1377.

[976] M. Harel-Raviv, S. Bernier, E. Raviv, M. Gornitsky, Oral epidermolysis bullosa in adults, Spec. Care Dentist. 15 (4) (1995) 144–148.

[977] J.T. Wright, J.D. Fine, L.B. Johnson, T.T. Steinmetz, Oral involvement of recessive dystrophic epidermolysis bullosa inversa, Am. J. Med. Genet. 47 (8) (1993) 1184–1188.

[978] J.T. Wright, J.D. Fine, L. Johnson, Hereditary epidermolysis bullosa: oral manifestations and dental management, Pediatr. Dent. 15 (4) (1993) 242–248.

[979] S.P. Travis, J.A. McGrath, A.J. Turnbull, O.M. Schofield, O. Chan, A.F. O'Connor, et al., Oral and gastrointestinal manifestations of epidermolysis bullosa, Lancet 340 (8834–8835) (1992) 1505–1506.

[980] B.K. Moghadam, R.E. Gier, Epidermolysis bullosa: oral management and case reports, ASDC J. Dent. Child. 59 (1) (1992) 66–69.

[981] J.V. Bagan Sebastian, M. Catala Pizarro, G. Gil Loscos, [Recessive dystrophic epidermolysis bullosa. Oral manifestations apropos of 3 cases], Rev. Eur. Odontoestomatol. 3 (5) (1991) 337–342.

[982] J.T. Wright, J.D. Fine, L.B. Johnson, Oral soft tissues in hereditary epidermolysis bullosa, Oral Surg. Oral Med. Oral Pathol. 71 (4) (1991) 440–446.

[983] V. Pilato, R. Serpico, G. Laino, [Focus of oral pathology: epidermolysis bullosa], Arch. Stomatol. (Napoli) 27 (2) (1986) 285–295.

[984] K. Weismann, Dystrophic epidermolysis bullosa treated unsuccessfully with oral zinc, Arch. Dermatol. Res. 277 (5) (1985) 404–405.

[985] S. Frohlich, E. Beetke, H. Heise, P. Skierlo, K. Buschatz, Epidermolysis bullosa and its changes in the oral cavity, Zahn Mund Kieferheilkd Zentralbl 73 (7) (1985) 711–720.

[986] M.S. Block, B.D. Gross, Epidermolysis bullosa dystrophica recessive: oral surgery and anesthetic considerations, J. Oral Maxillofac. Surg. 40 (11) (1982) 753–758.

[987] M. Cattabriga, S. Longhi, [Oral manifestations of epidermolysis bullosa], Minerva. Stomatol. 30 (5) (1981) 409–412.

[988] M. D'Angelo, Oral lesions in epidermolysis bullosa, Minerva Stomatol. 30 (3) (1981) 169–174.

[989] M.M. el-Khashab, A.M. Abdel-Aziz, Oral aspects in epidermolysis bullosa and systemic scleroderma, Egypt Dent. J. 24 (3) (1978) 235–248.

[990] R. Nilsen, J. Livden, S. Thunold, Oral lesions of epidermolysis bullosa acquisita, Oral Surg. Oral Med. Oral Pathol. 45 (5) (1978) 749–754.

[991] M.M. Album, A. Gaisin, K.W. Lee, B.E. Buck, W.G. Sharrar, F.M. Gill, Epidermolysis bullosa dystrophica polydysplastica. A case of anesthetic management in oral surgery, Oral Surg. Oral Med. Oral Pathol. 43 (6) (1977) 859–872.

[992] E.G. Crawford, Jr., E.J. Burkes, Jr., R.A. Briggaman, Hereditary epidermolysis bullosa: oral manifestations and dental therapy, Oral Surg. Oral Med. Oral Pathol. 42 (4) (1976) 490–500.

[993] S.B. Kahn, N. Trieger, Epidermolysis bullosa hereditaria letalis. A case report with special emphasis on oral manifestations, J. Oral. Med. 31 (2) (1976) 32–35.

[994] J.W. Gormley, C.E. Schow, Epidermolysis bullosa and associated problems in oral surgical treatment, J. Oral Surg. 34 (1) (1976) 45–52.

[995] J.S. Giansanti, Oral nodular excrescences in epidermolysis bullosa, Oral Surg. Oral Med. Oral Pathol. 40 (3) (1975) 385–390.

[996] W.P. Unger, J.R. Nethercott, Epidermolysis bullosa dystrophica treated with vitamin E and oral corticosteroids, Can. Med. Assoc. J. 108 (9) (1973) 1136–1138.

[997] H. Kinast, E. Schuh, Epidermolysis bullosa and its oral form, Osterr. Z Stomatol. 70 (5) (1973) 166–175.

[998] H. Kinast, E. Schuh, Epidermolysis bullosa and its oral manifestations, Osterr. Z Stomatol. 70 (5) (1973) 166–175.

[999] B.P. Levy, C.M. Reeve, R.R. Kierland, The oral aspects of epidermolysis bullosa dystrophica: a case report, J. Periodontol. 40 (7) (1969) 431–434.

[1000] T. Arwill, A. Bergenholtz, O. Olsson, Epidermolysis bullosa hereditaria. Iv. Histologic changes of the oral mucosa in the polydysplastic dystrophic and the letalis forms, Odontol Revy 16 (1965) 101–111.

[1001] T. Arwill, A. Bergenholtz, H. Thilander, Epidermolysis bullosa hereditaria. 5. The ultrastructure of oral mucosa and skin in four cases of the letalis form, Acta Pathol. Microbiol. Scand. 74 (3) (1968) 311–324.

[1002] A. Komori, W.A. Welton, E.E. Kelln, The behavior of the basement membrane of skin and oral lesions in patients with lichen planus, erythema multiforme, lupus erythematosus, pemphigus vulgaris, benign mucous membrane pemphigoid, and epidermolysis bullosa, Oral Surg. Oral Med. Oral Pathol. 22 (6) (1966) 752–763.

[1003] J.O. Andreasen, E. Hjorting-Hansen, M. Ulmansky, J.J. Pindborg, Milia formation in oral lesions in epidermolysis bullosa, Acta Pathol. Microbiol. Scand. 63 (1965) 37–41.

[1004] D. Winstock, Oral aspects of epidermolysis bullosa, Br. J. Dermatol. 74 (1962) 431–438.

[1005] J.T. Wright, K.I. Hall, T.G. Deaton, J.D. Fine, Structural and compositional alteration of tooth enamel in hereditary epidermolysis bullosa, Connect. Tissue Res. 34 (4) (1996) 271–279.
[1006] L.C. Silva, R.A. Cruz, L.R. Abou-Id, L.N. Brini, L.S. Moreira, Clinical evaluation of patients with epidermolysis bullosa: review of the literature and case reports, Spec. Care Dentist 24 (1) (2004) 22–27.
[1007] E. Sadler, M. Laimer, A. Diem, A. Klausegger, G. Pohla-Gubo, W. Muss, et al., Dental alterations in junctional epidermolysis bullosa--report of a patient with a mutation in the LAMB3-gene, J. Dtsch. Dermatol. Ges. 3 (5) (2005) 359–363.
[1008] J. Kirkham, C. Robinson, S.M. Strafford, R.C. Shore, W.A. Bonass, S.J. Brookes, et al., The chemical composition of tooth enamel in junctional epidermolysis bullosa, Arch. Oral Biol. 45 (5) (2000) 377–386.
[1009] J. Kirkham, C. Robinson, S.M. Strafford, R.C. Shore, W.A. Bonass, S.J. Brookes, et al., The chemical composition of tooth enamel in recessive dystrophic epidermolysis bullosa: significance with respect to dental caries, J. Dent. Res. 75 (9) (1996) 1672–1678.
[1010] M. De Benedittis, M. Petruzzi, G. Favia, R. Serpico, Oro-dental manifestations in Hallopeau-Siemens-type recessive dystrophic epidermolysis bullosa, Clin. Exp. Dermatol. 29 (2) (2004) 128–132.
[1011] O. Swensson, E. Christophers, Generalized atrophic benign epidermolysis bullosa in 2 siblings complicated by multiple squamous cell carcinomas, Arch. Dermatol. 134 (2) (1998) 199–203.
[1012] W.S. Pear, New roles for Notch in tuberous sclerosis, J. Clin. Invest. 120 (1) (2009) 84–87.
[1013] L. Meikle, D.M. Talos, H. Onda, K. Pollizzi, A. Rotenberg, M. Sahin, et al., A mouse model of tuberous sclerosis: neuronal loss of Tsc1 causes dysplastic and ectopic neurons, reduced myelination, seizure activity, and limited survival, J. Neurosci. 27 (21) (2007) 5546–5558.
[1014] L. Meikle, K. Pollizzi, A. Egnor, I. Kramvis, H. Lane, M. Sahin, et al., Response of a neuronal model of tuberous sclerosis to mammalian target of rapamycin (mTOR) inhibitors: effects on mTORC1 and Akt signaling lead to improved survival and function, J. Neurosci. 28 (21) (2008) 5422–5432.
[1015] D.J. Kwiatkowski, H. Zhang, J.L. Bandura, K.M. Heiberger, M. Glogauer, N. el-Hashemite, et al., A mouse model of TSC1 reveals sex-dependent lethality from liver hemangiomas, and up-regulation of p70S6 kinase activity in Tsc1 null cells, Hum. Mol. Genet. 11 (5) (2002) 525–534.
[1016] J.D. Sparling, C.H. Hong, J.S. Brahim, J. Moss, T.N. Darling, Oral findings in 58 adults with tuberous sclerosis complex, J. Am. Acad. Dermatol. 56 (5) (2007) 786–790.
[1017] M.A. Duran-Padilla, G.F. Paredes-Farreras, C. Torres-Gonzalez, [Enamel hypoplasia in tuberous sclerosis], Rev. Invest. Clin. 53 (2) (2001) 126–128.
[1018] A. Cutando, J.A. Gil, J. Lopez, Oral health management implications in patients with tuberous sclerosis, Oral Surg. Oral Med. Oral Pathol. Oral Radiol. Endod. 90 (4) (2000) 430–435.
[1019] B.G. Russell, M.B. Russell, F. Praetorius, C.A. Russell, Deciduous teeth in tuberous sclerosis, Clin. Genet. 50 (1) (1996) 36–40.
[1020] C.S. Ho, S.P. Lin, E.Y. Shen, N.C. Chiu, Enamel pitting in childhood tuberous sclerosis, J. Formos. Med. Assoc. 94 (7) (1995) 414–417.
[1021] G. Mlynarczyk, Enamel pitting. A common sign of tuberous sclerosis, Ann. N.Y. Acad. Sci. 615 (1991) 367–369.

[1022] G. Mlynarczyk, Enamel pitting: a common symptom of tuberous sclerosis, Oral Surg. Oral Med. Oral Pathol. 71 (1) (1991) 63–67.

[1023] J.R. Sampson, D. Attwood, A.S. al Mughery, J.S. Reid, Pitted enamel hypoplasia in tuberous sclerosis, Clin. Genet. 42 (1) (1992) 50–52.

[1024] D. Webb, J.P. Osborne, A. Clarke, Pitted enamel hypoplasia in tuberous sclerosis, Clin. Genet. 45 (5) (1994) 269.

[1025] K.S. Au, A.T. Williams, E.S. Roach, L. Batchelor, S.P. Sparagana, M.R. Delgado, et al., Genotype/phenotype correlation in 325 individuals referred for a diagnosis of tuberous sclerosis complex in the United States, Genet. Med. 9 (2) (2007) 88–100.

[1026] P. Curatolo, R. Bombardieri, S. Jozwiak, Tuberous sclerosis, Lancet 372 (9639) (2008) 657–668.

[1027] N. Flanagan, W.J. O'Connor, B. McCartan, S. Miller, J. McMenamin, R. Watson, et al., Developmental enamel defects in tuberous sclerosis: a clinical genetic marker? J. Med. Genet. 34 (8) (1997) 637–639.

[1028] O. Sancak, M. Nellist, M. Goedbloed, P. Elfferich, C. Wouters, A. Maat-Kievit, et al., Mutational analysis of the TSC1 and TSC2 genes in a diagnostic setting: genotype – phenotype correlations and comparison of diagnostic DNA techniques in tuberous sclerosis complex, Eur. J. Hum. Genet. 13 (6) (2005) 731–741.

[1029] R.A. Schwartz, G. Fernandez, K. Kotulska, S. Jozwiak, Tuberous sclerosis complex: advances in diagnosis, genetics, and management, J. Am. Acad. Dermatol. 57 (2) (2007) 189–202.

[1030] C. Scully, Orofacial manifestations in tuberous sclerosis, Oral Surg. Oral Med. Oral Pathol. 44 (5) (1977) 706–716.

[1031] C. Scully, Orofacial manifestations in the neurodermatoses, ASDC J. Dent. Child. 47 (4) (1980) 255–260.

[1032] C. Scully, Oral mucosal lesions in association with epilepsy and cutaneous lesions: the Pringle-Bourneville syndrome, Int. J. Oral Surg. 10 (1) (1981) 68–72.

[1033] D.C. Smith, S.R. Porter, C. Scully, Gingival and other oral manifestations in tuberous sclerosis: a case report, Periodontal Clin. Investig. 15 (2) (1993) 13–16.

[1034] G. Sekiguchi, [Dental enamel pitting in tuberous sclerosis complex], No To Hattatsu 37 (6) (2005) 512–516.

[1035] M. van Slegtenhorst, R. de Hoogt, C. Hermans, M. Nellist, B. Janssen, S. Verhoef, et al., Identification of the tuberous sclerosis gene TSC1 on chromosome 9q34, Science 277 (5327) (1997) 805–808.

[1036] J.T. Wright, Oral manifestations in the epidermolysis bullosa spectrum, Dermatol. Clin. 28 (1) (2010) 159–164.

[1037] L.E. Pereira, P. Bostik, A.A. Ansari, The development of mouse APECED models provides new insight into the role of AIRE in immune regulation, Clin. Dev. Immunol. 12 (3) (2005) 211–216.

[1038] R. Perniola, G. Tamborrino, S. Marsigliante, C. De Rinaldis, Assessment of enamel hypoplasia in autoimmune polyendocrinopathy-candidiasis-ectodermal dystrophy (APECED), J. Oral Pathol. Med. 27 (6) (1998) 278–282.

[1039] E. McGovern, P. Fleming, C. Costigan, M. Dominguez, D.C. Coleman, J. Nunn, et al., Oral health in autoimmune polyendocrinopathy candidiasis ectodermal dystrophy (APECED), Eur. Arch. Paediatr. Dent. 9 (4) (2008) 236–244.

[1040] R. Rautemaa, J. Hietanen, S. Niissalo, S. Pirinen, J. Perheentupa, Oral and oesophageal squamous cell carcinoma – a complication or component of autoimmune polyendocrinopathy-candidiasis-ectodermal dystrophy (APECED, APS-I), Oral Oncol. 43 (6) (2007) 607–613.

[1041] J. Uittamo, E. Siikala, P. Kaihovaara, M. Salaspuro, R. Rautemaa, Chronic candidiasis and oral cancer in APECED-patients: production of carcinogenic acetaldehyde from glucose and ethanol by *Candida albicans*, Int. J. Cancer 124 (3) (2009) 754–756.

[1042] P. Bjorses, J. Aaltonen, N. Horelli-Kuitunen, M.L. Yaspo, L. Peltonen, Gene defect behind APECED: a new clue to autoimmunity, Hum. Mol. Genet. 7 (10) (1998) 1547–1553.

[1043] Finnish-German APECED Consortium An autoimmune disease, APECED, caused by mutations in a novel gene featuring two PHD-type zinc-finger domains, Nat. Genet. 17 (4) (1997) 399–403.

[1044] K. Nagamine, P. Peterson, H.S. Scott, J. Kudoh, S. Minoshima, M. Heino, et al., Positional cloning of the APECED gene, Nat. Genet. 17 (4) (1997) 393–398.

[1045] P. Peterson, J. Pitkanen, N. Sillanpaa, K. Krohn, Autoimmune polyendocrinopathy candidiasis ectodermal dystrophy (APECED): a model disease to study molecular aspects of endocrine autoimmunity, Clin. Exp. Immunol. 135 (3) (2004) 348–357.

[1046] J. Villasenor, C. Benoist, D. Mathis, AIRE and APECED: molecular insights into an autoimmune disease, Immunol. Rev. 204 (2005) 156–164.

[1047] E. Proust-Lemoine, J.L. Wemeau, [Apeced syndrome or autoimmune polyendocrine syndrome Type 1], Presse Med. 37 (7–8) (2008) 1158–1171.

[1048] S. Myllarniemi, J. Perheentupa, Oral findings in the autoimmune polyendocrinopathy-candidiasis syndrome (APECS) and other forms of hypoparathyroidism, Oral Surg. Oral Med. Oral Pathol. 45 (5) (1978) 721–729.

[1049] P. Ahonen, S. Myllarniemi, A. Kahanpaa, J. Perheentupa, Ketoconazole is effective against the chronic mucocutaneous candidiasis of autoimmune polyendocrinopathy-candidiasis-ectodermal dystrophy (APECED), Acta Med. Scand. 220 (4) (1986) 333–339.

[1050] A. Pavlic, J. Waltimo-Siren, Clinical and microstructural aberrations of enamel of deciduous and permanent teeth in patients with autoimmune polyendocrinopathy-candidiasis-ectodermal dystrophy, Arch. Oral Biol. 54 (7) (2009) 658–665.

[1051] P.L. Lukinmaa, J. Waltimo, S. Pirinen, Microanatomy of the dental enamel in autoimmune polyendocrinopathy-candidiasis-ectodermal dystrophy (APECED): report of three cases, J. Craniofac. Genet. Dev. Biol. 16 (3) (1996) 174–181.

[1052] L.S. Weinstein, T. Xie, Q.H. Zhang, M. Chen, Studies of the regulation and function of the Gs alpha gene Gnas using gene targeting technology, Pharmacol. Ther. 115 (2) (2007) 271–291.

[1053] M.F. Gomes, A.M. Camargo, T.A. Sampaio, M.A. Graziozi, M.C. Armond, Oral manifestations of Albright hereditary osteodystrophy: a case report, Rev. Hosp. Clin. Fac. Med. Sao Paulo 57 (4) (2002) 161–166.

[1054] S. Izraeli, A. Metzker, G. Horev, D. Karmi, P. Merlob, Z. Farfel, et al., Albright hereditary osteodystrophy with hypothyroidism, normocalcemia, and normal Gs protein activity: a family presenting with congenital osteoma cutis, Am. J. Med. Genet. 43 (4) (1992) 764–767.

[1055] M.A. Levine, T.G. Ahn, S.F. Klupt, K.D. Kaufman, P.M. Smallwood, H.R. Bourne, et al., Genetic deficiency of the alpha subunit of the guanine nucleotide-binding protein Gs as the molecular basis for Albright hereditary osteodystrophy, Proc. Natl. Acad. Sci. U.S.A. 85 (2) (1988) 617–621.

[1056] V.V. Rao, S. Schnittger, I. Hansmann, G protein Gs alpha (GNAS 1), the probable candidate gene for Albright hereditary osteodystrophy, is assigned to human chromosome 20q12-q13.2, Genomics 10 (1) (1991) 257–261.

[1057] L.S. Weinstein, P.V. Gejman, E. Friedman, T. Kadowaki, R.M. Collins, E.S. Gershon, et al., Mutations of the Gs alpha-subunit gene in Albright hereditary osteodystrophy detected by denaturing gradient gel electrophoresis, Proc. Natl. Acad. Sci. U.S.A. 87 (21) (1990) 8287–8290.

[1058] S.J. Rickard, L.C. Wilson, Analysis of GNAS1 and overlapping transcripts identifies the parental origin of mutations in patients with sporadic Albright hereditary osteodystrophy and reveals a model system in which to observe the effects of splicing mutations on translated and untranslated messenger RNA, Am. J. Hum. Genet. 72 (4) (2003) 961–974.

[1059] F. Albright, A.P. Forbes, P.H. Henneman, Pseudo-pseudohypoparathyroidism, Trans. Assoc. Am. Physicians 65 (1952) 337–350.

[1060] F. Albright, C.H. Burnett, P.H. Smith, W. Parson, et al., Pseudo-hypoparathyroidism – an example of 'Seabright-Bantam syndrome': report of three cases, Endocrinology 30 (1942) 922–932.

[1061] M.A. Aldred, R.J. Bagshaw, K. Macdermot, D. Casson, S.H. Murch, J.A. Walker-Smith, et al., Germline mosaicism for a GNAS1 mutation and Albright hereditary osteodystrophy, J. Med. Genet. 37 (11) (2000) E35.

[1062] M.A. Aldred, S. Aftimos, C. Hall, K.S. Waters, R.V. Thakker, R.C. Trembath, et al., Constitutional deletion of chromosome 20q in two patients affected with albright hereditary osteodystrophy, Am. J. Med. Genet. 113 (2) (2002) 167–172.

[1063] L. De Sanctis, D. Romagnolo, M. Olivero, F. Buzi, M. Maghnie, G. Scire, et al., Molecular analysis of the GNAS1 gene for the correct diagnosis of Albright hereditary osteodystrophy and pseudohypoparathyroidism, Pediatr. Res. 53 (5) (2003) 749–755.

[1064] N. Fitch, Albright's hereditary osteodystrophy: a review, Am. J. Med. Genet. 11 (1) (1982) 11–29.

[1065] E.L. Germain-Lee, J. Groman, J.L. Crane, S.M. Jan de Beur, M.A. Levine, Growth hormone deficiency in pseudohypoparathyroidism type 1a: another manifestation of multihormone resistance, J. Clin. Endocrinol. Metab. 88 (9) (2003) 4059–4069.

[1066] A. Lagarde, L.M. Kerebel, B. Kerebel, Structural and ultrastructural study of the teeth in a suspected case of pseudohypoparathyroidism, J. Biol. Buccale 17 (2) (1989) 109–114.

[1067] S.B. Jensen, F. Illum, E. Dupont, Nature and frequency of dental changes in idiopathic hypoparathyroidism and pseudohypoparathyroidism, Scand. J. Dent. Res. 89 (1) (1981) 26–37.

[1068] L.K. Croft, C.J. Witkop, Jr., J.E. Glas, Pseudohypoparathyroidism, Oral Surg. Oral Med. Oral Pathol. 20 (6) (1965) 758–770.

[1069] G.M. Ritchie, Dental manifestations of pseudohypoparathyroidism, Arch. Dis. Child. 40 (213) (1965) 565–572.

[1070] L.S. Weinstein, J. Liu, A. Sakamoto, T. Xie, M. Chen, Minireview: GNAS: normal and abnormal functions, Endocrinology 145 (12) (2004) 5459–5464.

[1071] M.D. Brown, G. Aaron, Pseudohypoparathyroidism: case report, Pediatr. Dent. 13 (2) (1991) 106–109.

[1072] A.M. Flenniken, L.R. Osborne, N. Anderson, N. Ciliberti, C. Fleming, J.E. Gittens, et al., A Gja1 missense mutation in a mouse model of oculodentodigital dysplasia, Development 132 (19) (2005) 4375–4386.

[1073] R. Richardson, D. Donnai, F. Meire, M.J. Dixon, Expression of Gja1 correlates with the phenotype observed in oculodentodigital syndrome/type III syndactyly, J. Med. Genet. 41 (1) (2004) 60–67.

[1074] E. Eidelman, A. Chosack, M.L. Wagner, Orodigitofacial dysostosis and oculodentodigital dysplasia. Two distinct syndromes with some similarities, Oral Surg. Oral Med. Oral Pathol. 23 (3) (1967) 311–319.

[1075] S. Gunbay, B. Zeytinoglu, F. Ozkinay, C. Ozkinay, A. Oncag, N.J. Burzynski, et al., Orofaciodigital syndrome I: a case report oral-facial-digital syndrome. A family case report tooth anomalies in the oral-facial-digital syndrome mental retardation and dermatoglyphics in a family with the oral-facial-digital syndrome orodigitofacial dysostosis and oculodentodigital dysplasia. Two distinct syndromes with some similarities [On the association of dental, cervicofacial, cardiac and digital malformations], J. Clin. Pediatr. Dent. 20 (4) (1996) 329–332.

[1076] W.A. Paznekas, S.A. Boyadjiev, R.E. Shapiro, O. Daniels, B. Wollnik, C.E. Keegan, et al., Connexin 43 (GJA1) mutations cause the pleiotropic phenotype of oculodentodigital dysplasia, Am. J. Hum. Genet. 72 (2) (2003) 408–418.

[1077] A. Fenwick, R.J. Richardson, J. Butterworth, M.J. Barron, M.J. Dixon, Novel mutations in GJA1 cause oculodentodigital syndrome, J. Dent. Res. 87 (11) (2008) 1021–1026.

[1078] R.J. Richardson, S. Joss, S. Tomkin, M. Ahmed, E. Sheridan, M.J. Dixon, et al., A nonsense mutation in the first transmembrane domain of connexin 43 underlies autosomal recessive oculodentodigital syndrome, J. Med. Genet. 43 (7) (2006) e37.

[1079] E.I. Traboulsi, B.M. Faris, V.M. Der Kaloustian, Persistent hyperplastic primary vitreous and recessive oculo-dento-osseous dysplasia, Am. J. Med. Genet. 24 (1) (1986) 95–100.

[1080] R.J. van Es, D. Wittebol-Post, F.A. Beemer, Oculodentodigital dysplasia with mandibular retrognathism and absence of syndactyly: a case report with a novel mutation in the connexin 43 gene, Int. J. Oral Maxillofac. Surg. 36 (9) (2007) 858–860.

[1081] R.J. Gorlin, L.H. Miskin, G.J. St, Oculodentodigital dysplasia, J. Pediatr. 63 (1963) 69–75.

[1082] J.A. Dean, J.E. Jones, B.W. Vash, Dental management of oculodentodigital dysplasia: report of case, ASDC J. Dent. Child. 53 (2) (1986) 131–134.

[1083] A. Itro, A. Marra, V. Urciuolo, P. Difalco, A. Amodio, Oculodentodigital dysplasia. A case report, Minerva Stomatol. 54 (7–8) (2005) 453–459.

[1084] W. Chen, S. Yang, Y. Abe, M. Li, Y. Wang, J. Shao, et al., Novel pycnodysostosis mouse model uncovers cathepsin K function as a potential regulator of osteoclast apoptosis and senescence, Hum. Mol. Genet. 16 (4) (2007) 410–423.

[1085] C.Y. Li, K.J. Jepsen, R.J. Majeska, J. Zhang, R. Ni, B.D. Gelb, et al., Mice lacking cathepsin K maintain bone remodeling but develop bone fragility despite high bone mass, J. Bone Miner. Res. 21 (6) (2006) 865–875.

[1086] K.W. Fleming, G. Barest, O. Sakai, Dental and facial bone abnormalities in pyknodysostosis: CT findings, AJNR Am. J. Neuroradiol. 28 (1) (2007) 132–134.

[1087] J.P. Schmitz, C.J. Gassmann, A.M. Bauer, B.R. Smith, Mandibular reconstruction in a patient with pyknodysostosis, J. Oral Maxillofac. Surg. 54 (4) (1996) 513–517.

[1088] J.V. Francisco, T.J. Nicholoff, Jr., Pyknodysostosis: an unusual presentation in a denture wearer. A case report, Oral Surg. Oral Med. Oral Pathol. 72 (6) (1991) 693–695.

[1089] M. Talbi, M. Martinetti, P. Goudot, J.M. Vaillant, M. Auriol, G. Chomette, et al., [Pyknodysostosis. An unusual form of osteopetrosis], Rev. Stomatol. Chir. Maxillofac. 91 (4) (1990) 299–303.

[1090] K. Kawahara, M. Nishikiori, K. Imai, K. Kishi, Y. Fujiki, Radiographic observations of pyknodysostosis. Report of a case, Oral Surg. Oral Med. Oral Pathol. 44 (3) (1977) 476–482.

[1091] N.P. Hunt, S.J. Cunningham, N. Adnan, M. Harris, The dental, craniofacial, and biochemical features of pyknodysostosis: a report of three new cases, J. Oral Maxillofac. Surg. 56 (4) (1998) 497–504.

[1092] B.D. Gelb, K. Moissoglu, J. Zhang, J.A. Martignetti, D. Bromme, R.J. Desnick, et al., Cathepsin K: isolation and characterization of the murine cDNA and genomic sequence, the homologue of the human pycnodysostosis gene, Biochem. Mol. Med. 59 (2) (1996) 200–206.

[1093] B.D. Gelb, G.P. Shi, H.A. Chapman, R.J. Desnick, Pycnodysostosis, a lysosomal disease caused by cathepsin K deficiency, Science 273 (5279) (1996) 1236–1238.

[1094] B.D. Gelb, E. Spencer, S. Obad, G.J. Edelson, S. Faure, J. Weissenbach, et al., Pycnodysostosis: refined linkage and radiation hybrid analyses reduce the critical region to 2cM at 1q21 and map two candidate genes, Hum. Genet. 98 (2) (1996) 141–144.

[1095] B.D. Gelb, J.G. Edelson, R.J. Desnick, Linkage of pycnodysostosis to chromosome 1q21 by homozygosity mapping, Nat. Genet. 10 (2) (1995) 235–237.

[1096] H.D. Sedano, R.J. Gorlin, V.E. Anderson, Pycnodysostosis. Clinical and genetic considerations, Am. J. Dis. Child. 116 (1) (1968) 70–77.

[1097] D. Alves Pereira, L. Berini Aytes, C. Gay Escoda, Pycnodysostosis. A report of 3 clinical cases, Med. Oral Patol. Oral Cir. Bucal. 13 (10) (2008) E633–E635.

[1098] C.S. Fonteles, C.M. Chaves, Jr., A. Da Silveira, E.C. Soares, J.L. Couto, F. de Azevedo Mde, et al., Cephalometric characteristics and dentofacial abnormalities of pycnodysostosis: report of four cases from Brazil, Oral Surg. Oral Med. Oral Pathol. Oral Radiol. Endod. 104 (4) (2007) e83–e90.

[1099] C.M. Jones, J.S. Rennie, A.S. Blinkhorn, Pycnodysostosis. A review of reported dental abnormalities and a report of the dental findings in two cases, Br. Dent. J. 164 (7) (1988) 218–220.

[1100] O. Dardenne, J. Prud'homme, A. Arabian, F.H. Glorieux, R. St-Arnaud, Targeted inactivation of the 25-hydroxyvitamin D(3)-1(alpha)-hydroxylase gene (CYP27B1) creates an animal model of pseudovitamin D-deficiency rickets, Endocrinology 142 (7) (2001) 3135–3141.

[1101] O. Dardenne, J. Prudhomme, S.A. Hacking, F.H. Glorieux, R. St-Arnaud, Rescue of the pseudo-vitamin D deficiency rickets phenotype of CYP27B1-deficient mice by treatment with 1,25-dihydroxyvitamin D3: biochemical, histomorphometric, and biomechanical analyses, J. Bone Miner. Res. 18 (4) (2003) 637–643.

[1102] S. Kato, T. Yoshizazawa, S. Kitanaka, A. Murayama, K. Takeyama, Molecular genetics of vitamin D-dependent hereditary rickets, Horm. Res. 57 (3–4) (2002) 73–78.

[1103] F.H. Glorieux, Pseudo-vitamin D deficiency rickets, J. Endocrinol. 154 (Suppl.) (1997) S75–S78.

[1104] F.H. Glorieux, R. St-Arnaud, Molecular cloning of (25-OH D)-1 alpha-hydroxylase: an approach to the understanding of vitamin D pseudo-deficiency, Recent Prog. Horm. Res. 53 (1998) 341–349, discussion 350.

[1105] W.L. Miller, A.A. Portale, Genetics of vitamin D biosynthesis and its disorders, Best Pract. Res. Clin. Endocrinol. Metab. 15 (1) (2001) 95–109.

[1106] M. Zambrano, N.G. Nikitakis, M.C. Sanchez-Quevedo, J.J. Sauk, H. Sedano, H. Rivera, et al., Oral and dental manifestations of vitamin D-dependent rickets type I: report of a pediatric case, Oral Surg. Oral Med. Oral Pathol. Oral Radiol. Endod. 95 (6) (2003) 705–709.

[1107] S. Kitanaka, K. Takeyama, A. Murayama, S. Kato, The molecular basis of vitamin D-dependent rickets type I, Endocr. J. 48 (4) (2001) 427–432.

[1108] M. Nishino, K. Kamada, K. Arita, T. Takarada, [Dentofacial manifestations in children with vitamin D-dependent rickets type II], Shoni Shikagaku Zasshi 28 (2) (1990) 346–358.

[1109] K. Kikuchi, T. Okamoto, M. Nishino, E. Takeda, Y. Kuroda, M. Miyao, et al., Vitamin D-dependent rickets type II: report of three cases, ASDC J. Dent. Child. 55 (6) (1988) 465–468.

[1110] E. Decker, A. Stellzig-Eisenhauer, B.S. Fiebig, C. Rau, W. Kress, K. Saar, et al., PTHR1 loss-of-function mutations in familial, nonsyndromic primary failure of tooth eruption, Am. J. Hum. Genet. 83 (6) (2008) 781–786.

[1111] A. Wolff, M.J. Koch, S. Benzinger, H. van Waes, N.I. Wolf, E. Boltshauser, et al., Rare dental peculiarities associated with the hypomyelinating leukoencephalopathy 4H syndrome/ADDH, Pediatr. Dent. 32 (5) (2010) 386–392.

[1112] G. Bernard, E. Chouery, M.L. Putorti, M. Tetreault, A. Takanohashi, G. Carosso, et al., Mutations of POLR3A encoding a catalytic subunit of RNA polymerase pol III cause a recessive hypomyelinating leukodystrophy, Am. J. Hum. Genet. 89 (3) (2011) 415–423.

[1113] E. Reinstein, R.Y. Wang, L. Zhan, D.L. Rimoin, W.R. Wilcox, Ehlers-Danlos type VIII, periodontitis-type: further delineation of the syndrome in a four-generation pedigree, Am. J. Med. Genet. A 155A (4) (2011) 742–747.

[1114] G.E. Wise, Cellular and molecular basis of tooth eruption, Orthod. Craniofac. Res. 12 (2) (2009) 67–73.

[1115] W. Beertsen, T. VandenBos, V. Everts, Root development in mice lacking functional tissue non-specific alkaline phosphatase gene: inhibition of acellular cementum formation, J. Dent. Res. 78 (6) (1999) 1221–1229.

[1116] A. Reibel, M.C. Maniere, F. Clauss, D. Droz, Y. Alembik, E. Mornet, et al., Orodental phenotype and genotype findings in all subtypes of hypophosphatasia, Orphanet J. Rare Dis. 4 (2009) 6.

[1117] G.V. Rayasam, O. Wendling, P.O. Angrand, M. Mark, K. Niederreither, L. Song, et al., NSD1 is essential for early post-implantation development and has a catalytically active SET domain, EMBO J. 22 (12) (2003) 3153–3163.

[1118] R.E. Ward, P.L. Jamison, J.E. Allanson, Quantitative approach to identifying abnormal variation in the human face exemplified by a study of 278 individuals with five craniofacial syndromes, Am. J. Med. Genet. 91 (1) (2000) 8–17.

[1119] J.M. Gomes-Silva, D.B. Ruviere, R.A. Segatto, A.M. de Queiroz, A.C. de Freitas, Sotos syndrome: a case report, Spec. Care Dentist. 26 (6) (2006) 257–262.

[1120] M. Inokuchi, J. Nomura, Y. Mtsumura, M. Sekida, T. Tagawa, Sotos syndrome with enamel hypoplasia: a case report, J. Clin. Pediatr. Dent. 25 (4) (2001) 313–316.

[1121] A. Berio, R. Trucchi, M. Meliota, [Maxillofacial and dental abnormalities in some multiple abnormality syndromes. 'Cri du chat' syndrome, Wilms' tumor-aniridia syndrome; Sotos syndrome; Goldenhar syndrome], Minerva Pediatr. 44 (5) (1992) 223–229.

[1122] K. Takei, K. Sueishi, H. Yamaguchi, Y. Ohtawa, Dentofacial growth in patients with Sotos syndrome, Bull. Tokyo Dent. Coll. 48 (2) (2007) 73–85.

[1123] G. Baujat, V. Cormier-Daire, Sotos syndrome, Orphanet J. Rare Dis. 2 (2007) 36.

[1124] S.W. Park, M.S. Park, J.S. Hwang, Y.S. Shin, S.H. Yoon, A case of Sotos syndrome with subduroperitoneal shunt, Pediatr. Neurosurg. 42 (3) (2006) 174–179.

[1125] A.P. Callanan, P. Anand, E.C. Sheehy, Sotos syndrome with hypodontia, Int. J. Paediatr. Dent. 16 (2) (2006) 143–146.

[1126] N. Staffolani, S. Belcastro, M. Guerra, [Maxillofacial and dental anomalies in multiple-abnormality syndromes. The clinical and therapeutic aspects in Sotos' syndrome], Minerva Stomatol. 43 (11) (1994) 525–529.

[1127] S. Scaglioni, R. Besana, M.T. Ortisi, L. Calcagni, M. Giovannini, [Sotos' syndrome. Presentation of a clinical case], Minerva Pediatr. 34 (15–16) (1982) 669–674.

[1128] R.R. Welbury, H.J. Fletcher, Cerebral gigantism (Sotos syndrome) – two case reports, J. Paediatr. Dent. 4 (1) (1988) 41–44.

[1129] M.M. Villaverde, J.A. Da Silva, Sotos' syndrome: hypertelorism, antimongoloid slant of eye, and high arched palate complex, J. Med. Soc. N.J. 68 (10) (1971) 805–808.

[1130] F. Faravelli, NSD1 mutations in Sotos syndrome, Am. J. Med. Genet. C Semin. Med. Genet. 137C (1) (2005) 24–31.

[1131] N. Niikawa, Molecular basis of Sotos syndrome, Horm. Res. 62 (Suppl. 3) (2004) 60–65.

[1132] G. Leventopoulos, S. Kitsiou-Tzeli, K. Kritikos, S. Psoni, A. Mavrou, E. Kanavakis, et al., A clinical study of Sotos syndrome patients with review of the literature, Pediatr. Neurol. 40 (5) (2009) 357–364.

[1133] P. Lapunzina, Risk of tumorigenesis in overgrowth syndromes: a comprehensive review, Am. J. Med. Genet. C Semin. Med. Genet. 137C (1) (2005) 53–71.

[1134] K. Tatton-Brown, J. Douglas, K. Coleman, G. Baujat, T.R. Cole, S. Das, et al., Genotype-phenotype associations in Sotos syndrome: an analysis of 266 individuals with NSD1 aberrations, Am. J. Hum. Genet. 77 (2) (2005) 193–204.

[1135] K. Tatton-Brown, J. Douglas, K. Coleman, G. Baujat, K. Chandler, A. Clarke, et al., Multiple mechanisms are implicated in the generation of 5q35 microdeletions in Sotos syndrome, J. Med. Genet. 42 (4) (2005) 307–313.

[1136] T. Ogawa, T. Onishi, T. Hayashibara, S. Sakashita, R. Okawa, T. Ooshima, et al., Dentinal defects in Hyp mice not caused by hypophosphatemia alone, Arch. Oral Biol. 51 (1) (2006) 58–63.

[1137] L. Wang, L. Du, B. Ecarot, Evidence for Phex haploinsufficiency in murine X-linked hypophosphatemia, Mamm. Genome 10 (4) (1999) 385–389.

[1138] C.R. Scriver, H.S. Tenenhouse, X-linked hypophosphataemia: a homologous phenotype in humans and mice with unusual organ-specific gene dosage, J. Inherit. Metab. Dis. 15 (4) (1992) 610–624.

[1139] L. Ye, S. Zhang, H. Ke, L.F. Bonewald, J.Q. Feng, Periodontal breakdown in the Dmp1 null mouse model of hypophosphatemic rickets, J. Dent. Res. 87 (7) (2008) 624–629.

[1140] B. Yuan, M. Takaiwa, T.L. Clemens, J.Q. Feng, R. Kumar, P.S. Rowe, et al., Aberrant Phex function in osteoblasts and osteocytes alone underlies murine X-linked hypophosphatemia, J. Clin. Invest. 118 (2) (2008) 722–734.

[1141] G.I. Baroncelli, M. Angiolini, E. Ninni, V. Galli, R. Saggese, M.R. Giuca, et al., Prevalence and pathogenesis of dental and periodontal lesions in children with X-linked hypophosphatemic rickets, Eur. J. Paediatr. Dent. 7 (2) (2006) 61–66.

[1142] I.A. Holm, A.E. Nelson, B.G. Robinson, R.S. Mason, D.J. Marsh, C.T. Cowell, et al., Mutational analysis and genotype-phenotype correlation of the PHEX gene in X-linked hypophosphatemic rickets, J. Clin. Endocrinol. Metab. 86 (8) (2001) 3889–3899.

[1143] C.M. Pereira, C.R. de Andrade, P.A. Vargas, R.D. Coletta, O.P. de Almeida, M.A. Lopes, et al., Dental alterations associated with X-linked hypophosphatemic rickets, J. Endod. 30 (4) (2004) 241–245.

[1144] T. Yamamoto, Diagnosis of X-linked hypophosphatemic vitamin D resistant rickets, Acta Paediatr. Jpn. 39 (4) (1997) 499–502.

[1145] C. Gaucher, T. Boukpessi, D. Septier, F. Jehan, P.S. Rowe, M. Garabedian, et al., Dentin noncollagenous matrix proteins in familial hypophosphatemic rickets, Cells Tissues Organs 189 (1–4) (2009) 219–223.

[1146] C. Chaussain-Miller, C. Sinding, D. Septier, M. Wolikow, M. Goldberg, M. Garabedian, et al., Dentin structure in familial hypophosphatemic rickets: benefits of vitamin D and phosphate treatment, Oral Dis. 13 (5) (2007) 482–489.
[1147] T. Boukpessi, D. Septier, S. Bagga, M. Garabedian, M. Goldberg, C. Chaussain-Miller, et al., Dentin alteration of deciduous teeth in human hypophosphatemic rickets, Calcif. Tissue Int. 79 (5) (2006) 294–300.
[1148] C. Chaussain-Miller, C. Sinding, M. Wolikow, J.J. Lasfargues, G. Godeau, M. Garabedian, et al., Dental abnormalities in patients with familial hypophosphatemic vitamin D-resistant rickets: prevention by early treatment with 1-hydroxyvitamin D, J. Pediatr. 142 (3) (2003) 324–331.
[1149] P. Batra, Z. Tejani, M. Mars, X-linked hypophosphatemia: dental and histologic findings, J. Can. Dent. Assoc. 72 (1) (2006) 69–72.
[1150] D. Douyere, C. Joseph, C. Gaucher, C. Chaussain, F. Courson, Familial hypophosphatemic vitamin D-resistant rickets – prevention of spontaneous dental abscesses on primary teeth: a case report, Oral Surg. Oral Med. Oral Pathol. Oral Radiol. Endod. 107 (4) (2009) 525–530.
[1151] S.L. Abboud, K. Woodruff, C. Liu, V. Shen, N. Ghosh-Choudhury, Rescue of the osteopetrotic defect in op/op mice by osteoblast-specific targeting of soluble colony-stimulating factor-1, Endocrinology 143 (5) (2002) 1942–1949.
[1152] W.C. Dougall, M. Glaccum, K. Charrier, K. Rohrbach, K. Brasel, T. De Smedt, et al., RANK is essential for osteoclast and lymph node development, Genes Dev. 13 (18) (1999) 2412–2424.
[1153] H. Ida-Yonemochi, T. Noda, H. Shimokawa, T. Saku, Disturbed tooth eruption in osteopetrotic (op/op) mice: histopathogenesis of tooth malformation and odontomas, J. Oral Pathol. Med. 31 (6) (2002) 361–373.
[1154] B. Jalevik, A. Fasth, G. Dahllof, Dental development after successful treatment of infantile osteopetrosis with bone marrow transplantation, Bone Marrow Transplant. 29 (6) (2002) 537–540.
[1155] Z. Wang, L.K. McCauley, Osteoclasts and odontoclasts: signaling pathways to development and disease, Oral Dis. 17 (2) (2011) 129–142.
[1156] D.V. Novack, Role of NF-kappaB in the skeleton, Cell Res. 21 (1) (2011) 169–182.
[1157] H.C. Anderson, D. Harmey, N.P. Camacho, R. Garimella, J.B. Sipe, S. Tague, et al., Sustained osteomalacia of long bones despite major improvement in other hypophosphatasia-related mineral deficits in tissue nonspecific alkaline phosphatase/nucleotide pyrophosphatase phosphodiesterase 1 double-deficient mice, Am. J. Pathol. 166 (6) (2005) 1711–1720.
[1158] S. Di Mauro, T. Manes, L. Hessle, A. Kozlenkov, J.M. Pizauro, M.F. Hoylaerts, et al., Kinetic characterization of hypophosphatasia mutations with physiological substrates, J. Bone Miner. Res. 17 (8) (2002) 1383–1391.
[1159] K.N. Fedde, L. Blair, J. Silverstein, S.P. Coburn, L.M. Ryan, R.S. Weinstein, et al., Alkaline phosphatase knock-out mice recapitulate the metabolic and skeletal defects of infantile hypophosphatasia, J. Bone Miner. Res. 14 (12) (1999) 2015–2026.
[1160] D. Harmey, K.A. Johnson, J. Zelken, N.P. Camacho, M.F. Hoylaerts, M. Noda, et al., Elevated skeletal osteopontin levels contribute to the hypophosphatasia phenotype in Akp2(-/-) mice, J. Bone Miner. Res. 21 (9) (2006) 1377–1386.
[1161] J.L. Millan, S. Narisawa, I. Lemire, T.P. Loisel, G. Boileau, P. Leonard, et al., Enzyme replacement therapy for murine hypophosphatasia, J. Bone Miner. Res. 23 (6) (2008) 777–787.

[1162] S. Narisawa, N. Frohlander, J.L. Millan, Inactivation of two mouse alkaline phosphatase genes and establishment of a model of infantile hypophosphatasia, Dev. Dyn. 208 (3) (1997) 432–446.

[1163] D. Hotton, N. Mauro, F. Lezot, N. Forest, A. Berdal, Differential expression and activity of tissue-nonspecific alkaline phosphatase (TNAP) in rat odontogenic cells in vivo, J. Histochem. Cytochem. 47 (12) (1999) 1541–1552.

[1164] E. Mornet, B. Simon-Bouy, [Genetics of hypophosphatasia], Arch. Pediatr. 11 (5) (2004) 444–448.

[1165] E. Mornet, Hypophosphatasia, Orphanet J. Rare Dis. 2 (2007) 40.

[1166] I. Brun-Heath, A. Taillandier, J.L. Serre, E. Mornet, Characterization of 11 novel mutations in the tissue non-specific alkaline phosphatase gene responsible for hypophosphatasia and genotype-phenotype correlations, Mol. Genet. Metab. 84 (3) (2005) 273–277.

[1167] C. Stoll, M. Fischbach, J. Terzic, Y. Alembik, M.O. Vuillemin, E. Mornet, Severe hypophosphatasia due to mutations in the tissue-nonspecific alkaline phosphatase (TNSALP) gene, Genet. Couns. 13 (3) (2002) 289–295.

[1168] A. Taillandier, S.L. Sallinen, I. Brun-Heath, P. De Mazancourt, J.L. Serre, E. Mornet, Childhood hypophosphatasia due to a de novo missense mutation in the tissue-nonspecific alkaline phosphatase gene, J. Clin. Endocrinol. Metab. 90 (4) (2005) 2436–2439.

[1169] D. Fauvert, I. Brun-Heath, A.S. Lia-Baldini, L. Bellazi, A. Taillandier, J.L. Serre, et al., Mild forms of hypophosphatasia mostly result from dominant negative effect of severe alleles or from compound heterozygosity for severe and moderate alleles, BMC Med. Genet. 10 (2009) 51.

[1170] P. Moulin, F. Vaysse, E. Bieth, E. Mornet, I. Gennero, S. Dalicieux-Laurencin, et al., Hypophosphatasia may lead to bone fragility: don't miss it, Eur. J. Pediatr. 168 (7) (2009) 783–788.

[1171] T. van den Bos, G. Handoko, A. Niehof, L.M. Ryan, S.P. Coburn, M.P. Whyte, et al., Cementum and dentin in hypophosphatasia, J. Dent. Res. 84 (11) (2005) 1021–1025.

[1172] C.R. Greenberg, J.A. Evans, S. McKendry-Smith, S. Redekopp, J.C. Haworth, R. Mulivor, et al., Infantile hypophosphatasia: localization within chromosome region 1p36.1-34 and prenatal diagnosis using linked DNA markers, Am. J. Hum. Genet. 46 (2) (1990) 286–292.

[1173] D.E. Cole, Hypophosphatasia update: recent advances in diagnosis and treatment, Clin. Genet. 73 (3) (2008) 232–235.

[1174] M. Herasse, M. Spentchian, A. Taillandier, K. Keppler-Noreuil, A.N. Fliorito, J. Bergoffen, et al., Molecular study of three cases of odontohypophosphatasia resulting from heterozygosity for mutations in the tissue non-specific alkaline phosphatase gene, J. Med. Genet. 40 (8) (2003) 605–609.

[1175] J.K. Hartsfield, Jr., Premature exfoliation of teeth in childhood and adolescence, Adv. Pediatr. 41 (1994) 453–470.

[1176] M.D. McKee, Y. Nakano, D.L. Masica, J.J. Gray, I. Lemire, R. Heft, et al., Enzyme replacement therapy prevents dental defects in a model of hypophosphatasia, J. Dent. Res. 90 (4) (2011) 470–476.

[1177] E. Mornet, C. Beck, A. Bloch-Zupan, H. Girschick, M. Le Merrer, Clinical utility gene card for: hypophosphatasia, Eur. J. Hum. Genet. 19 (3) (2011).

[1178] E. Mornet, A. Yvard, A. Taillandier, D. Fauvert, B. Simon-Bouy, A molecular-based estimation of the prevalence of hypophosphatasia in the European population, Ann. Hum. Genet. 75 (3) (2011) 439–445.

[1179] M.P. Whyte, Physiological role of alkaline phosphatase explored in hypophosphatasia, Ann. N.Y. Acad. Sci. 1192 (2010) 190–200.

[1180] S.F. de Haar, W. Tigchelaar-Gutter, V. Everts, W. Beertsen, Structure of the periodontium in cathepsin C-deficient mice, Eur. J. Oral. Sci. 114 (2) (2006) 171–173.
[1181] U.M. Nadkarni, A.M. Jawdekar, S.G. Damle, S.G. Sujan, Papillon–Lefevre syndrome: management of a case, J. Indian Soc. Pedod. Prev. Dent. 19 (2) (2001) 61–66.
[1182] C. Ullbro, S. Twetman, Review paper: dental treatment for patients with Papillon–Lefevre syndrome (PLS), Eur. Arch. Paediatr. Dent. 8 (Suppl. 1) (2007) 4–11.
[1183] R.J. Gorlin, H. Sedano, V.E. Anderson, The syndrome of palmar-plantar hyperkeratosis and premature periodontal destruction of the teeth. A clinical and genetic analysis of the Papillon–Lefevre syndrome, J. Pediatr. 65 (1964) 895–908.
[1184] P.J. Dhanrajani, Re: 'Dental implants in a young patient with Papillon–Lefevre syndrome: a case report' (Implant Dent. 2003, 12 (2) 140–144), Implant Dent. 13 (4) (2004) 280.
[1185] B. Noack, H. Gorgens, U. Hempel, J. Fanghanel, T. Hoffmann, A. Ziegler, et al., Cathepsin C gene variants in aggressive periodontitis, J. Dent. Res. 87 (10) (2008) 958–963.
[1186] B. Noack, H. Gorgens, T. Hoffmann, J. Fanghanel, T. Kocher, P. Eickholz, et al., Novel mutations in the cathepsin C gene in patients with pre-pubertal aggressive periodontitis and Papillon–Lefevre syndrome, J. Dent. Res. 83 (5) (2004) 368–370.
[1187] B. Noack, H. Gorgens, B. Schacher, M. Puklo, P. Eickholz, T. Hoffmann, et al., Functional cathepsin C mutations cause different Papillon–Lefevre syndrome phenotypes, J. Clin. Periodontol. 35 (4) (2008) 311–316.
[1188] T. Lundgren, R.S. Parhar, S. Renvert, D.N. Tatakis, Impaired cytotoxicity in Papillon–Lefevre syndrome, J. Dent. Res. 84 (5) (2005) 414–417.
[1189] T. Lundgren, S. Renvert, Periodontal treatment of patients with Papillon–Lefevre syndrome: a 3-year follow-up, J. Clin. Periodontol. 31 (11) (2004) 933–938.
[1190] C. Ullbro, C.G. Crossner, T. Lundgren, P.A. Stalblad, S. Renvert, Osseointegrated implants in a patient with Papillon–Lefevre syndrome. A 4 1/2-year follow up, J. Clin. Periodontol. 27 (12) (2000) 951–954.
[1191] Y. Zhang, T. Lundgren, S. Renvert, D.N. Tatakis, E. Firatli, C. Uygur, et al., Evidence of a founder effect for four cathepsin C gene mutations in Papillon–Lefevre syndrome patients, J. Med. Genet. 38 (2) (2001) 96–101.
[1192] C. Hewitt, D. McCormick, G. Linden, D. Turk, I. Stern, I. Wallace, et al., The role of cathepsin C in Papillon–Lefevre syndrome, prepubertal periodontitis, and aggressive periodontitis, Hum. Mutat. 23 (3) (2004) 222–228.
[1193] S. Chen, S. Rani, Y. Wu, A. Unterbrink, T.T. Gu, J. Gluhak-Heinrich, et al., Differential regulation of dentin sialophosphoprotein expression by Runx2 during odontoblast cytodifferentiation, J. Biol. Chem. 280 (33) (2005) 29717–29727.
[1194] J.J. Pacheco, C. Coelho, F. Salazar, A. Contreras, J. Slots, C.H. Velazco, Treatment of Papillon–Lefevre syndrome periodontitis, J. Clin. Periodontol. 29 (4) (2002) 370–374.
[1195] I. Woo, D.P. Brunner, D.D. Yamashita, B.T. Le, Dental implants in a young patient with Papillon–Lefevre syndrome: a case report, Implant Dent. 12 (2) (2003) 140–144.
[1196] V.F. Cury, R.S. Gomez, J.E. Costa, E. Friedman, W. Boson, L. De Marco, A homozygous cathepsin C mutation associated with Haim–Munk syndrome, Br. J. Dermatol. 152 (2) (2005) 353–356.
[1197] T.C. Hart, P.S. Hart, M.D. Michalec, Y. Zhang, E. Firatli, T.E. Van Dyke, et al., Haim–Munk syndrome and Papillon–Lefevre syndrome are allelic mutations in cathepsin C, J. Med. Genet. 37 (2) (2000) 88–94.

[1198] S.A. Janjua, N. Iftikhar, I. Hussain, A. Khachemoune, Dermatologic, periodontal, and skeletal manifestations of Haim–Munk syndrome in two siblings, J. Am. Acad. Dermatol. 58 (2) (2008) 339–344.

[1199] M. Lidar, A. Zlotogorski, P. Langevitz, N. Tweezer-Zaks, G. Zandman-Goddard, Destructive arthritis in a patient with Haim–Munk syndrome, J. Rheumatol. 31 (4) (2004) 814–817.

[1200] M.A. Van Steensel, M. Van Geel, P.M. Steijlen, New syndrome of hypotrichosis, striate palmoplantar keratoderma, acro-osteolysis and periodontitis not due to mutations in cathepsin C, Br. J. Dermatol. 147 (3) (2002) 575–581.

[1201] I. Berdowska, Cysteine proteases as disease markers, Clin. Chim. Acta 342 (1–2) (2004) 41–69.

[1202] P. Pahwa, A.K. Lamba, F. Faraz, S. Tandon, Haim–Munk syndrome, J. Indian Soc. Periodontol. 14 (3) (2010) 201–203.

[1203] S.H. Ralston, Juvenile Paget's disease, familial expansile osteolysis and other genetic osteolytic disorders, Best Pract. Res. Clin. Rheumatol. 22 (1) (2008) 101–111.

[1204] I. Marik, A. Marikova, E. Hyankova, K. Kozlowski, Familial expansile osteolysis – not exclusively an adult disorder, Skeletal Radiol. 35 (11) (2006) 872–875.

[1205] A. Daneshi, Y. Shafeghati, M.H. Karimi-Nejad, A. Khosravi, F. Farhang, Hereditary bilateral conductive hearing loss caused by total loss of ossicles: a report of familial expansile osteolysis, Otol. Neurotol. 26 (2) (2005) 237–240.

[1206] T.L. Johnson-Pais, F.R. Singer, H.G. Bone, C.T. McMurray, M.F. Hansen, R.J. Leach, Identification of a novel tandem duplication in exon 1 of the TNFRSF11A gene in two unrelated patients with familial expansile osteolysis, J. Bone Miner. Res. 18 (2) (2003) 376–380.

[1207] L. Palenzuela, C. Vives-Bauza, I. Fernandez-Cadenas, A. Meseguer, N. Font, E. Sarret, et al., Familial expansile osteolysis in a large Spanish kindred resulting from an insertion mutation in the TNFRSF11A gene, J. Med. Genet. 39 (10) (2002) E67.

[1208] M.P. Whyte, W.R. Reinus, M.N. Podgornik, B.G. Mills, Familial expansile osteolysis (excessive RANK effect) in a 5-generation American kindred, Medicine (Baltimore) 81 (2) (2002) 101–121.

[1209] M.P. Whyte, A.E. Hughes, Expansile skeletal hyperphosphatasia is caused by a 15-base pair tandem duplication in TNFRSF11A encoding RANK and is allelic to familial expansile osteolysis, J. Bone Miner. Res. 17 (1) (2002) 26–29.

[1210] A.E. Hughes, S.H. Ralston, J. Marken, C. Bell, H. MacPherson, R.G. Wallace, et al., Mutations in TNFRSF11A, affecting the signal peptide of RANK, cause familial expansile osteolysis, Nat. Genet. 24 (1) (2000) 45–48.

[1211] G.H. Esselman, J.A. Goebel, F.J. Wippold, 2nd, Conductive hearing loss caused by hereditary incus necrosis: a study of familial expansile osteolysis, Otolaryngol. Head Neck Surg. 114 (4) (1996) 639–641.

[1212] A.E. Hughes, A.M. Shearman, J.L. Weber, R.J. Barr, R.G. Wallace, Genetic linkage of familial expansile osteolysis to chromosome 18q, Hum. Mol. Genet. 3 (2) (1994) 359–361.

[1213] G.R. Dickson, P.V. Shirodria, J.A. Kanis, M.N. Beneton, K.E. Carr, R.A. Mollan, Familial expansile osteolysis: a morphological, histomorphometric and serological study, Bone 12 (5) (1991) 331–338.

[1214] C.A. Mitchell, J.G. Kennedy, R.G. Wallace, Dental abnormalities associated with familial expansile osteolysis: a clinical and radiographic study, Oral Surg. Oral Med. Oral Pathol. 70 (3) (1990) 301–307.

[1215] C.A. Mitchell, J.G. Kennedy, P.D. Owens, Dental histology in familial expansile osteolysis, J. Oral Pathol. Med. 19 (2) (1990) 65–70.
[1216] M.D. Crone, R.G. Wallace, The radiographic features of familial expansile osteolysis, Skeletal Radiol. 19 (4) (1990) 245–250.
[1217] R.G. Wallace, R.J. Barr, P.H. Osterberg, R.A. Mollan, Familial expansile osteolysis, Clin. Orthop. Relat. Res. November (248) (1989) 265–277.
[1218] P.H. Osterberg, R.G. Wallace, D.A. Adams, R.S. Crone, G.R. Dickson, J.A. Kanis, et al., Familial expansile osteolysis. A new dysplasia, J. Bone Joint Surg. Br. 70 (2) (1988) 255–260.
[1219] C.B. Olsen, K. Tangchaitrong, I. Chippendale, H.K. Graham, H.M. Dahl, J.R. Stockigt, Tooth root resorption associated with a familial bone dysplasia affecting mother and daughter, Pediatr. Dent. 21 (6) (1999) 363–367.
[1220] Y. Ueki, V. Tiziani, C. Santanna, N. Fukai, C. Maulik, J. Garfinkle, et al., Mutations in the gene encoding c-Abl-binding protein SH3BP2 cause cherubism, Nat. Genet. 28 (2) (2001) 125–126.
[1221] M.T. Cobourne, G.M. Xavier, M. Depew, L. Hagan, J. Sealby, Z. Webster, et al., Sonic hedgehog signalling inhibits palatogenesis and arrests tooth development in a mouse model of the nevoid basal cell carcinoma syndrome, Dev. Biol. 331 (1) (2009) 38–49.
[1222] A.E. Oro, K.M. Higgins, Z. Hu, J.M. Bonifas, E.H. Epstein, Jr., M.P. Scott, et al., Basal cell carcinomas in mice overexpressing sonic hedgehog, Science 276 (5313) (1997) 817–821.
[1223] H. Hahn, L. Wojnowski, G. Miller, A. Zimmer, The patched signaling pathway in tumorigenesis and development: lessons from animal models, J. Mol. Med. 77 (6) (1999) 459–468.
[1224] R.B. Corcoran, M.P. Scott, A mouse model for medulloblastoma and basal cell nevus syndrome, J. Neurooncol. 53 (3) (2001) 307–318.
[1225] K. Kimi, K. Ohki, H. Kumamoto, M. Kondo, Y. Taniguchi, A. Tanigami, et al., Immunohistochemical and genetic analysis of mandibular cysts in heterozygous ptc knockout mice, J. Oral Pathol. Med. 32 (2) (2003) 108–113.
[1226] S. Pazzaglia, Ptc1 heterozygous knockout mice as a model of multi-organ tumorigenesis, Cancer Lett. 234 (2) (2006) 124–134.
[1227] L. Lo Muzio, Nevoid basal cell carcinoma syndrome (Gorlin syndrome)., Orphanet J. Rare Dis. 3 (2008) 32.
[1228] R.J. Gorlin, Nevoid basal cell carcinoma (Gorlin) syndrome, Genet. Med. 6 (6) (2004) 530–539.
[1229] R.J. Gorlin, Nevoid basal cell carcinoma (Gorlin) syndrome: unanswered issues, J. Lab. Clin. Med. 134 (6) (1999) 551–552.
[1230] R.J. Gorlin, Nevoid basal cell carcinoma syndrome, Dermatol. Clin. 13 (1) (1995) 113–125.
[1231] R.J. Gorlin, Nevoid basal-cell carcinoma syndrome, Medicine (Baltimore) 66 (2) (1987) 98–113.
[1232] R.J. Gorlin, H.O. Sedano, The multiple nevoid basal cell carcinoma syndrome revisited, Birth Defects Orig. Artic. Ser. 7 (8) (1971) 140–148.
[1233] R.J. Gorlin, The multiple nevoid basal cell carcinoma syndrome, Contact Point 48 (6) (1970) 212–214.
[1234] R.J. Gorlin, R.A. Vickers, E. Kellen, J.J. Williamson, Multiple basal-cell nevi syndrome. An analysis of a syndrome consisting of multiple nevoid basal-cell carcinoma, jaw cysts, skeletal anomalies, medulloblastoma, and hyporesponsiveness to parathormone, Cancer 18 (1965) 89–104.

[1235] L. Pastorino, P. Ghiorzo, S. Nasti, L. Battistuzzi, R. Cusano, C. Marzocchi, et al., Identification of a SUFU germline mutation in a family with Gorlin syndrome, Am. J. Med. Genet. A 149A (7) (2009) 1539–1543.

[1236] L.C. Wilson, E. Ajayi-Obe, B. Bernhard, S.M. Maas, Patched mutations and hairy skin patches: a new sign in Gorlin syndrome, Am. J. Med. Genet. A 140 (23) (2006) 2625–2630.

[1237] M.M. Cajaiba, A.E. Bale, M. Alvarez-Franco, J. McNamara, M. Reyes-Mugica, Rhabdomyosarcoma, Wilms tumor, and deletion of the patched gene in Gorlin syndrome, Nat. Clin. Pract. Oncol. 3 (10) (2006) 575–580.

[1238] S. Levanat, R.J. Gorlin, S. Fallet, D.R. Johnson, J.E. Fantasia, A.E. Bale, A two-hit model for developmental defects in Gorlin syndrome, Nat. Genet. 12 (1) (1996) 85–87.

[1239] W. Zedan, P.A. Robinson, A.F. Markham, A.S. High, Expression of the Sonic Hedgehog receptor 'PATCHED' in basal cell carcinomas and odontogenic keratocysts, J. Pathol. 194 (4) (2001) 473–477.

[1240] L. Lo Muzio, L. Pastorino, S. Levanat, V. Musani, M. Situm, G.B. Scarra, Clinical utility gene card for: Gorlin syndrome, Eur. J. Hum. Genet. 19 (2011) 8.

[1241] N. Levaot, P.D. Simoncic, I.D. Dimitriou, A. Scotter, J. La Rose, A.H. Ng, et al., 3BP2-deficient mice are osteoporotic with impaired osteoblast and osteoclast functions, J. Clin. Invest. 121 (8) (2011) 3244–3257.

[1242] B. Baskin, S. Bowdin, P.N. Ray, Cherubism, in: R.A. Pagon, T.D. Bird, C.R. Dolan, K. Stephens, (Eds.), GeneReviews [Internet], University of Washington, Seattle (WA), 1993.

Printed in the United States
By Bookmasters